THE WAYSIDE
AND WOODLAND
SERIES

FLIES OF THE BRITISH ISLES

FRONTISPIECE

Volucella zonaria Poda ♀

FLIES
OF THE
BRITISH ISLES

By CHARLES N. COLYER, F.R.E.S., F.Z.S.

IN COLLABORATION WITH
CYRIL O. HAMMOND, F.R.E.S.

★

WITH
25 PLATES IN COLOUR, 28 HALF-TONE PLATES
AND NUMEROUS TEXT FIGURES AND DIAGRAMS DEPICTING
286 REPRESENTATIVE SPECIES FROM
ORIGINAL ENLARGED MICROSCOPE DRAWINGS
BY
CYRIL O. HAMMOND, F.R.E.S.

FREDERICK WARNE & CO. LTD.
LONDON · NEW YORK

First Edition . . . 1951

7232 0004 1

PRINTED IN GREAT BRITAIN BY
LOWE AND BRYDONE (PRINTERS) LIMITED, LONDON

615 . 567

ACKNOWLEDGEMENTS—First Edition

The authors desire to acknowledge with grateful thanks the courteous and kindly assistance vouchsafed to them at all times by Mr. H. Oldroyd, Mr. P. Freeman and Mr. R. L. Coe, of the Dept. of Entomology, British Museum (Nat. Hist.), and by Dr. F. Van Emden of the Commonwealth Institute of Entomology.

They wish to record their appreciation of useful advice and practical assistance in dipterological matters given by Mr. E. A. Fonseca, Mr. H. Audcent, Mr. H. Britten, Jun., and Mr. M. Niblett; their thanks are also due to Mr. A. E. LeGros and Mr. R. D. Weal of the London Natural History Society, and Mr. Tom Spence, of the Veterinary Laboratory, Ministry of Agriculture and Fisheries, for specimens presented or loaned and observation-notes furnished by them.

To Mr. L. Parmenter especially, they are indebted for many hours of congenial company whilst collecting and for much valuable practical knowledge imparted to them, particularly in the field.

C. N. C.
C. O. H.

ACKNOWLEDGEMENTS—Second Edition

Such amendments and additions as have been considered necessary for the second edition have been made mainly as a result of the invaluable help of Mr. K. G. V. Smith and his colleagues, Messrs. R. L. Coe, B. H. Cogan, A. M. Hutson, H. Oldroyd and A. C. Pont at the Department of Entomology, British Museum (Natural History), whose constant courtesy has been a great encouragement to the authors.

C. N. C.
C. O. H.

CONTENTS

INTRODUCTORY

TRUE Flies, or Diptera, are distinguished from nearly all other winged insects by the possession, in the typical adult form, of only one pair of wings; the pair corresponding to the hind pair in other insects being represented by small, knobbed organs known as "balancers" or "halteres" (Fig. 1). The Order, both in aggregate population and numbers of species, is vast by comparison with most other kinds of living creatures. In Britain alone, more than 5,000 species have been recorded, varying in size from almost a pin-point to one and a quarter inches in body-length and two and a half inches in wing-span.

Flies have not captured the popular fancy in the same way as Butterflies, Moths, Beetles and Dragonflies; yet, to anyone who is prepared to take the trouble to use pocket-lens and microscope, they will reveal themselves as objects of great beauty and interest. In their manner of living and the rôle they play in Nature, they are of much significance to the naturalist.

Many of our friends and acquaintances, some already entomologists, have expressed a desire to know more of the Flies, but have had their enthusiasm damped when they have learned that the literature relating to Diptera (in English, at all events) is relatively sparse, mostly technical and scattered among entomological periodicals and the publications of learned societies, where it is not always easy of access.

In some other Orders, however, through the agency of well-illustrated "popular" books, it is possible to acquire a very substantial basis of knowledge, which serves not only to heighten interest in Natural History and ecological study generally, but stimulates a desire for further knowledge of the chosen Order. That such a facility should not, so far, be available in respect of Diptera has seemed to the authors most unfortunate; they have, therefore, attempted to make good the deficiency.

This book, then, is intended to provide in simple fashion such illustrations and descriptions as will serve to give a general introduction to the British Flies, together with brief accounts of some of the more interesting life-histories and habits. For the assistance of those readers who have the opportunity and intention of pursuing the study of flies further than the scope of this book will allow, the names of authors of interesting and useful works are cited in appropriate places, followed by numbers which refer to the bibliography at the end of each chapter. Many of the works quoted themselves contain extensive further references. No attempt has been made to provide scientific keys for accurate identification down to the level of species; such keys may not necessarily furnish the field-naturalist with the kind of assistance he requires at the outset of his excursion into Dipterology. The Royal Entomological Society of London is in course of publishing an excellent series of keys for the identification of British insects, which will provide what is required in this direction; as regards Diptera, reference may be made to the Introductory Handbook by Oldroyd.[1] Since it will be impracticable for the beginner to attempt to identify, in the majority of cases, live, free specimens in the field, resort must be had to the notebook and net in order to associate the results of observation with dead specimens subsequently classified; by this means the student will gradually acquire the ability to refer specimens in the field at least to their correct family, and very often to their correct genus. Thus it will be increasingly possible to collate the results of individual observation and those recorded by others. To assist in the more rapid running-down of mounted specimens to families, therefore, keys are included in appropriate places.

The difficulty which most readers will, in all probability, experience when using keys, particularly at first, is a feeling of uncertainty as to whether the determination is correct, and a desire for corroboration, which is best accomplished by reference to a reliably-named collection. This is a facility which will not be readily available to many; suitable illustrations, however, will assist a great deal in resolving doubt; thus the Family Keys bear references to illustrations and figures as well as to the text.

In considering the question of suitable illustrations, the authors have decided that, owing to the risk of important structural characters being obscured or lost in photographic reproduction, drawings would be preferable, with an accentuation, where considered necessary, of characters peculiar to the group or species under consideration. Every drawing has been made direct from a binocular microscope and subsequently reduced in reproduction so as to give the greatest possible sharpness the process allows; the specimens illustrated, with a few exceptions, form part of the authors' personal collections.

It is hoped that, armed with the contents of these pages and with those indispensable qualities of the true field-naturalist, an alert eye, a receptive mind and a capacity for appreciating minute detail, the reader may be encouraged to discover in the field fresh knowledge of what is, as yet, a relatively little-explored part of the insect world. Opportunities for this are not lacking for the amateur, who may often, by patient field-work, supplement the researches of the professional entomologist. In the search, the seeker will find that the feet will be led into many new places, the body encouraged to new forms of effort and that, inevitably, the eyes will perceive many strange, fascinating and unexpected things apart from the special objects of the search. The dividend of physical health and mental satisfaction to be derived from such endeavour is a thing enjoyed by all too few in this harassed world of today.

As with all kinds of study, before one may take advantage of the knowledge accumulated by previous students and recorded in the literature, it is necessary to become familiar with the language peculiar to the selected subject; that is to say, with the more "technical" terms, which, although capable of wider interpretation in other fields, have by usage in a particular field acquired a special significance. In this connection, Dipterology is no exception, not only in so far as the structure and characteristics of flies are concerned, but also due to the fact that only a few flies have English or "popular" names. Such names, where they exist, must be treated with reserve, since in many cases they were bestowed in days when scanty or even mistaken ideas about the nature and habits of insects prevailed. Sometimes the same "popular" name is applied to widely different species in various parts of the country, especially by rustic folk; so that, for instance, the "thunder fly" of Herefordshire may be quite different from its namesake in Hampshire. Consequently, it is advisable in most cases to refer to flies by their scientific names; the reader who is unfamiliar with such names should not be deterred; if doubt is felt as to the pronunciation usually adopted by entomologists, assistance may be sought from Longfield.[2] After all, when there is occasion to refer to something new or unfamiliar, one name will serve as well as another; for our present purpose, it will suffice to recognise a name or term and associate it with the particular species, group or structure to which it refers. For ease of reference, a glossary of dipterological terms and a general index will be found on pages 353 and 369 respectively.

GENERAL STRUCTURE OF THE ADULT FLY. The shape or build of adult flies varies greatly through the Order. It will be convenient for us to recognise six main sections and we must first acquire a general impression of these by a preliminary examination of some representative examples. The diagrammatic figures we shall consider will indicate the more prominent parts which can be made out with the naked eye or a good pocket-lens, but it will be necessary to use a compound

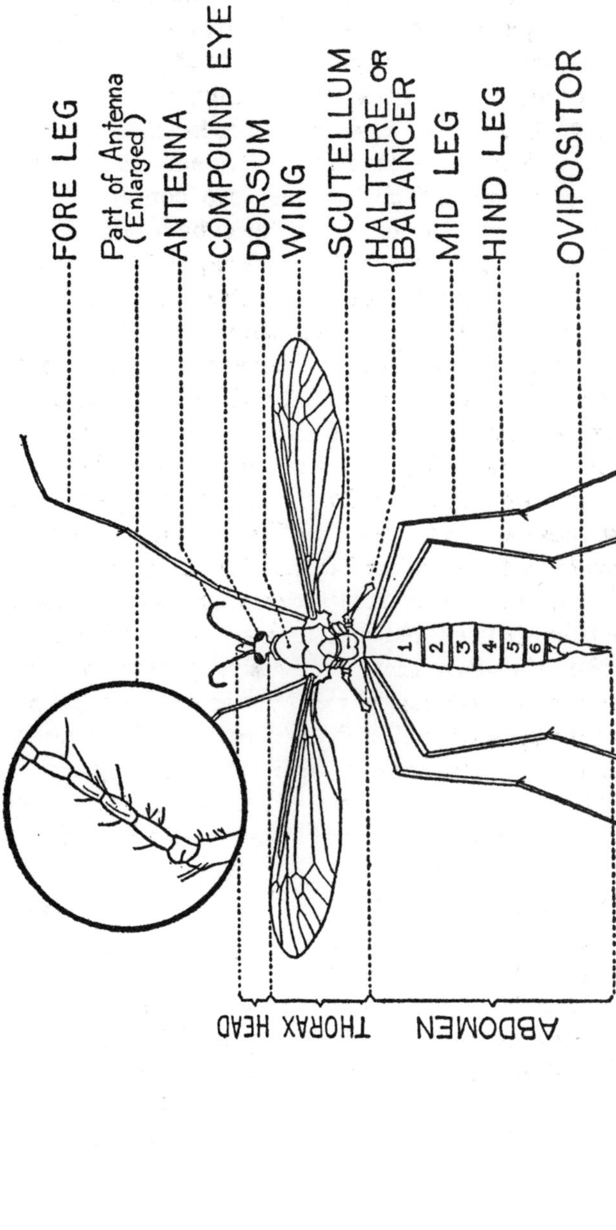

FIG. 1.—Representative fly of Sub-Order NEMATOCERA (*Tipula* sp.), p. 16.

microscope to appreciate some of the smaller features. Hints on the use of equipment for the examination of flies in the field or at home, with simple techniques for mounting, are to be found in the Appendix at p. 319.

Fig. 1 represents a Crane Fly or " Daddy-long-legs ", one of the more primitive kinds of fly in the Sub-Order NEMATOCERA (Thread-horns), which will constitute our Section I. The name had its origin in one of the principal characters upon which the Sub-Order is based, namely, the long antennae or " feelers " which consist of six or more practically uniform segments. In this fly it will be noticed that the venation, or vein-arrangement of the wings, is complex and that the body is long and slender. These two characters together, however, are not invariable throughout the Sub-Order, as will be appreciated by reference to Pls. **7** and **10**.

Fig. 2 shows a representative of the Sub-Order BRACHYCERA (Short-horns), which forms our Section II. Here it will be seen that the antennae are shorter, more robust and composed of fewer segments ; there are indications that some of the terminal segments have been fused together. The wing-venation is less complex, some of the veins found in more primitive flies having disappeared in the course of evolution. Moreover, the body is of a decidedly more robust appearance.

Nematocera and Brachycera together form the group Orthorrhapha, of which more will be said later.

The group Cyclorrhapha includes the remaining four Sections and constitutes a large and complex Sub-Order. Fig. 3 illustrates a representative Cyclorrhaphous fly, the well-known Blue Bottle. Here the antennae are restricted to three principal segments, the third usually being enlarged or dilated. Segments of structures such as the antennae, palpi and legs are numbered from the basal segment to the apical one ; the basal segment is that nearest to the point of attachment to the body and the apical or distal segment that which is furthest away. In the abdomen, the basal segment is that nearest

ASILIDAE-I
Asilus crabroniformis L., ♀, p. 120.

TIPULIDAE-I
Tipula maxima Poda, ♀, p. 47.

Plate I

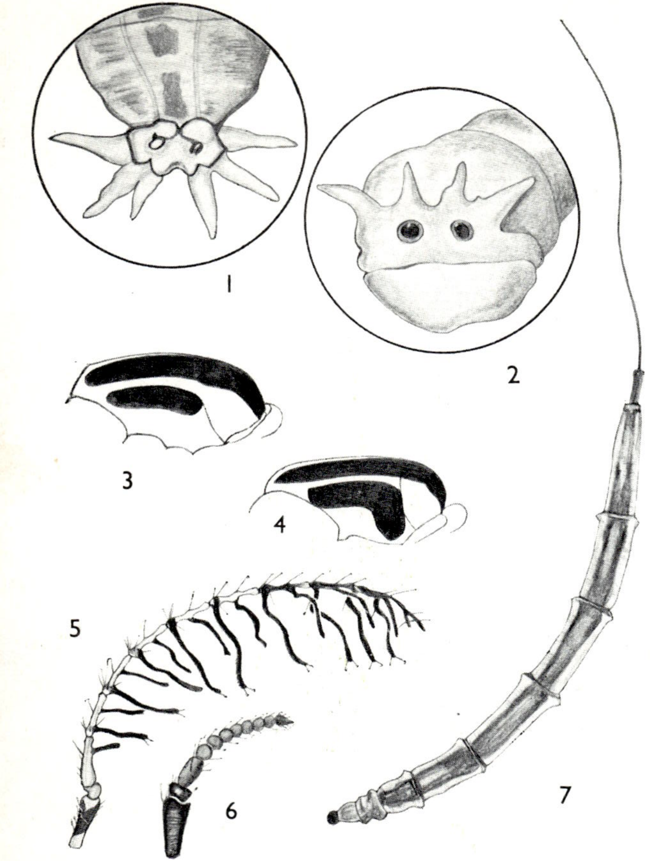

TIPULIDAE-II

1. *Tipula* sp., aquatic type larva, p. 47. 2. *Nephrotoma* sp., terrestrial type larva, p. 47. 3. *Nephrotoma* sp., dorsum, profile, p. 49. 4. *Nephrotoma* sp., dorsum, profile, p. 49. 5. *Dictenidia*, ♂ antenna. 6. ♀ antenna, p. 50.

PTYCHOPTERIDAE-I 7. *Ptychoptear*, larva, p. 58.

PLATE 2

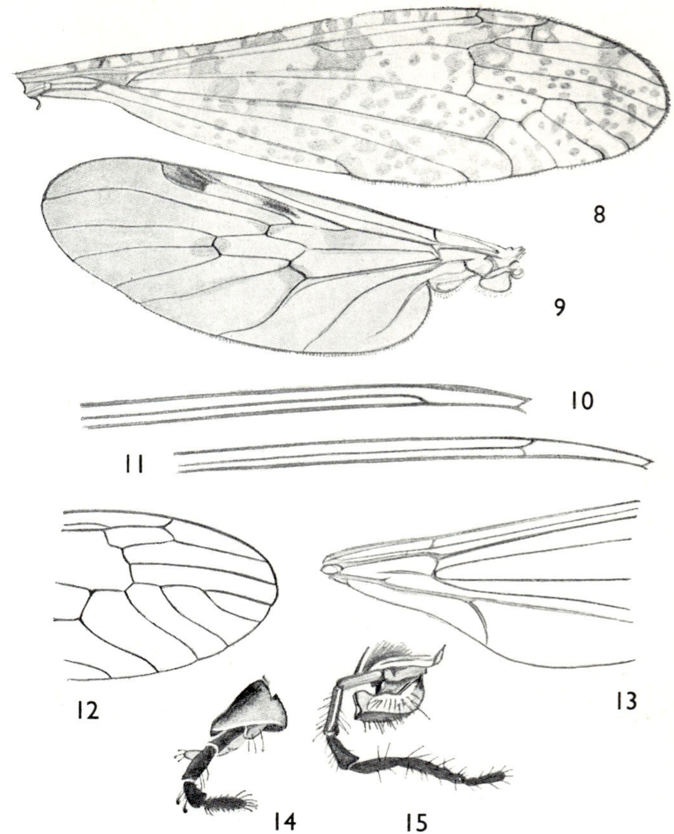

TIPULIDAE-III

8. *Limnophila punctata* Schrank, wing, p. 56. 10. Tipulinae, sub-costa, p. 46. 11. Short-palped Crane Fly, sub-costa, p. 46. 12. *Dolichopeza albipes* Stroem, apex of wing, p. 49. 14. Short-palped Crane Fly, palp, p. 46. 15. Tipulinae, palp, p. 46.

ANISOPODIDAE 9. *Sylvicola fenestralis* Scop., wing, p. 59.

TRICHOCERIDAE 13. *Trichocera* sp., base of wing, p. 57.

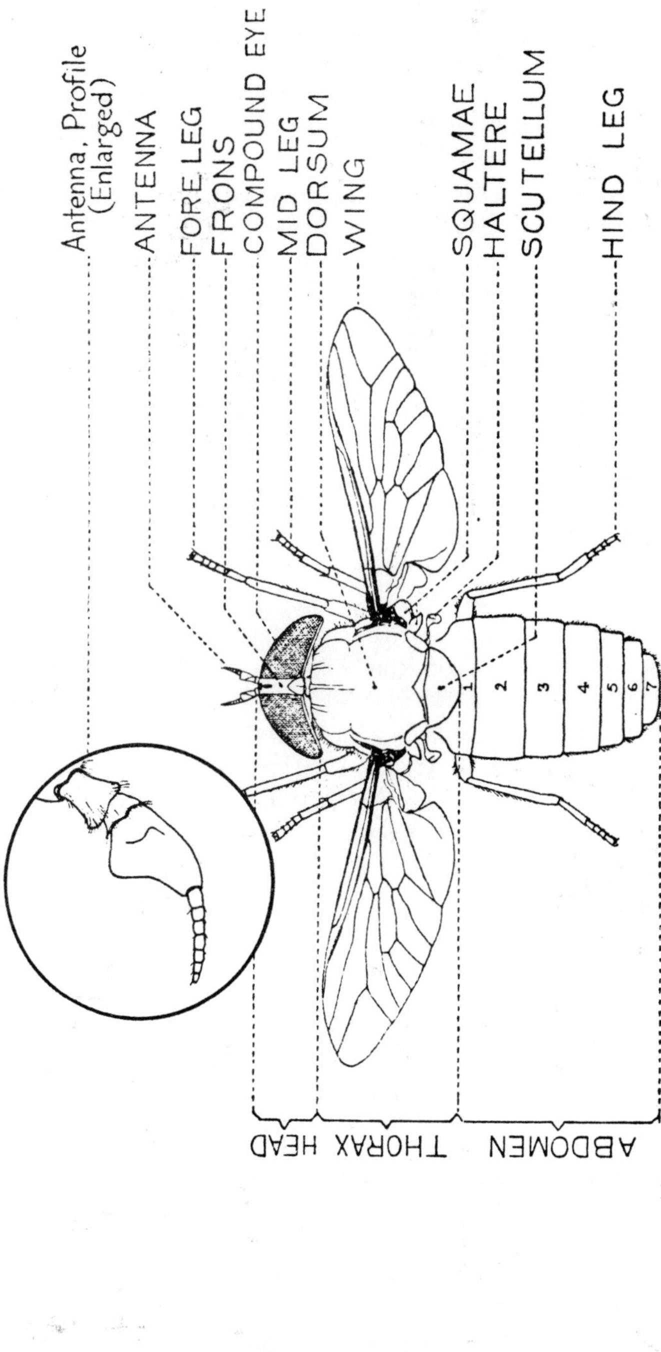

FIG. 2.—Representative fly of Sub-Order BRACHYCERA (*Tabanus* sp.), p. 16.

the thorax. Wing-venation in the Cyclorrhapha is further simplified or reduced, and the body generally much larger in relation to the size of the wings. It will be seen that the number of abdominal segments visible from above is also considerably reduced.

In resolving the Sub-Order Cyclorrhapha into its component Sections, it will be convenient to separate first the PUPIPARA, the species of which comprise our Section VI. These consist of a few aberrant, highly-specialised flies, with a tough, leathery integument and more or less flattened body, which have adapted themselves to a parasitic mode of existence, their hosts being principally mammals and birds. In some cases, wings are absent, the need for them having disappeared owing to the manner of living; the legs and claws, on the other hand, have developed in a way which makes them admirably suited to clinging tightly to the hairs or feathers of their hosts. Some examples of the Section will be found on Pls. **44**, 2, and **50**.

Fig. 4 demonstrates the differentiation of the remainder of the Cyclorrhapha between the ASCHIZA (Section III) and the SCHIZOPHORA (Sections IV and V). The latter are characterised by a facial depression in which the bases of the antennae are situated and which is bounded by a groove or ptilinal suture (cf. page 38). There is also a clearly defined lunule or crescent-shaped mark above the bases of the antennae. The depression and suture are absent or very faintly indicated in Aschiza; if a lunule is present, it is often indistinct.

SCHIZOPHORA may now be separated into the ACALYPTERATES (Section IV) and the CALYPTERATES (Section V). Fig. 5 shows that the Calypterates have a cleft or fissure in the second antennal segment, which is absent in the Acalypterates. Furthermore, the Calypterates have, in many cases but not invariably, well-developed squamae, or flap-like appendages at the base of the wing, often, but again by no means invariably, concealing the halteres; whereas in the Acalypterates,

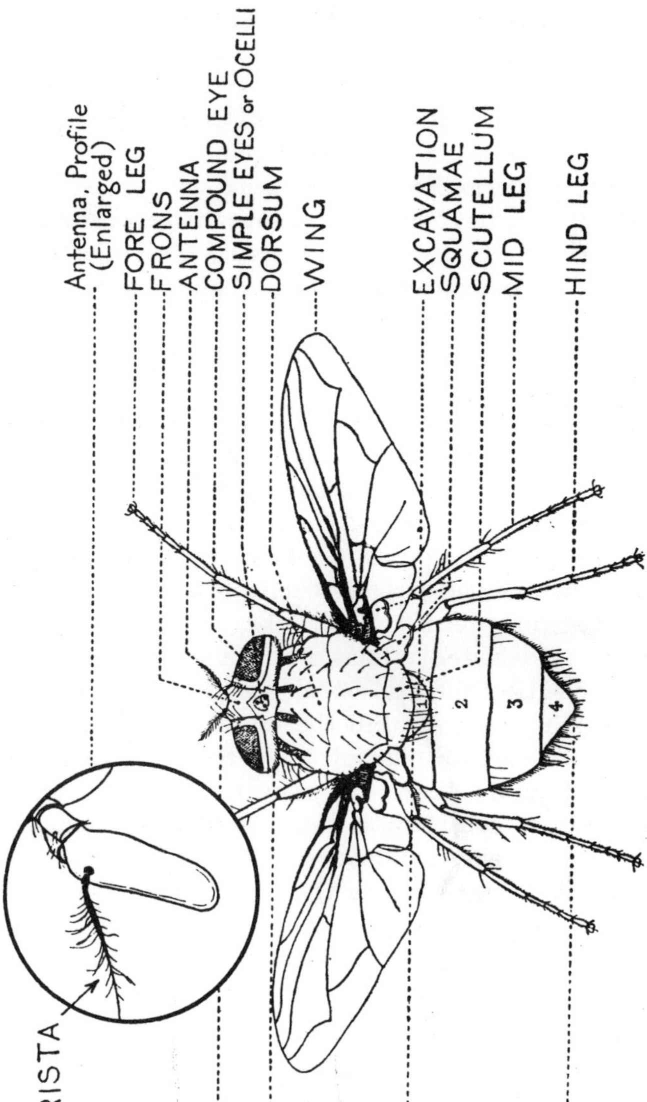

FIG. 3.—Representative fly of Sub-Order CYCLORRHAPHA (*Calliphora* sp.), p. 16.

squamae are mostly vestigial and usually leave the halteres revealed. The dorsum also provides distinguishing characters; in the Acalypterates, the transverse suture is incomplete and the posterior calli barely noticeable, while in the Calypterates, the suture is extended right across the dorsum and prominent posterior calli are present.

The foregoing will have merely conveyed a general impression of the limits and principal features of the Order. It is now necessary to examine the structure of the adult fly in greater detail; for this purpose, it will be methodical to commence at the front and work progressively to the rear.

THE HEAD. In Fig. 4 it will be seen that the main features consist of the antennae, the compound eyes, the ocelli or simple eyes, the mouthparts, the areas of the surface not taken up by the foregoing organs, and various bristles. Although the areas are not sharply delimited, in broad terms it may be said that the area at the top of the head, upon which stands the ocellar triangle, is the vertex; below it, as far as the bases of the antennae, lies the frons; beneath the antennae is the face, the sides of which are known as the genae, or cheeks; the areas bounded by the oral opening and the lower margin of the compound eyes are called the jowls, while the back of the head is designated the occiput. The head is joined to the succeeding part of the body, the thorax, by a very narrow neck and possesses considerable latitude of movement, even to the extent of almost complete rotation.

It has already been seen how the general structure of the antennae may vary; in addition there may be, as in some species of Brachycera, a stout terminal "style", or process arising from the apex of the last segment (Pl. **19**, 2, 4, 5); or, as is general in Cyclorrhapha, a bristle-like process, or "arista" (Fig. 4) on the dorsal surface of the last segment, i.e. the upper surface, if the antennae are imagined to be stretched out horizontally. Various bristles or circlets of hairs may adorn some or all of the segments.

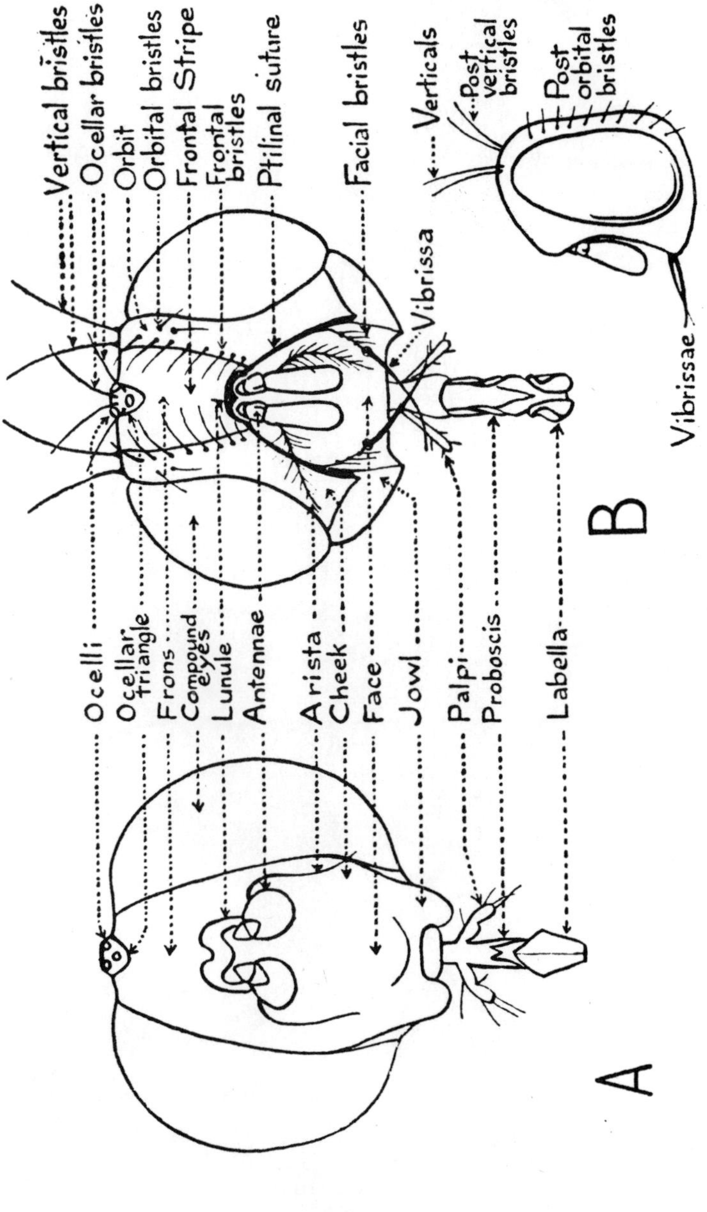

FIG. 4.—CYCLORRHAPHOUS HEADS, p. 18.
A. Aschiza. B. Schizophora.

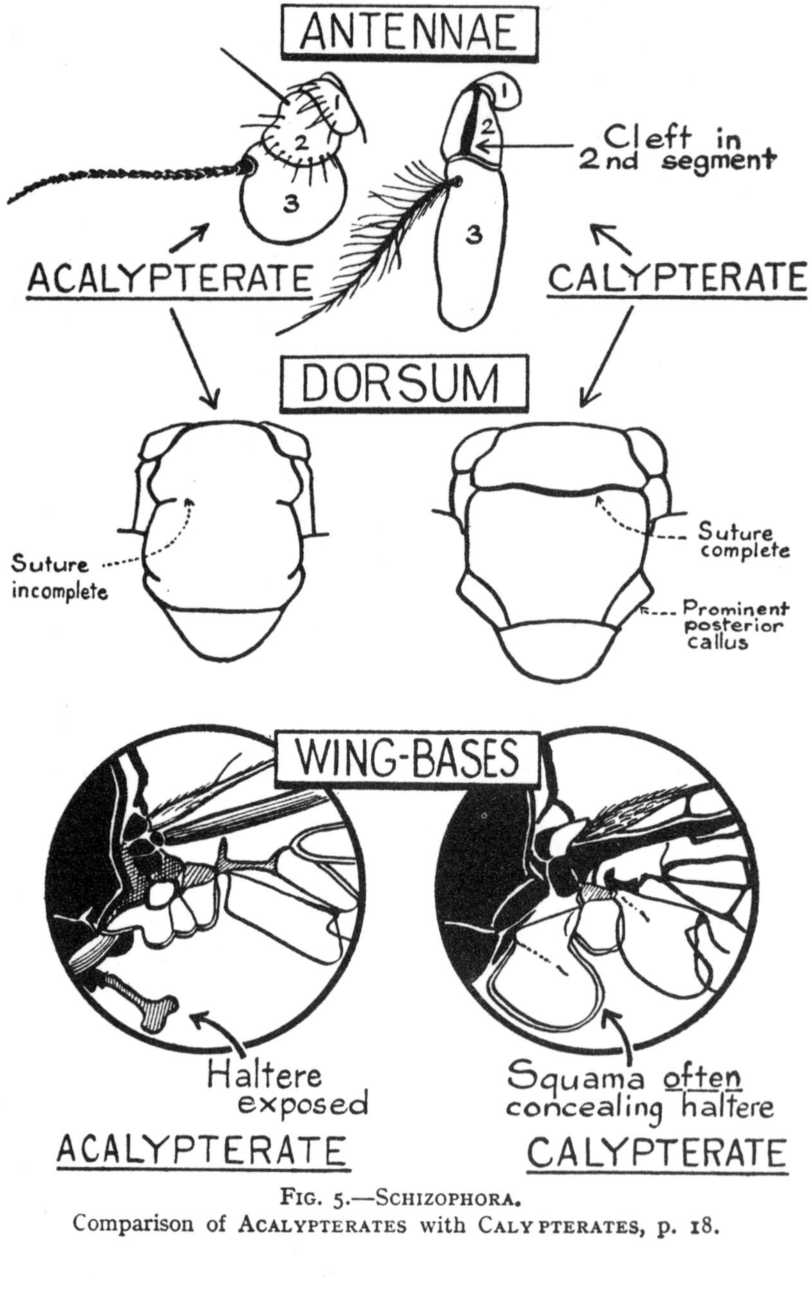

FIG. 5.—SCHIZOPHORA.
Comparison of ACALYPTERATES with CALYPTERATES, p. 18.

A fly's compound eyes consist of a very large number of individual facets, which may vary in size in different parts of the eye-surface ; hairs of various lengths may occur between the facets. When present, the ocelli are usually arranged in the form of a triangle, one in front and two at the rear, but sometimes only the front one is present. In the males of many species, the compound eyes are so far developed that the frons is reduced to a very narrow strip or almost completely obliterated, a condition known as holoptic, while the females of the same species have a much broader frons, leaving the eyes clearly separated ; here the eyes are said to be dichoptic.

Mouthparts of adult flies exhibit considerable variation in form. In essence, the primitive insect biting mouthparts have, in Diptera, undergone a change into a kind of suctorial tube, some of the original movable parts having been fused together and generally modified. The proboscis, in blood-feeding and predatory flies, is normally much strengthened and capable of being made rigid for the effective perforation of the skin of their hosts. In some cases, it is in the form of a hollow, piercing needle ; in others, of a broad, dagger-shaped weapon enclosing the feeding-channel ; in yet others, of a tube with abrasive apparatus at the tip. Only the females of some bloodsucking species practise the habit, the mouthparts of the males being different and not adapted for the purpose. In non-bloodsucking flies, the proboscis, which is usually retractable within the oral opening, bears at its tip a kind of pad capable of being used as a filter ; in the closed position, only liquids and very tiny solid particles are absorbed, but the opening of the feeding-channel can be voluntarily opened to permit the ingress of larger items. The mouthparts of certain flies are non-functional, no food being taken in the adult stage, which is usually of short duration and occupied mainly with reproduction. Situated on either side of the proboscis are the palpi, which may be many-segmented as in Nematocera

(Pl. **5**, 7, 8), or few-segmented and mostly porrect (forwardly projecting) in the other Sections (Fig. 4).

Bristles or hairs are much used for purposes of classification. The location of the largest and most robust may still be traced after they have been broken off, since they arise from clearly marked depressions or sockets. Others vary from stiff, stout hairs down to a very short " pile " or pubescence. Sometimes small hairs in certain parts of the structure are referred to as cilia.

As regards the head, Fig. 4 shows that the principal bristles are named in relation to the area on which they occur, thus: ocellar bristles occur on, or adjacent to the ocellar triangle; vertical bristles on the vertex; behind the ocellar triangle, the postocellars or postverticals; orbitals on the orbits, or eye-borders (which are often differentiated from the centra portion or " frontal stripe "); and frontals on the inner edge of the orbits along the stripe. The term fronto-orbitals is often applied to orbital bristles.

Very prominent bristles which occur at the lower end of the lateral ridge-borders of the face in many Cyclorrhaphous flies are known as vibrissae, and above these may be an ascending row of facial bristles. Occipital bristles are found on the occiput, and postorbitals or postoculars on the posterior orbits, or hinder eye-margins.

THE THORAX. In profile (Fig. 6A), the thorax is divided into more or less clearly-defined areas or pleural plates, called pleurites. Collectively they are termed the pleura and consist of the notopleuron, propleuron, mesopleuron, pteropleuron, metapleuron, sternopleuron and hypopleuron. The wings arise from the pteropleuron; spiracles or apertures of the tracheae (breathing-tubes) are located before the mesopleuron and behind the hypopleuron respectively. The basal segments of the legs are joined to the lower pleurites. Areas on the ventral or lower surface of the thorax are known as sternites; the upper surface, as a whole, is referred to as the

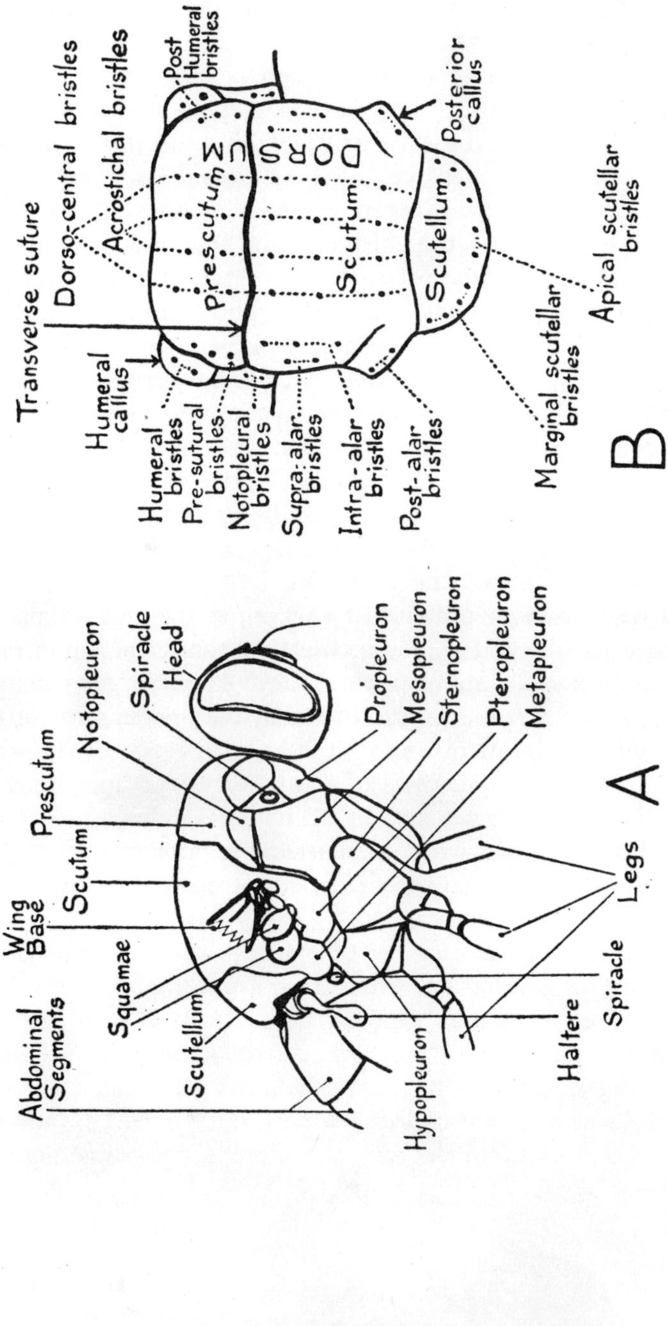

FIG. 6.—THE THORAX, P. 24.
A. Profile. B. Dorsal aspect.

dorsum (Fig. 6B); it exhibits three main divisions, the prescutum, scutum and scutellum. The two former are divided by a transverse suture, which is terminated on each side by the notopleura. This suture may be complete, as in the Calypterates; visible only at the sides, as in the Acalypterates; or, in some cases, vestigial or absent.

Thoracic bristles are often of considerable importance. The names of those on the sides usually bear reference to the pleurite on which they occur, thus: sternopleurals on the sternopleuron, hypopleurals on the hypopleuron, etc. Those on the dorsum have special names as shown in the figure. It should be realised, however, that not all the bristles indicated will necessarily be found in one particular fly; some flies are almost completely devoid of bristles.

THE WINGS. The three features of the wings with which we need concern ourselves are the veins or nervures, the membrane, and the small bristles or hairs which may occur on either. The membrane may be clear, tinged with colour, or marked with spots or patches of varying shape and intensity; sometimes small hyaline (transparent) patches are found on a tinged membrane. Microscopic hairs, some larger (macrotrichia) and some smaller (microtrichia) may occur, separately or together, on the membrane; the anterior edge of the wing often bears one or more rows of small bristles or hairs, usually called cilia; there may be, in addition, fairly stout spines near the base of the wing or distributed along the anterior edge; cilia also occur on the veins, either on the upper surface or the lower.

There have been a number of systems of reference to the parts of the wing-venation, which is of great importance. The one which, with various modifications, is in general use for scientific purposes today is known as the Comstock-Needham system and is capable of application to all winged insects. Since we, however, are concerned here only with Diptera, we shall find it less involved and therefore more convenient to adopt a simple method, much used by Dipterists, which is, in essence, based on numbering the longitudinal veins. We have already seen that, as flies have evolved from the more primitive Nematocerous types toward the highly-developed Cyclorrhapha, a reduction of the wing-venation has taken place, i.e. certain of the veins have atrophied and disappeared. It will be necessary, therefore, to bear in mind that what, for example, may be Vein 7 in a primitive wing may not be present in the wing of a fly which has evolved further. Fig. 7 demonstrates how the system is applied. The usually stout nervure which forms the leading edge of the wing is known as the costa (*c*) and the usually short auxiliary vein lying directly behind it and mostly ending in it is called the subcosta (*sc*). Thence the long veins are numbered 1, 2, 3, etc. Veins 6 and 7 are often referred to as the first and second anal veins respectively and Vein 7 not infrequently as the axillary vein. Some veins fork into branches before reaching the edge of the wing; these may be referred to as the 1st, 2nd, 3rd branches, etc., of a particular vein. It will be noticed in the figure that, in some wings, certain long veins do not arise from the base of the wing but spring from, or have a common stem with another; this has come about through the origins having become fused. Consequently, doubt may be experienced as to whether an apparent branch is really a long vein, or vice versa; or, to which of the long veins the various branches properly belong. On studying Fig. 7, however, it will be seen that two of the transverse veins, or "cross-veins", have been emphasised, namely, the anterior (*a*), sometimes called the middle or small cross-vein, which lies between the 3rd and 4th long veins; and the posterior (*p*), sometimes called the large cross-vein, which lies between the 4th and 5th long veins. After a scrutiny of various kinds of wing, it will be found that the location of Veins 3, 4 and 5, in conjunction with these cross-veins, becomes progressively easier.

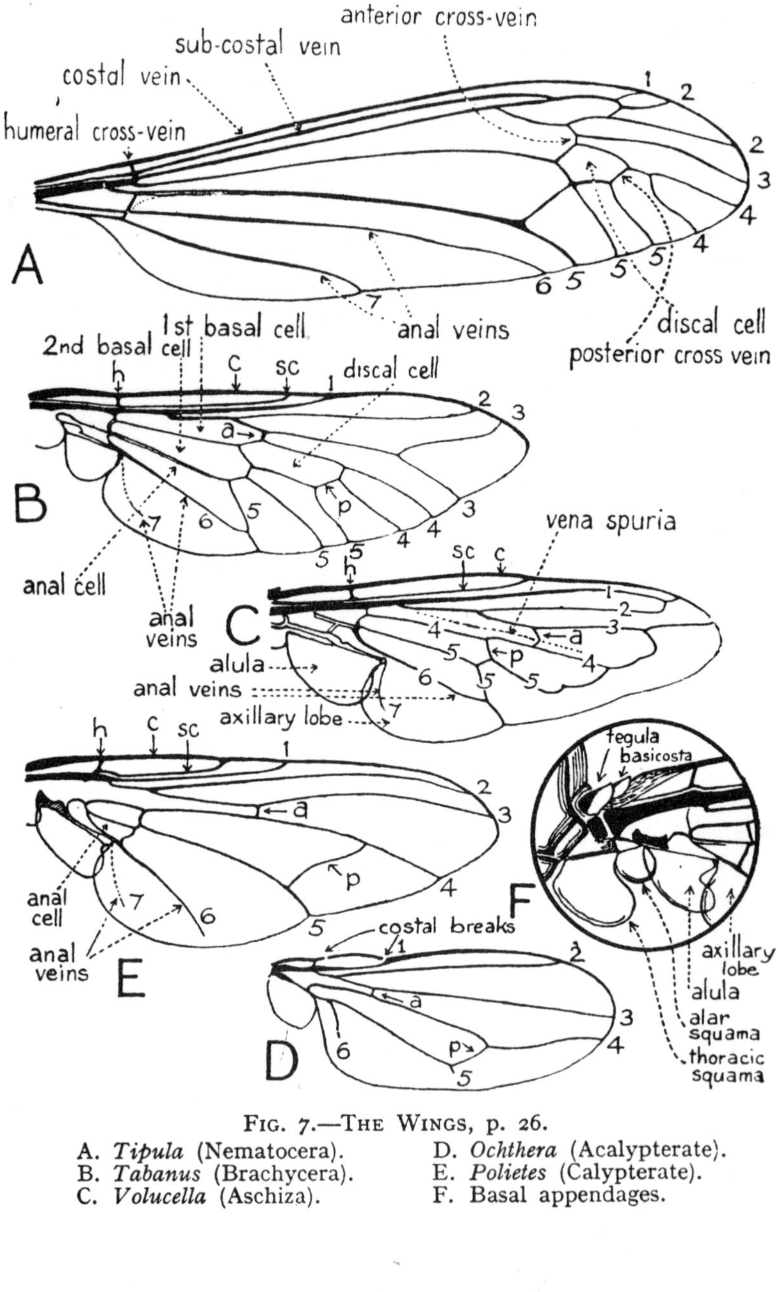

FIG. 7.—THE WINGS, p. 26.

A. *Tipula* (Nematocera).
B. *Tabanus* (Brachycera).
C. *Volucella* (Aschiza).
D. *Ochthera* (Acalypterate).
E. *Polietes* (Calypterate).
F. Basal appendages.

Sometimes the sub-costa may be absent or so weak as to be almost invisible ; it may appear from the dorsal view, which is the normal aspect envisaged by the system, to be almost fused with Vein 1. Its presence, in these doubtful cases, is frequently more easily detected by regarding the wing from below against the light. Either Vein 6 or 7, or both, may be vestigial or absent, as may also be the case with Vein 2 ; but once the location of Veins 3, 4 and 5 has been established, the diagnosis of such cases becomes more simple.

The humeral cross-vein (*h*) lies between the costa and sub-costa, near the base of the wing ; other cross-veins occur in various wings, particularly the primitive forms, and are accorded special names, with which we need not concern ourselves unless a special study of a particular group is to be undertaken.

Areas bounded by the long veins, cross-veins or the wing-margin are called cells and are accorded names by various systems. If a cell is entirely bounded by veins, it is said to be " closed ", e.g. the " discal " cell (Fig. 7A, B) ; a closed cell which is terminated apically before the wing-margin by the fusion of two long veins is said to be " stalked " (Fig. 7B). The particular cell instanced, which is bounded by Veins 5 and 6, is called the anal cell ; the two cells preceding it, near the base of the wing and bounded by Veins 3–4 and 4–5 respectively, are known as 1st and 2nd basal cells.

Often an incision is present in the hind margin of the wing near the base (Fig. 7B–F) and separates an area of membrane, called the alula ; the area between it and Vein 6 is sometimes designated the axillary lobe since it bears upon it the axillary vein (Vein 7) or its vestiges. On the other side of the alula, at the base of the wing, is the alar squama, or upper calypter ; the lower calypter is attached to the thorax and is consequently known as the thoracic squama.

In some wing-plans, there are to be seen one or two interruptions in the costal vein, at or near the humeral cross-vein, or the apex of *sc* or Vein 1, or both (Fig. 7D). These "breaks" are of much assistance in identification, but require considerable care in their interpretation.

THE LEGS. The general structure of the legs may be appreciated by reference to Fig. 8. Attached to the thorax is the basal segment or coxa; next comes a small one known as the trochanter, followed by the femur, or "thigh", which is usually robust and frequently dilated; the tibia, or "shin", which sometimes possesses apical "spurs", comes next, bearing at its apex the five-segmented tarsus, or "foot". Of these five segments, paradoxically, the first is often called the metatarsus and is generally the largest; at the apex of the fifth, or last segment, are to be found the claws and when present, the pad-like pulvilli, of which there may be two or three (Fig. 8B, C). If there are three, the middle one is properly known as an empodium; sometimes this is merely in the form of a bristle or hair.

Various modifications and adornments of the leg-segments occur in the males of some species and are thought to have some significance as regards courtship; some very efficient-looking modifications occur in both sexes of certain predatory flies and are obviously adapted for the purpose of securing and retaining prey.

References to the bristle-arrangement of the legs, which is of considerable importance in identification, may prove a little confusing at first. Longitudinally, descriptions of the location of individual bristles or groups of bristles present little difficulty. Where the segment is long enough to permit differentiation, e.g. femur or tibia, one may find basal, subbasal, mid, preapical or apical bristles (Fig. 8A). In the interpretation of descriptions of the surface upon which a bristle stands, however, the student may experience some doubt, since what might be the dorsal surface of a segment when the leg is in one position may cease to be such if the position of the leg were substantially different. Grimshaw's

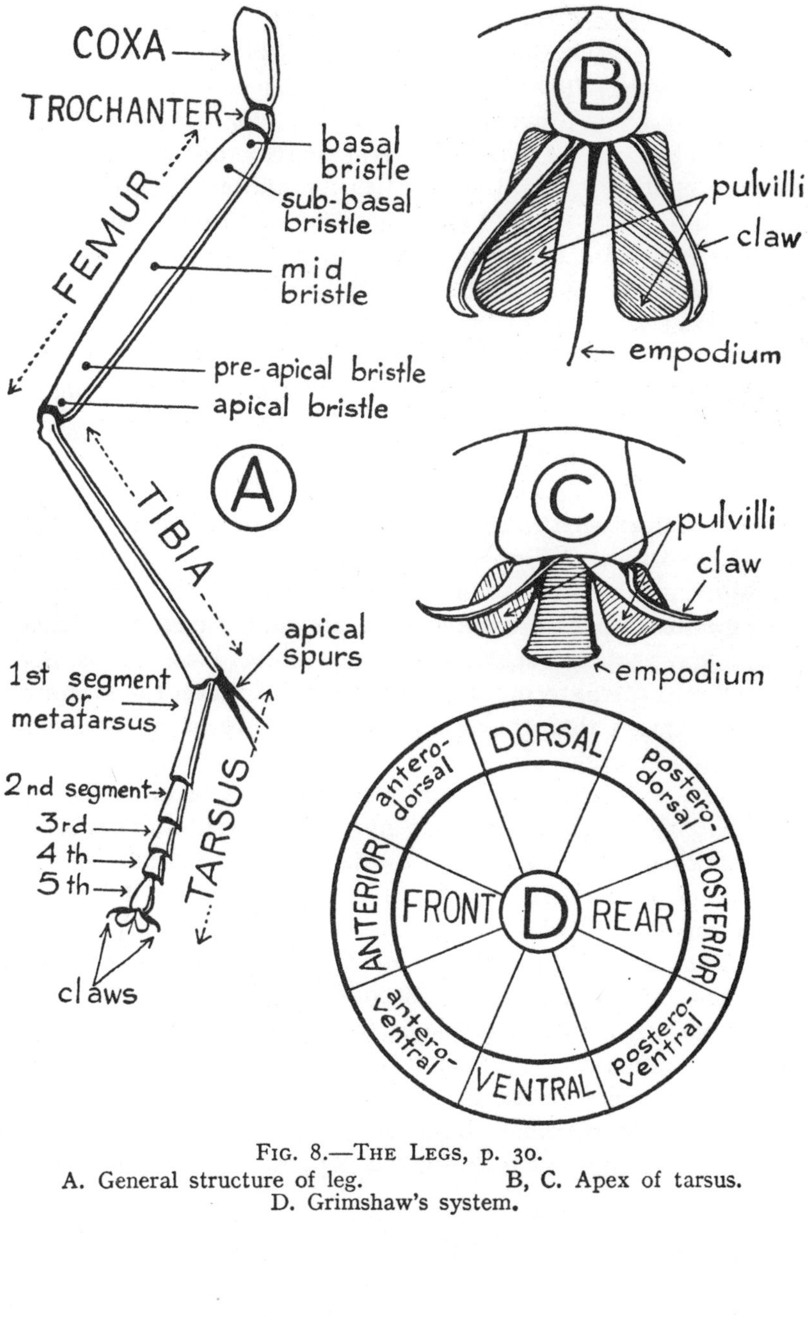

FIG. 8.—THE LEGS, p. 30.
A. General structure of leg. B, C. Apex of tarsus.
D. Grimshaw's system.

system regularises the matter by referring to the surfaces of the segments as if the whole leg were stretched horizontally from the body. The periphery of a transverse section of any segment is then divided into eight imaginary parts, which, reading clockwise, are called dorsal, postero-dorsal, posterior, postero-ventral, ventral, antero-ventral, anterior and antero-dorsal (Fig. 8D). It will now be seen that the position of leg bristles may be accurately described. Thus, a reference such as: " hind tibia with mid antero-dorsal bristle ", would leave little room for doubt. The presence or absence of a dorsal preapical bristle on all or any of the tibiae is a character much used in differentiating the Acalypterate families. The phrase " tibia with preapical bristle " may refer to the fact that such a bristle is present on at least one of the tibiae, not necessarily all; the phrase " tibia without preapical bristle ", however, indicates that such a bristle is absent from all of the tibiae.

THE ABDOMEN. Figs. 1–3 indicate that a variable number of segments may be visible from a dorsal aspect; as might be expected, there are fewer in more highly-developed flies. Generally, each segment consists of a dorsal plate, or tergite, and a ventral, or sternite; in some advanced flies, the sternites are so reduced that the tergites curve round on each side to the ventral surface and may even, in some cases, conceal the sternite. The first tergite may bear upon its surface a hollow or excavation adjusted to the contour of the scutellum (Fig. 3).

Bristles of the abdomen are named by reference to the particular segment on which they occur and to their location upon the surface of that segment; thus, lateral bristles are found on the sides, marginals on the fore or hind (generally the hind) margins, and discals on the disc or central area.

THE GENITALIA. When considering the eyes, mouthparts and legs, allusion was made to certain differences which may occur between the sexes. Whilst these secondary sexual characters are of much value in determining sex (and, very often, the identity of species), much more general use is made

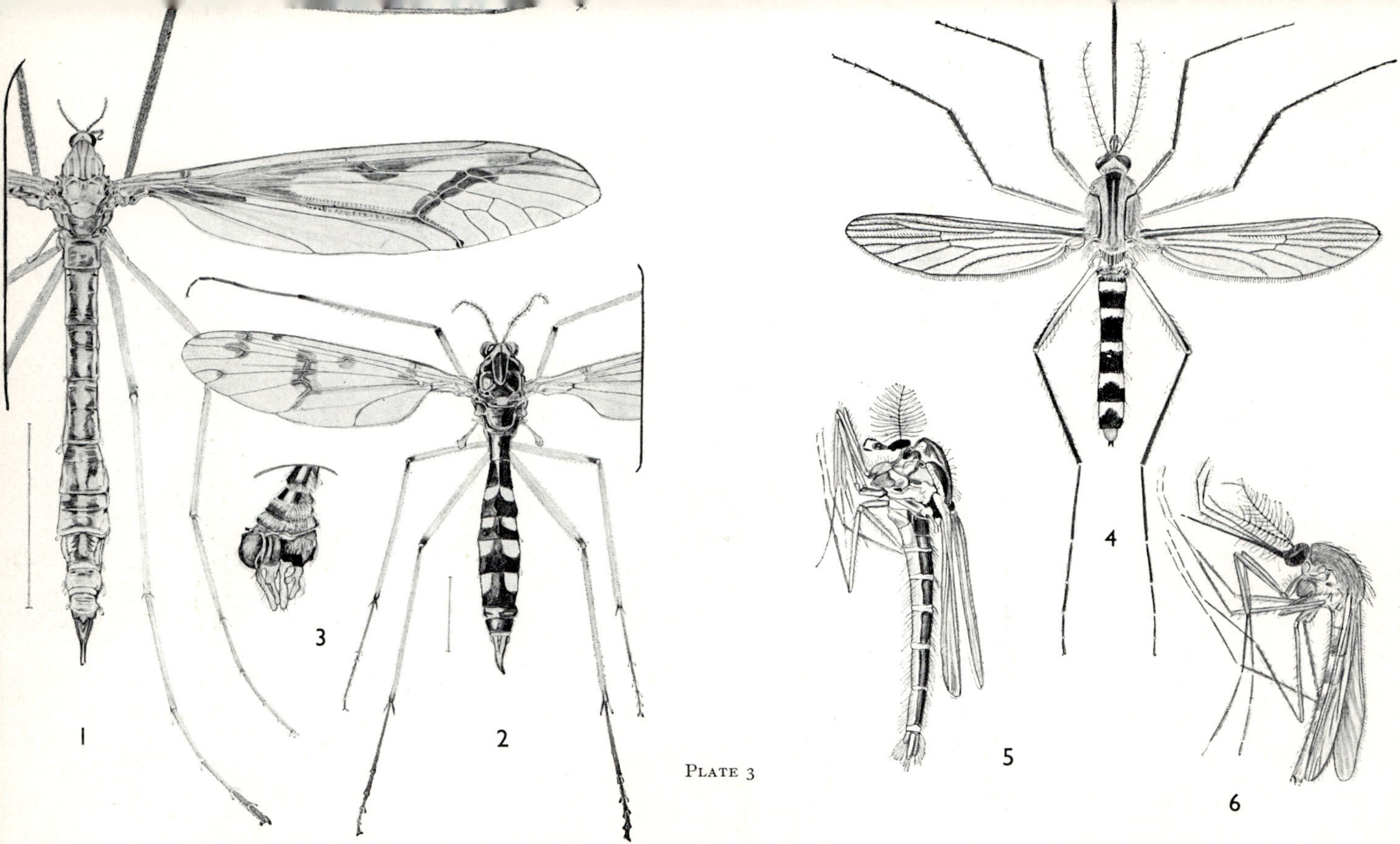

PLATE 3

TIPULIDAE-IV

1. *Pedicia rivosa* L., ♀, p. 56.

PTYCHOPTERIDAE-II

2. *Ptychoptera contaminata* L., ♀., p. 58. 3. *Ptychoptera* sp., ♂ genitalia, p. 58.

CULICIDAE-I

4. *Aëdes punctor* Kirby, ♀, p. 63. 5. *Chaoborus crystallinus* Deg., ♂, p. 66. 6. *Culex pipiens* L., ♂, p. 62.

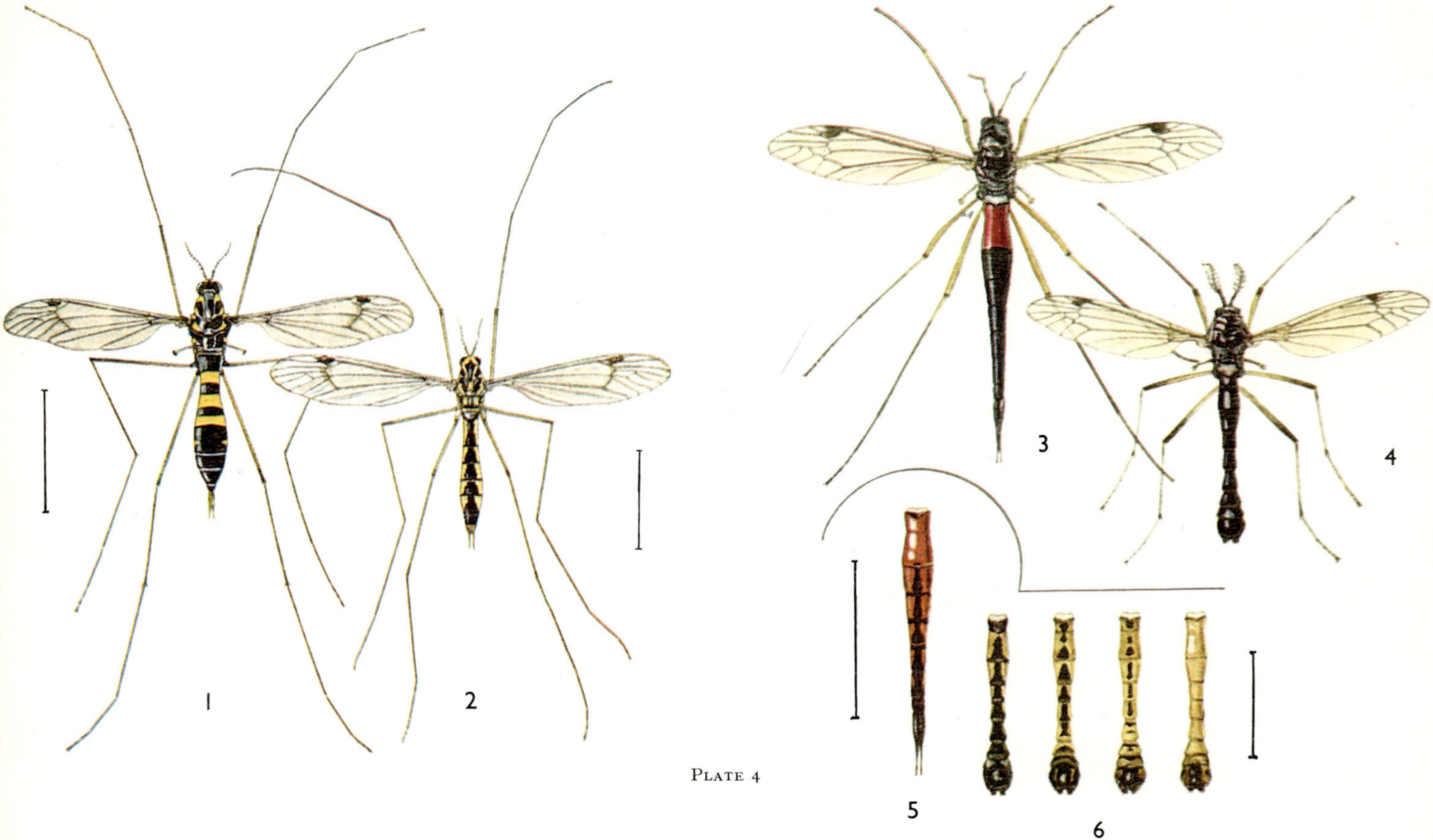

PLATE 4

TIPULIDAE-V

1. *Nephrotoma crocata* L., ♀, p. 49.
2. *Nephrotoma quadrifaria* Mg., ♀, p. 49.

3. *Tanyptera atrata* L., ♀, p. 53. 4. *Tanyptera atrata* L., ♂. 5, 6. Abdominal colour forms.

of characters provided by the genitalia. These organs are located at the hind end of the abdomen. In the male, the principal parts with which we shall be concerned are various paired appendages, which are often exposed; an important function of these is to clasp the female during copulation. The shape of these structures is modified in many ways and often provides the only satisfactory means of separating otherwise indistinguishable species. The externally-visible organs of the female consist of an aperture from which the eggs pass; this may be on the surface of the abdomen or at the end of a tubular egg-laying apparatus, or ovipositor, which is sometimes strengthened and adapted for piercing and sometimes pliant and capable of being protruded or retracted. A pair of small appendages, known as "cerci" may be visible, which are probably tactile in function.

Fig. 9 depicts various forms of externally-visible genitalia of both sexes. In some species the genitalia are sufficiently visible without preparation or may be easily withdrawn when mounting, but in others it will be found necessary to go to the trouble of dissecting and preparing microscopic mounts. Some hints on simple techniques for the latter, if found necessary, are given in the Appendix at pp. 340–2.

For further reading on the structure and physiology of adult flies, the student may consult Imms [3] and other works cited by that author. Tuxen[5] deals with genitalia.

REPRODUCTION AND METAMORPHOSES. Whilst there are, in a few cases, forms of asexual reproduction in Diptera, to which reference will be made later, most species reproduce themselves sexually by the adult-egg-larva-pupa cycle. Some, however, are viviparous, that is to say, the females deposit living larvae, which, in certain cases, are so far developed that they immediately pupate; the earlier stages being passed within the parental body.

Egg-forms are often very beautiful and interesting, as in other Orders, on account of their external sculpture and

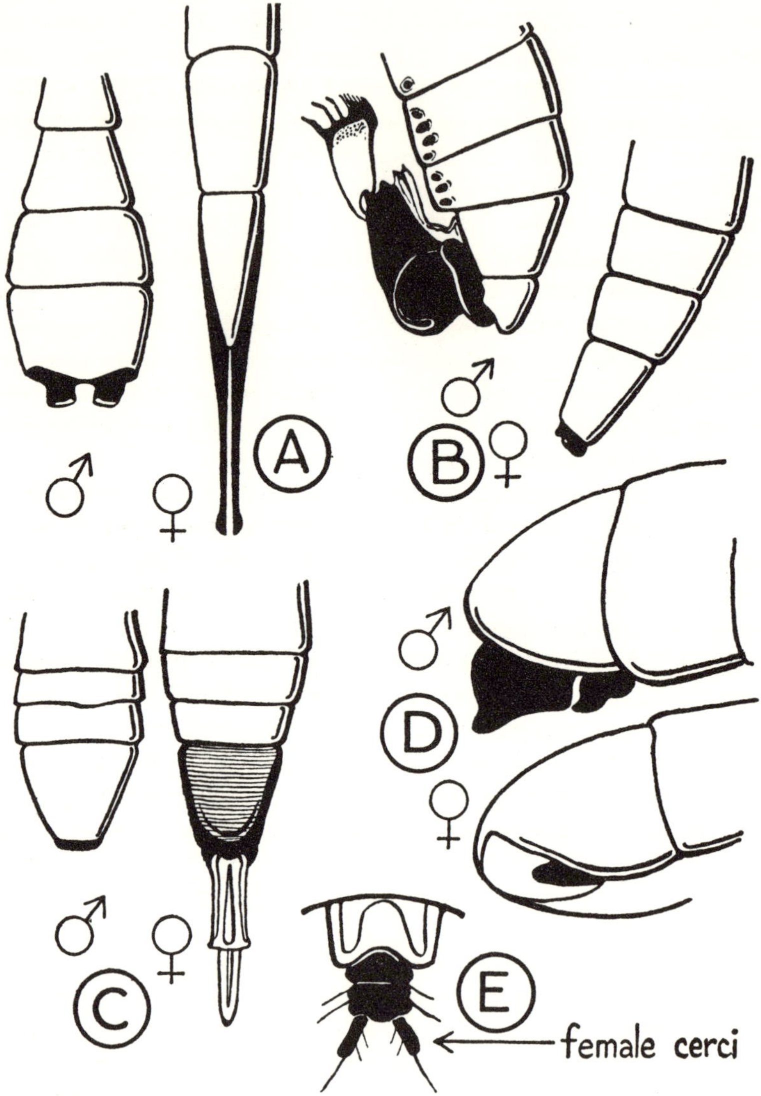

FIG. 9.—SOME FORMS OF EXTERNALLY-VISIBLE GENITALIA, p. 33.

A. Tipulidae (*Tanyptera*).
B. Dolichopodidae (*Dolichopus*).
C. Tephritidae (*Xyphosia*).
D. Tachinidae (*Servillia*).
E. Phoridae (*Triphleba*).

peculiar modifications, adapted to meet the needs imposed by their environment.

THE LARVA. Primitive types of Dipterous larvae exhibit much variation in form, especially those which lead an aquatic existence; external features such as specialised respiratory apparatus are frequently in evidence (Fig. 10B–D). Some very primitive types found among Nematocera have a well-developed head, with biting mouthparts (Fig. 10A); there is a gradual transition through Brachycera (Fig. 10E) and onwards to the typical maggot of the Cyclorrhapha (Fig. 10F), which is virtually headless, tapering anteriorly and truncate posteriorly. Only vestigial mouthparts are present, the main feature being a pair of hooks; the body is devoid of noticeable external appendages in most cases.

Respiration is normally effected through a pair of spiracles near the head and another more prominent pair opening at or close to the apex of the terminal segment (Fig. 10A); in aquatic forms, ingenious provisions for the exclusion of water from the spiracles are found.

The larval skin is cast two or three times, so that three or four larval stages are passed before pupation.

THE PUPA. Important characteristics are displayed by the pupae. In the group Orthorrhapha, already alluded to, the adult fly, or imago, escapes through a longitudinal split at one end of the puparium, or pupal integument. The outward form in Nematocera is usually marked by various external appendages, such as the respiratory apparatus and the sheaths of the developing wings (Fig. 11A–D), which are continued to some extent in Brachycerous pupae; backwardly-directed spines on the segments and hind extremity are also very characteristic of Brachycera (Fig. 11E). In Cyclorrhapha (Fig. 11F) emergence of the adult is effected through a more or less circular orifice, a piece of the puparium corresponding to it being forced off; in Schizophora, this is accomplished by the inflation of a sac on the fly's face; this instrument is also

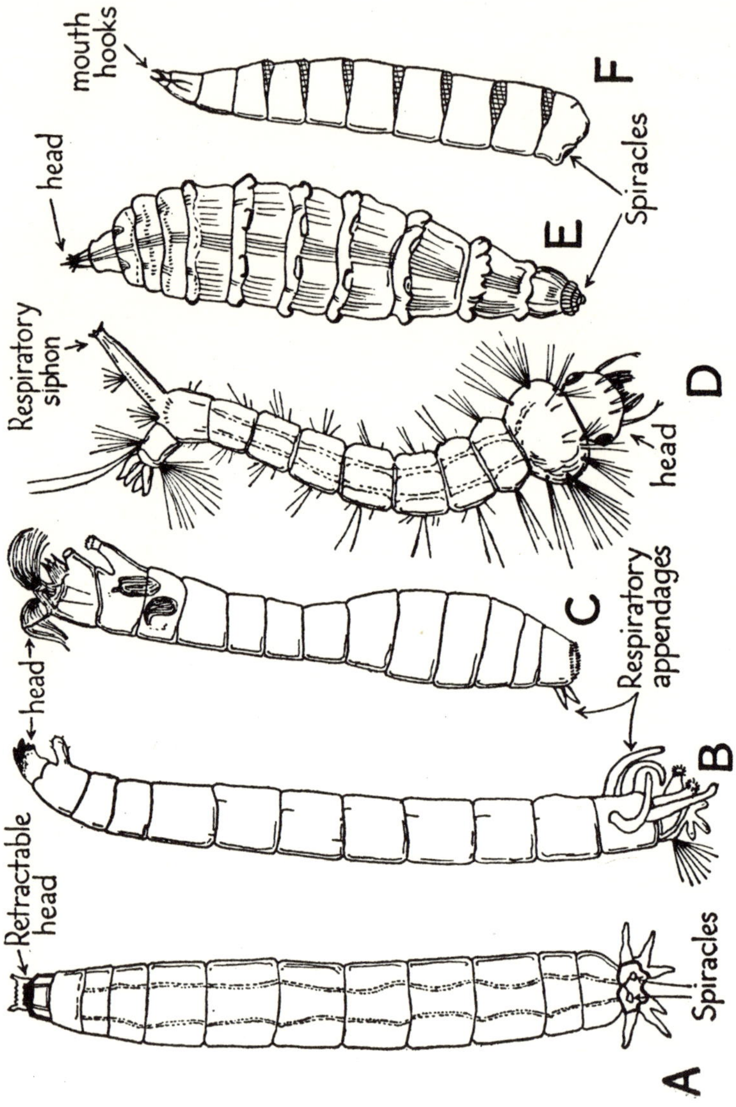

FIG. 10.—DIPTEROUS LARVAE, p. 35.

A. *Tipula* (Nematocera).
B. *Chironomus* (Nematocera).
C. *Simulium* (Nematocera).
D. *Culex* (Nematocera).
E. *Tabanus* (Brachycera).
F. Typical Cyclorrhaphous larva.

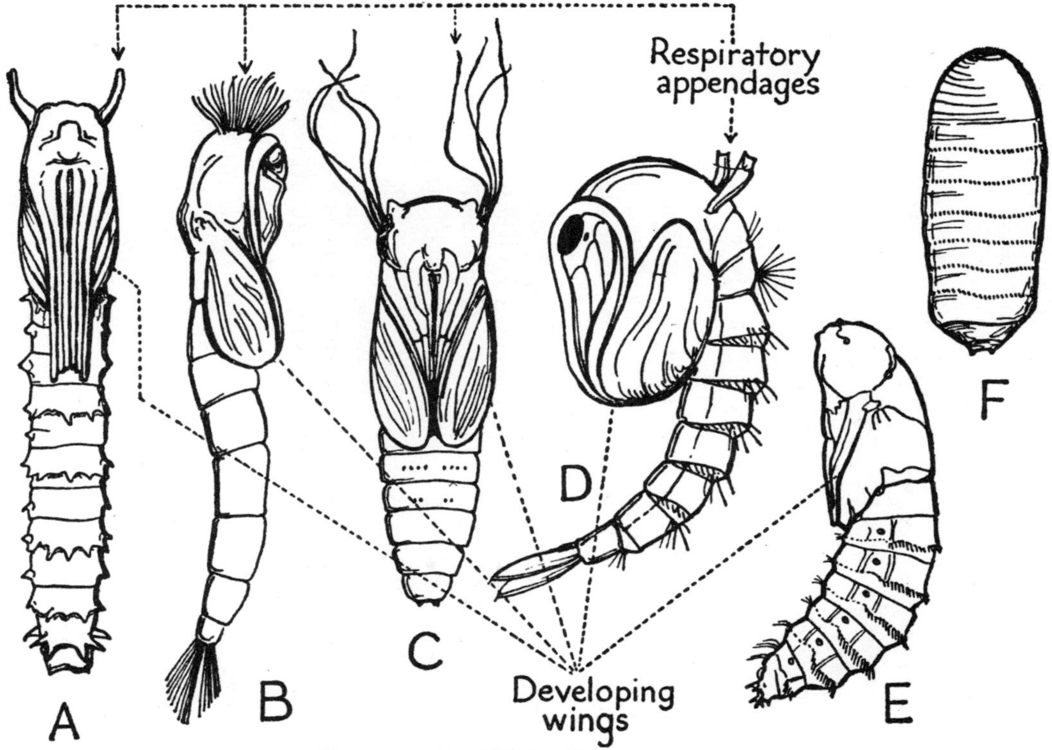

FIG. 11.—DIPTEROUS PUPAE, p. 35.

A. *Tipula* (Nematocera).
B. *Chironomus* (Nematocera).
C. *Simulium* (Nematocera).
D. *Culex* (Nematocera).
E. *Tabanus* (Brachycera).
F. Typical Cyclorrhaphous pupa.

used, in species which pupate in the earth, for the purpose of gaining access to the open air ; to permit the inflation of the sac, the face falls forward as a sort of hinged flap (cf. Pl. **46,** 7). The sac, or ptilinum, is deflated once its function has been fulfilled and shrinks to leave the ptilinal suture, to which reference has already been made (Fig. 4B). Cyclorrhaphous pupae are more or less ovoid in shape and without external adornment (Fig. 11F), the puparium consisting of the skin of the last-stage larva. Hennig[6] deals with immature stages.

LIFE-HISTORIES AND HABITATS. Flies have adapted themselves to living in most places where life can be supported ; the larva of one foreign species even succeeds in maintaining existence in pools of crude petroleum ! Some flies, in the immature stages, are completely aquatic ; some have adapted themselves to semi-liquid media such as mud, wet vegetable matter or animal refuse ; while others live in the earth, in the tissues of plants, in rotten wood, or are internal parasites of mammals, insects or spiders ; others act as scavengers in the nests of insects and are tolerated as guests. In the adult stage, some very specialised forms of bloodsucking flies spend their existence on the bodies of mammal or bird hosts, while others are free-flying and attack various mammals, or are predatory upon insects ; certain flies are said to attach themselves to predatory Diptera for transport and to join in feasting upon the juices exuding from the wounded prey. Non-bloodsucking flies may be nectar-feeders, or subsist upon liquid or semi-liquid products of decaying vegetable or animal matter, excrement, or sap flowing from tree-wounds.

THE RÔLE OF THE FLIES. A little study in the field will cause the observer, should he not already be conscious of the fact, to realise the existence of enormous numbers of individual flies, apart from the vast array of different species. A comparatively small area systematically worked with a sweep-net will sometimes yield a prodigious haul. Observation of flowers and comparative counts of the kinds of insects visiting them over a given period will lead to the conclusion that flies play a very prominent part in pollination. Another very significant function which they fulfil is that of scavengers; carrion, excrement and various deposits of decaying animal or vegetable matter all receive the attention of flies, whose larvae assist materially in their disposal.

From the medical standpoint, flies cause harm to Man as vectors of malaria, typhoid, yellow fever, "sand-fly fever" and other diseases; conversely, the larvae of certain species have been successfully used in the treatment of wounds (cf. *Lucilia*, p. 284). From the economic standpoint, Man is adversely affected by agricultural pests, such as the Hessian Fly, the Fruit Flies, the Cabbage Root Fly, the Narcissus Flies and many others, which destroy cereals, fruits, vegetables and flower bulbs; also by the Gad Flies, Warble Flies and Bot Flies, which attack livestock. On the other hand, the larvae of various Hover Flies consume large numbers of destructive Plant Lice (Aphids), while the predatory flies also help, in some measure, to control insect pests.

CLASSIFICATION. Fig. 12 gives, in diagrammatic form, an outline of the classification-structure of Diptera. A fly is usually alluded to by two names, a generic, which denotes the genus to which it belongs and a specific, or trivial, the name of the species. These are followed by the name of the author who originated the specific name, thus: *Tipula oleracea* Linnaeus. Generic and specific names are customarily printed in italics, the former commencing with a capital letter; the author's name is in roman type and is often abbreviated if such name is well-known and in frequent use, thus: *Tipula oleracea* L. Where varieties of a species have been accorded names, they are expressed thus: *Rhagio tringaria* L. v. (or var.) *nigriventris* Loew. Males are denoted by the sign ♂ and females by ♀.

After a careful study of the plates, the reader will almost certainly remark the striking superficial resemblance that

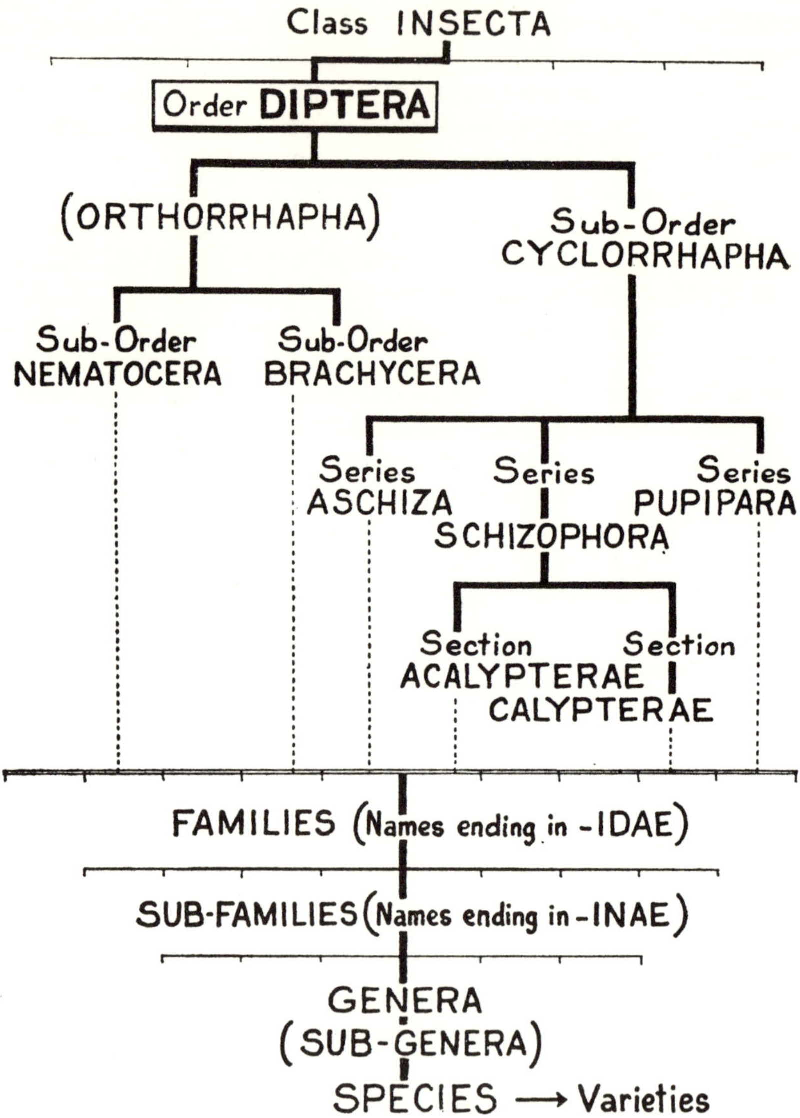

FIG. 12.—CLASSIFICATION OF DIPTERA, p. 39.

some flies bear to others in far-removed families. The groupings of early authors were often based on superficial resemblances of outward form and colour; modern research on fossil forms, the metamorphoses and other aspects has demonstrated that such groupings were, in many cases, quite unnatural. Colour is an extremely variable character, even in live specimens, and may change considerably after death. Size of the adult is also subject to much variation and is largely dependent upon degree of nutrition and type of environment in the larval stages. When sizes of adults are quoted, without any indication to the contrary, average or typical body-length is inferred. This is usually expressed in millimetres and taken to be the distance between the front of the head and the tip of the abdomen, from a dorsal aspect. It is, of course, assumed that the body is not abnormally contorted; features such as the antennae and proboscis, which might be subject to variation in position in a dead specimen, are excluded; abnormally-extruded genitalia would likewise be discounted. In the plates, where the life-size of the specimen illustrated is indicated by adjacent lines, no mention of size is made in the text.

The arrangement in the following chapters is broadly based on that of Kloet and Hincks [4] in their check list of British insects, hereafter referred to as the Check List. Nomenclature adopted therein, the validity of which is in certain cases disputed by other authorities, has been followed (except in the family Syrphidae *), not necessarily because the authors are in agreement, but in order to provide the reader who is desirous of so doing, with a means of quickly ascertaining the more important alternative names for a species, genus, family, etc., which have been considered invalid for various

*For this family, the names used by Verrall, in "British Flies", VIII, (1901), have been followed. A completely revised edition of the Check List is in preparation and the nomenclature to be used in this work has been used in all families where recent research renders this desirable. Some of this research is already published in the American catalogue[7].

reasons, e.g. when it is found that a species has already been described by another author. Unless the student is aware of these alternatives (synonyms), some of the literature relating to a particular species may be missed. Certain names shown in the Check List as synonyms have been used for many years and are familiar to dipterists ; these are, in appropriate places in the following pages, quoted in brackets, thus: Micropezidae (=Tylidae), Rhagionidae(=Leptidae). This should not be confused with an indication of a sub-genus in brackets, which is quoted thus: *Erioptera* (*Symplecta*) *stictica* Mg. The numbers of genera and species in a family, quoted in the summaries, are revised up to December 1966 and are intended to convey an idea of the relative size and complexity of each family.

LITERATURE CITED

1. Oldroyd, H. Handbooks for the Identification of British Insects. Vol. IX. Pt. 1. Diptera. Introductory. Royal Entomological Society of London. 1949.
2. Longfield, C. The Dragonflies of the British Isles (2nd Ed.). Warne, London. 1949.
3. Imms, A. D. A General Textbook of Entomology (9th Ed.). Methuen, London. 1957. (Revised by O. W. Richards and R. G. Davies).
4. Kloet, G. S., and Hincks, W. D. A Check List of British Insects. Stockport. 1945. (Under revision by Royal Entomological Society of London).
5. Tuxen, S. L. Taxonomist's Glossary of genitalia in Insects. Ejnar Munksgaard, Copenhagen. 1956.
6. Hennig, W. Die Larvenformen der Dipteren. I-III. Akademie-Verlag, Berlin. 1948–52.
7. Catalogue of the Diptera of America N. of Mexico. U.S. Dept. Agriculture, Washington. 1965.

OTHER USEFUL WORKS

Grimshaw, P. H. A Guide to the Literature of British Diptera. *Proc. roy. Phys. Soc. Edin.* **20**: 78–117. 1917.

Grimshaw, P. H. Introduction to the Study of Diptera, with a Key for the Identification of Families. *Proc. roy. Phys. Soc. Edin.* **22**: 187–215. 1934.

Oldroyd, H. The Natural History of Flies. Weidenfeld & Nicholson, London. 1964.
Brindle, A. Terrestrial Diptera larvae. *Ent. Rec.* **75**: 47–62. 1963.
Clausen, C. P. Entomophagous Insects. McGraw-Hill, New York. 1940.
Smart, J. Bibliography of Key Works for the Identification of the British Fauna and Flora. Adlard & Son, Dorking. 1967.

KEY TO SECTIONS

ILLUSTRATIONS are cited in the keys to assist in resolving doubts (see pp. 12–13) and do not always relate to the characters used; determinations should be checked with the family summaries on the pages indicated. Having decided to which of the following groups the specimen belongs, turn to the separate key on the page referred to after the name of the section, in order to determine the family. If the specimen does not run down satisfactorily or check with the family summary, the steps should be carefully retraced and alternatives in the couplets explored. Sections II and III are grouped together for convenience.

1. Abnormal[1] flies, often leathery and flattened, parasitic on warm-blooded animals; head closely joined to thorax, fitting into an excavation or capable of being folded back into a groove on dorsum; wings may be of odd shape, reduced, or absent; legs and claws peculiarly developed for clinging to hairs or feathers of hosts (Pl. **44**, 2: **50**)
Section VI. Pupipara, p. 313

Normal flies, not as above 2

2. Antennae usually long and thin, always consisting of not less than six free (i.e. not fused) and more or less uniform segments (Fig. 1); palpi with several visible segments, drooping (Pl. **2**, 14, 15) .... **Section I. Nematocera**, p. 89

Antennae with less than six free segments; terminal segment enlarged and may show traces of fusion of primitive segments (Figs. 2 and 3); palpi usually with only one visible segment, porrect (Fig. 4) 3

[1] But see *Braula*, p. 242, which is included in the Acalypterates (p. 244).

3. Ptilinal suture present (Fig. 4B)[1] 4
Ptilinal suture absent (Fig. 4A)
Sections II and III.
Brachycera and Aschiza, p. 175

4. A cleft or fissure in second antennal segment[2]; transverse suture of dorsum continuous; posterior calli differentiated from rest of dorsum; squamae often conceal halteres[3] (Fig. 5) . . . . **Section V. Calypterates**, p. 311

No cleft or fissure in second antennal segment; transverse suture of dorsum not visible in middle; posterior calli not differentiated; squamae do not usually conceal halteres (Fig. 5) . . . . **Section IV. Acalypterates**, p. 244

[1] To appreciate this character, refer to p. 277 and Pl. **46**, **7**, 8.
[2] Except *Gasterophilus*, Gasterophilidae, p. 311, and Pl. **44**, **1**.
[3] By no means invariably; the halteres are exposed, for instance, in Scatophagidae and the sub-families Anthomyiinae, Fucellinae and Coenosiinae (Muscidae).

CHAPTER I

SECTION I. NEMATOCERA. THREAD-HORNS

TIPULIDAE . . . . . . Crane Flies.
TRICHOCERIDAE . . . . . Winter Gnats.
PTYCHOPTERIDAE.
ANISOPODIDAE (= RHYPHIDAE). Window midges.

TIPULIDAE.

37 genera. Nearly 300 species. Slender-bodied, narrow-winged flies with long, delicate legs which break off easily; species varying greatly in size; **V**-shaped suture on dorsum; wing-venation complex; Veins 6 and 7 long, reaching margin; ocelli absent; genitalia prominent in both sexes. (Pls. **1**–**4**.)

"LEATHERJACKETS", the tough-skinned grey or greyish-brown grubs of the Crane Flies or "Daddy-long-legs", are well known to farmers, gardeners and groundsmen. These destructive pests belong to the primitive type of Dipterous larvae which are equipped with biting mouthparts. They usually feed during the day just under the surface of the soil and attack the roots and lower stems of a wide range of plants, including cereals, grasses, seedling crops of various kinds and even seedling forest trees. Often, when the weather is warm and damp, they will feed at night above ground, cutting through the stems at surface level. Advantage of this propensity is frequently taken by groundsmen, who thoroughly water their greens, spread tarpaulins, thus encouraging the grubs to come to the surface, and gather them in large numbers for destruction. Certain species are present in varying force in most farm land, especially pastures and meadows, which seem to be their natural haunt. When they are in considerable numbers in cornfields and grassland, their presence is betrayed by the appearance of large, bare patches where they have destroyed the plants, or in gardens by the wilting, collapse or total disappearance, on the surface, of large numbers of seedlings.

On farm land the most frequent offenders seem to be the larvae of *Tipula paludosa* Mg., a very common drab-coloured species, about 25 mm., while those of *Nephrotoma maculosa* Mg., a yellow and black insect, about 15–20 mm., of which we shall have more to say later, are addicted mainly to garden produce. These species belong to the sub-family Tipulinae, of which there are nearly 90 British species, mostly large. Tipulines differ from the members of the other sub-families, which are collectively known as the short-palped Crane Flies (Pl. **2,** 14), in having the palpi long, with apical segment much longer than the preceding one (Pl. **2,** 15). Another difference may be noted in the sub-costa, which in the Tipulinae ends in Vein 1 (Pl. **2,** 10), whereas, in the short-palps, it finishes, albeit sometimes indistinctly, in the costa (Pl. **2,** 11).

The life-history of *paludosa* may be considered as representative of the species which spend their immature stages in the earth. When laying their eggs, the females insert their long, pointed ovipositors almost vertically into the ground, the body suspended, as it were, in a cradle formed by the legs. Then the abdomen is revolved back and forth on its longitudinal axis until the ovipositor has penetrated sufficiently for the fly's requirements. Each female may lay several hundred eggs from about June onwards; they are long-oval shaped, roughly 1 mm. in length, and look very much like little black seeds. After a fortnight or so they hatch, the resulting tiny grey grubs going their ways in search of food. Since they require a cool, moist environment at this stage of their career, dry conditions arising from prolonged sunny periods and warm winds may kill off large numbers. Some are consumed by rooks, gulls, lapwings and other birds, while parasitic worms, fungi and insects, including flies (cf. *Siphona*,

p. 263), also take their toll. Enough usually survive, however, to constitute a factor to be reckoned with, feeding through the autumn, winter and spring to reach, when fully grown, as much as 40 mm. in length. Pupation takes place in late spring or early summer, the adult flies emerging about a fortnight later.

In shape, the larvae are elongated-cylindrical, a little tapered anteriorly and truncate at the rear (Fig. 10A). The body, although enclosed in a tough, corrugated skin, is capable of considerable contraction and the head can be withdrawn into it. Distinct mandibles and antennae are present. The rear aspect of the terminal segment reveals the two hind spiracles surrounded by fleshy, pointed processes ; in *paludosa* there are six, but the number may vary in other Tipulid larvae, while the shape of the processes appears to be related to the type of habitat favoured by the larva. Generally, the lobes are blunt in earth or wood-dwelling species (cf. Pl. **2**, 2) and attenuated in semi-aquatic or aquatic species (cf. Pl. **2**, 1).

The pupae bear a number of external appendages which give an indication of the developing wings and legs (Fig. 11A). There are two horn-like projections at the anterior end, which have a respiratory function ; downwardly-directed spines on the segments and hind extremity enable the pupa to work its way up through the earth so that the anterior part protrudes sufficiently for the fly to escape. This it does through a longitudinal rupture on the dorsal surface ; the process of emergence will be considered in some detail a little later (cf. *Dictenidia*, p. 50).

TIPULINAE.

Tipula is a large genus which includes species whose immature stages are spent under different circumstances. In contrast to the undistinguished-looking *paludosa*, *T. maxima* Poda (Pl. **1**) is not only a handsome insect, but has the added distinction of a larger wing-expanse, at about 65 mm., than any other British fly, and a body-length of about 32 mm. The larva is semi-aquatic, frequenting saturated margins of streams and ponds, or actually living under water near the edge. Pupation takes place ashore, the larva leaving the water to seek a drier location in which to bury itself. It will be interesting to examine the arrangement by means of which aquatic or semi-aquatic Tipulid larvae contrive to remain submerged without water entering the spiracles. Comparison of Pl. **2,** 1 and **2,** 2 shows that the short lobes of the terrestrial type are replaced by long finger-like structures in the aquatic form; these surround the spiracles, which are recessed. The larva, when submerged, avoids drowning by folding in the lobes, thus sealing off the orifices. When sufficiently near the surface, the tail end is allowed to ascend at intervals and the lobes to open, forming a buoyant float with the spiracles exposed to the air. The lobes are often referred to as anal gills; they assist in respiration beneath the surface by the absorption of oxygen dissolved in the water.

An example of another type of larval habitat is provided by *T. irrorata* Macq. This species, in the adult stage, is about 13 mm. long, with a bluish-grey bloom on the thorax and very prominent male genitalia; the immature stages are associated with the bark, decaying wood or mossy coverings of fallen trees. We have found almost fully-developed larvae in early May in cavities in prostrate trunks, where they were submerged in rain-water containing dead leaves, earth and fragments of twigs. Some of these larvae were removed with debris and reared in a jar, the water having been drained off and the debris kept moist from day to day with rain-water applied sparingly by means of an ordinary scent-spray. Pupation took place after two or three days and the adults emerged three weeks later.

Most of the adults in this genus, when at rest, carry their wings opened at roughly a right-angle to one another, but in a few cases, they are folded, one superimposed upon the other.

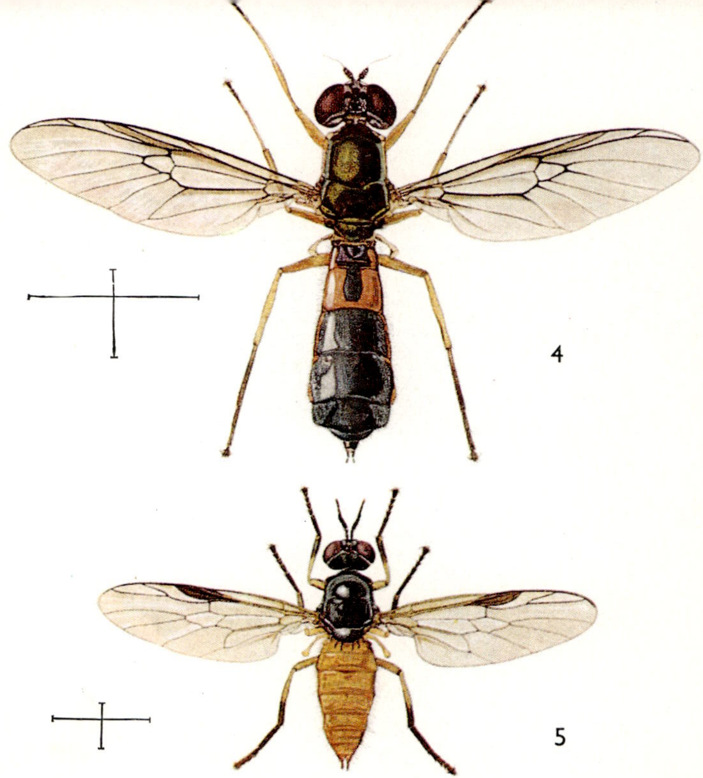

Plate 5

STRATIOMYIDAE-I

1. *Chloromyia formosa* Scop., ♂, p. 91. 2. *Microchrysa polita* L., ♀, p. 92. 3. *Chorisops tibialis* Mg., ♀, p. 93.
4. *Sargus bipunctatus* Scop., ♀, p. 92.
5. *Beris vallata* Forster, ♀, p. 92.

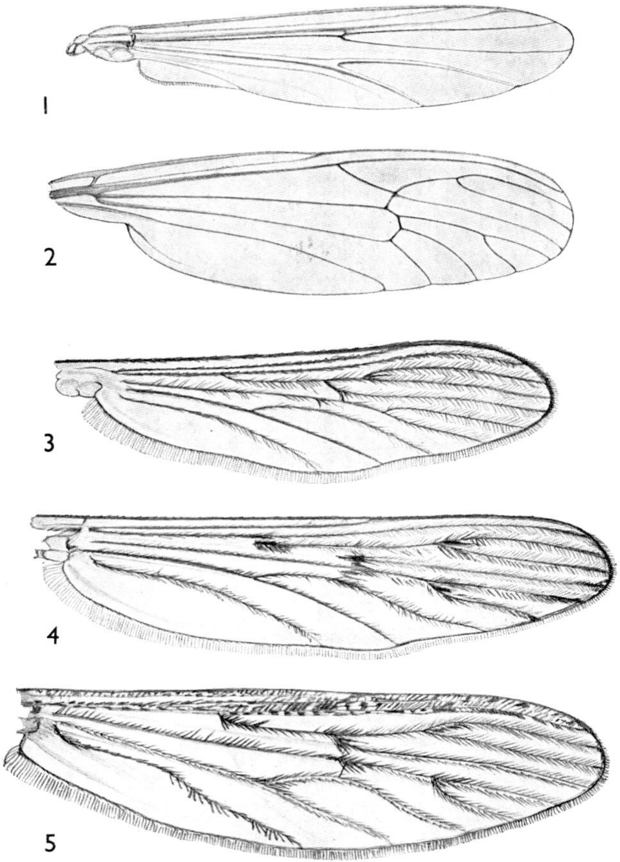

CULICIDAE-II

2. *Dixa* sp., p. 67. 3. *Culex pipiens* L., p. 62. 4. *Anopheles maculipennis* Mg., p. 63. 5. *Culiseta annulata* Schrank, p. 63.

CHIRONOMIDAE-I

PLATE 6

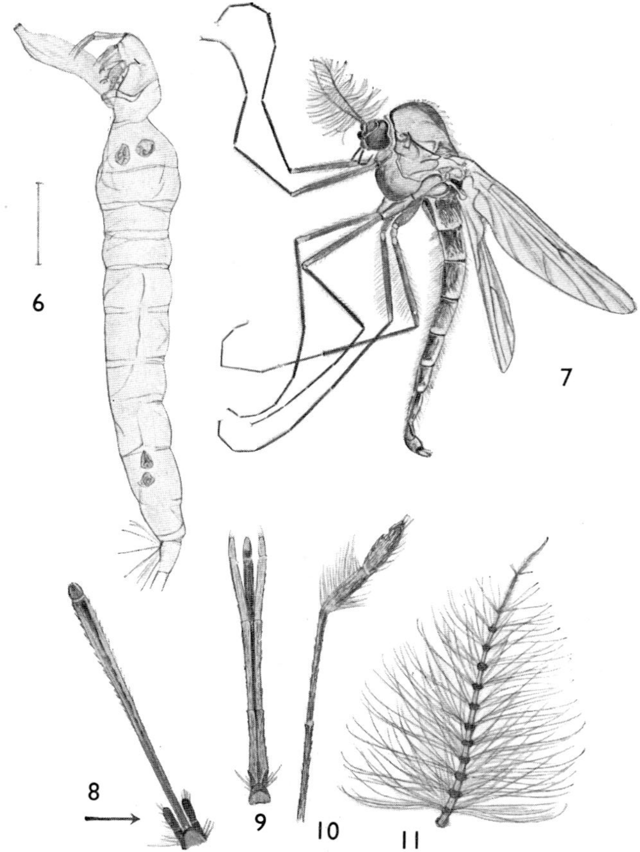

CULICIDAE-III

6. *Chaoborus*, larva, p. 66. 8. Culicine ♀, proboscis and palps, p. 62. 9. Anopheline ♀, proboscis, p. 62. 10. Anopheline ♂, palp, p. 62. 11. *Culex* ♂, antenna, p. 61.

CHIRONOMIDAE-II

There are examples of what is known as sexual dimorphism, or differences in appearance between the sexes; the females of two British species of *Tipula* are subapterous, i.e. with only rudimentary wings, while the males are normal.

Nephrotoma species are yellow and black and may be divided into two main types; one has bold black and yellow bands on the abdomen, while the other has black spots or a black median longitudinal stripe on a yellow abdomen. *N. crocata* L. (Pl. **4,** 1) may be taken as an example of the first type; it provides an elegant contrast to the less colourful species of *Tipula.* Not a great deal appears to be known about its life-history, but the larvae are said to be found in moist vegetable debris, while the adults occur in damp woodland bordering streams or ponds. Representative of the second type is *N. quadrifaria* Mg. (Pl. **4,** 2); it will be noticed that there is a small darkened area, or stigma, near the apex of the wing, below which a smoky marking extends over the cross-veins and along the apical portion of Vein 5. In certain species the stigma is faint and the marking absent, examples being *N. maculosa* Mg. and *N. flavescens* L. All of these three species are common and are characterised by the lateral stripes on the dorsum being curved downward (Pl. **2,** 4). Unfortunately for the gardener, *maculosa,* to which reference has already been made, is the commonest and constitutes a pest of some importance; its life-history is very similar to that of *T. paludosa,* the adults being on the wing from May to September. In *N. scurra* Mg., a larger species of about 22 mm., associated with sandy terrain, the lateral stripes on the dorsum are straight (Pl. **2,** 3) and the stigma faint.

In the next genus we shall touch upon, *Dolichopeza albipes* Stroem. is the only British species and is distinguishable from the rest of the Tipulines by the absence of a discal cell (Pl. **2,** 12); its conspicuous white tarsi will also attract attention. The fly is not uncommon in damp woods in summer; its larvae are said to live, during the day, beneath moss on tree-trunks and in similar situations, whence they emerge at night to feed on the moss. They are green in colour, as also are the pupae, and have five lobes on the spiracular disc. When resting, the adults usually suspend themselves by the front and mid-legs from some such support as the roof of a culvert or underside of a bridge, the hind legs trailing and the wings opened. According to Alexander,[1] this fly and certain other Tipulid species have a habit of resting on spiders' webs.

We come now to three genera which are somewhat sharply demarcated from the remainder of the Tipulinae ; the larvae live in wood and the adults are generally more colourful and robust in form, particularly as regards the legs. The male antennae show a curious modification, the segments bearing excrescences which cause them to appear pectinated, or comb-like ; the nature of these excrescences varies according to genus. In *Dictenidia bimaculata* L., the sole British species, the arrangement consists of two processes on each segment from the fourth onwards, the apical process shorter than the basal (Pl. **2**, 5). The female antennae, however, are quite different (Pl. **2**, 6). A shining black body, in the female with brown abdominal markings, orange legs and wings with two dark apical patches go to make the adult a distinguished-looking insect of about 15 mm. in length. We are able to give an eye-witness account of the emergence of this species. In early May, we discovered some of the larvae in damp, powdery debris inside the rotten trunk of a fallen beech. These were taken home with some of the debris and kept in a jar, the top of which was covered with butter-muslin retained in position with an elastic band. The debris was kept moist with rain-water applied with a scent-spray at intervals of two or three days and the jar kept in a cool, shaded situation. Adult flies began to emerge about three weeks later and one of them was obliging enough to commence its emergence during one of our periodical inspections of the jar ; we were thus privileged to witness the whole process from beginning to end. Our attention was attracted to a miniature upheaval going on apparently just beneath the surface of the debris; this resulted in the formation of a tiny hummock from the sides of which small wood-particles rolled down. The eruption subsided for a few minutes and then recommenced, attended by a slight rise of the hummock and the dislodgement of more particles. After this had been repeated once or twice, a distinct projection from the top of the hummock became visible which, viewed in silhouette, resolved itself into the respiratory horns and upper part of a Tipulid pupa. This projected, after a few more convulsions, nearly vertically from the debris to the extent of about $\frac{3}{8}$ in., exposing the part bearing the developing wings and legs. The pupa now remained quiescent for about ten minutes and then commenced a series of jerky, wriggling movements, bending the exposed part sharply back and forth. After a further ten minutes had elapsed, it was possible to discern a split on the ventral surface of the puparium bearing the sheaths of the wings and legs and another on the dorsal surface. The wriggles continued intermittently and at last there became perceptible a limp, shapeless greyish-white mass protruding upwards from the dorsal split. As this emerged further, the use of a lens made it possible to discern the shape of the humped thorax, with the head tucked underneath, while the limp wings, legs, antennae and palpi were still trailing inside the puparium. The upward and outward jerks continued until, ultimately, the antennae, palpi and front legs were free, the head then being moved and stretched forward to some extent, while the front legs clawed feebly at the surrounding debris as if attempting to assist in the emergence of the remainder of the body. Gradually the limp, trailing wings were withdrawn; the mid- and hind legs became free and assisted in the final extraction of the abdomen. The bedraggled creature which now rested somewhat helplessly on the surface of the debris, close to the puparium, had a pale, milky-green, semi-transparent appearance, with the head of a rather darker hue. Its exhausted aspect reminded us strongly of a newly-born foal, perched awkwardly and precariously upon its long, tottering legs. The insect, after a short respite, crawled feebly over the litter to the side of the jar, secured a foothold upon tiny pieces of damp wood adhering to the glass and reached upward to a ledge formed by the indented neck of the jar. There it suspended itself by the front legs, the others braced against some of the small particles which had formed its ladder ; the body and wings hung downwards, limp and motionless. At this point, the wings were crumpled in appearance and extended only as far as the middle of the abdomen. Within three minutes, the fly had begun to darken visibly, the legs having become a pale orange ; the thorax and demarcation of the abdominal segments brownish, and the head quite black. After a further three minutes, there was a violent contraction of the abdomen, coupled with the ejection of a large drop of clear liquid from the anus. Ten minutes later, or forty-three minutes from the rupture of the puparium, the wings were extended almost to the tip of the abdomen ; the wrinkles were disappearing and traces of the dark markings becoming evident. The abdomen, especially the dorsal region, was now much darker and more opaque. Almost exactly two hours after the pupa first emerged from the wood-litter, the insect, a fine female, was in full colour and started to fly somewhat agitatedly round the jar, resting from time to time by hanging from the ledge.

Individual flies continued to emerge for about three weeks and then, about the middle of June, we were fortunate in having a male appear shortly after a female. Not long after the male had fully developed, the pair mated, attempts at flight in the jar being made while still in copula. The next day, the female was observed probing into the debris at the side of the jar with her ovipositor, the body upheld by the partly-bent legs, with the abdomen and ovipositor pointed almost vertically downward. It was not possible to see, even with a lens, exactly what was going on, so the fly was transferred to another jar containing several layers of damp, white blotting-paper. She declined, however, to carry on whilst under observation; during the following night she must have decided otherwise, since the morning disclosed some fifty or more little black eggs round the edge of the blotting-paper, close to the glass. These were kept moist by an occasional fine mist applied by the scent-spray and stored in a cool place away from direct light. Just over a fortnight later, patience was rewarded by the spectacle of tiny grubs wandering slowly and aimlessly over the blotting-paper. Attempts at removal with a fine sable brush appeared to be fatal to these frail specks of life, so after their removal to alcohol for future examination, the blotting-paper, with live grubs and unhatched eggs, was transferred with the aid of forceps to the original wood-debris, there to form a new generation.

In the species of *Ctenophora*, our next genus, the males have the fourth to apical segments of the antennae bearing four process on each, two long basal and two short apical. They are handsome flies, of a yellowish or golden-brown colour with black markings and 20–25 mm. in length; somewhat scarce and recorded from decayed beech, birch, willow and cherry in old-established woodland, such as Epping Forest and the New Forest.

The last of the three wood-dwelling genera with pectinated antennae is *Tanyptera*, in which the male antennae bear, on each of the fourth to apical segments, two long basal and one short, knob-like, apical projections. *T. atrata* L. (Pl. **4**) is a lovely insect in which the genitalia of both sexes are especially pronounced. In the typical form shown in the plate, the antennae are black; below are shown the abdomens of a series of intermediate forms between this and v. *ruficornis* Mg., which is the palest form with some of the apical segments of the antennae reddish. The species has been bred from oak, beech, birch, poplar, lime and alder; we have found a wide range of varying forms emerging from a pile of sapling birch logs lying in a swampy patch in old forest land.

Next to be considered are the sub-families Cylindrotominae and Limoniinae, which comprise the short-palped Crane Flies; it will be remembered that they are characterised generally by the sub-costa ending in the costa, instead of Vein 1 as in the Tipulinae. Cylindrotominae may be distinguished by Vein 1 which curves downward to end in Vein 2 well before its apex, while Vein 5 is abruptly curved before it ends in the margin; in Limoniinae, Vein 1 carries on in more or less a straight line to end in the costa (Pl. **2,** 8 and **3,** 1), while Vein 5 is much less abruptly curved or hardly at all (Pl. **3,** 1). The curve of Vein 5 in Pl. **5,** 1 is exceptional.

CYLINDROTOMINAE.

Four genera, each with one species, make up the British Cylindrotominae, which we may pause to notice on account of the unusual larval forms. These are usually bright green in colour, resembling caterpillars, and feed on terrestrial or aquatic mosses, liverworts or flowering plants. The larva of *Cylindrotoma distinctissima* Mg. is said to bear a striking superficial resemblance to a Sawfly larva and has been recorded as feeding openly, like a caterpillar, on the upper sides of the leaves of Burdock and Marsh Marigold, in this country. A large number of long, thin processes clothe the body of the larva of *Phalacrocera replicata* L. and give it a centipede-like appearance; it feeds on aquatic mosses. An account of this insect is given by Miall and Shelford.[2]

LIMONIINAE.

Over 200 species are included in the Limoniinae; the habits of some of these are varied and interesting. Few Nature observers can have failed to notice the swarms of gnat-like insects which occur so commonly in damp situations just before sunset or just before sunrise; they dance up and down, or gyrate, always in a monotonously characteristic fashion, with now and again a sudden and seemingly co-ordinated ascent or descent of a foot or so by the whole swarm. If investigated, these swarms may prove to consist of species of other Nematocerous families, but sometimes will be found to be Limoniinines. Females, when present, always seem to be very much in a minority. At intervals, a female will enter the swarm, to be rapidly appropriated by a male, both thereafter dropping rapidly to the ground, where copulation is completed. It will be noticed that such a descent is often the signal for the apparently co-ordinated partial descent of the whole swarm. This interesting mating-dance provides one of the possible answers to a problem which must exercise the mind of nearly every entomologist at some time or other; whether or not the meetings of males and females, especially of the very tiny species, for the purpose of mating, amid the relatively vast areas around them and the myriad other forms of insect life, are governed by chance; or why, if such meetings are fortuitous, species with a low degree of fecundity do not speedily become extinct. As we progress, we shall come to the conclusion that meetings are far from being governed by chance. In an instance such as the mating-dance, it may well be that the stimulus which causes the males to swarm in a given situation at a given time is a combination of such factors as degree of light, or atmospheric temperature, pressure and humidity. The females may also be stimulated by these factors to seek similar situations, where the prominent swarm of males would speedily attract attention. The larvae of some Tipulidae are said to pupate gregariously, the males appearing first and lying in wait for the females to emerge; pairing takes place almost immediately, even before the females are sufficiently mature to fly. Males of some other kinds of flies frequent places where the females may come to feed or lay their eggs.

Another odd habit on the part of certain Limoniinines,

particularly species of *Limonia* (*Dicranomyia*) occurring in Britain, is that of swinging the body rapidly up and down by flexing the long legs, or "bobbing". According to Barnes,[3] this bobbing is associated with egg-laying; the abdomen, which is normally carried horizontally, being inclined to the vertical after a certain number of "bobs", thus assisting the ovipositor to penetrate into the soil. Some foreign species are said to take nectar from flowers while bobbing, the proboscis being forced into the florets on the downward strokes. Cuthbertson [4] and others have recorded the occurrence of bobbing while the insects were apparently at rest; we have seen the motion carried on for a long period without pause on window-panes and walls, where the possibility of feeding or oviposition could be ruled out.

Most species of Limoniinae rest with their wings folded, one above the other, over the abdomen, unlike the majority of Tipulines. *Pedicia rivosa* L. (Pl. **3,** 1) is an exception to this; moreover, its size and pictured wings make it liable, at first sight, to be classified as a Tipuline. However, the wing-markings are sufficiently distinctive to enable it to be recognised quite readily; examination of the palpi with a lens will confirm. This fine insect is not common; the larvae live in boggy places near streams and springs, where there is decaying vegetation.

Limnophila punctata Schrank (Pl. **2,** 8) is recorded from both wet earth and decaying wood. We have found the adults, which are about 10–13 mm. long, in considerable numbers flying over saturated areas adjacent to springs in the New Forest, in April, and in similar locations. The fully-grown larvae are said to be about 15 mm. long; reddish-brown in colour, with a number of spines on the body; and active in habits. Surrounding the hind spiracles are lobes bearing long fringes of fine hairs.

The larvae of *Dicranota bimaculata* Schummel, of which an account is given by Miall,[5] are of carnivorous habit. They lead an active existence, crawling through mud and gravel on the borders of streams and ponds, or readily progressing through the water by alternately bending and straightening the body. Their food is stated to consist mainly of small, red, aquatic worms, recognisable traces of which may be revealed, by the microscope, in the gut. In shape, the larvae are cylindrical, tapered at both ends, with one pair of the terminal lobes elongated so as to give a forked-tail appearance; they are up to 18 mm. in length. As might be expected from the carnivorous habit, the mandibles are robust and operated by powerful muscles.

Larval habitats of Limoniinae, generally, are varied; some live in wet earth, or decaying vegetation immersed in water or mud; some on partly-submerged stones in shallow streams, or on piles and bridge-piers, where they feed on algae; others live in fungi, or under the bark of trees. One foreign species is recorded as a leaf-miner.

Identification of the British species of Crane Flies may be essayed with the assistance of Coe[6] For Larvae, Brindle[9], Brindle and Bryce[9] and Chiswell[10] are useful.

TRICHOCERIDAE.

2 genera. 8 species. Strongly resembling some of the smaller Crane Flies; **V**-shaped suture on dorsum; Vein 7 characteristically short and curved (Pl. **2**, 13); ocelli present.

The popular name of Winter Gnats has been earned for the species of this family by their habit of congregating in dancing swarms in winter. They may appear in quite cold weather and are not infrequently seen when the winter sun is setting in the late afternoon and the atmosphere is still and damp; they may even be in evidence when snow is on the ground, particularly during a thaw, and may sometimes be found indoors. Their appearance is not confined to winter, however; they may be encountered throughout the remainder of the year, being abundant in spring and autumn but more sparingly in evidence, usually in cool, shady places, in summer. Normally, the larvae are to be found in leaf-mould, fungi and decaying vegetable matter; there is a record of considerable damage being caused by them to stored potatoes in Ireland. Some foreign species of the family are said to occur in caves and mines, sometimes at considerable depths, where their larvae are thought to live in animal excretions. For identification of British species, see Freeman[7], Laurence[11], and Dahl [14]. For the immature stages, see Keilin and Tate[12] and Brindle[8].

PTYCHOPTERIDAE.

1 genus. 7 species. Superficially resembling Tipulids; generally shining black, with yellow or brown markings on abdomen and legs; spotted wings; characteristic **U**-shaped suture on dorsum; Vein 7 absent; ocelli absent; male genitalia exceptionally prominent. (Pl. **3**, 2, 3.)

These flies may be distinguished from the other families of Nematocera, by the shape of the suture on the dorsum, which will be seen to differ from the **V**-shaped furrow of the Tipulidae and Trichoceridae and is less deeply incised. In addition to the characters mentioned above, apical spurs are present on the tibiae and there are peculiar creases between Veins 3–4 and 5–6 respectively; some species have a silvery bloom on the pleura; at rest, the wings are opened.

Ptychoptera contaminata L., the species figured, and *P. albimana* F. are very common and usually to be found flying or resting on low vegetation near water, or in damp situations generally; both species fly from May to October. *P. albimana* differs from *contaminata* mainly in lacking the small dark spot near the base of the wing.

The narrow, elongated larvae (Pl. **2**, 7) are aquatic, living in mud and debris beneath very shallow water, where they are able to breathe, while submerged, by means of a long, thin "tail" or tube which carries an extension of the tracheae. The device is partly reproduced in the pupa, one of the respiratory horns being greatly elongated, while the other is normal. According to Miall,[5] air is absorbed, in this case, not through spiracles, but through the membrane of the tube, which is attenuated in places. To assist in crawling through mud and debris, the hind margins of the segments in the larva project outward beyond the anterior margins of the succeeding segments and are armed with backwardly-directed spines; it is easy to appreciate how this arrangement, in conjunction with alternate expansion and contraction of the body, results in forward movement.

A key to the British species of Ptychopteridae is provided by Freeman[7]. Brindle[13] keys the larvae.

ANISOPODIDAE (= RHYPHIDAE).

2 genera. 5 species. No suture on dorsum; Vein 7 not reaching margin; ocelli present; tibiae with apical spurs; fork of Veins 2 and 3 at or before anterior cross-vein. (Pl. **2**, 9.)

Four of the five British species in this family belong to the genus *Sylvicola* (=*Anisopus* =*Rhyphus*) and are characterised by the presence of a discal cell, which is absent in the fifth species *Mycetobia pallipes* Mg. *Sylvicola* species are gnat-like flies with swarming habits similar to those of *Trichocera*. We have seen *S. punctatus* F., a not uncommon species about 8 mm. long, swarming in large numbers in the lee of coastal sand-dunes just before sunset, the flies rising and falling gently, exerting just enough effort to maintain their position facing the occasional slight breezes which reached them. *S. fenestralis* Scop. (Pl. **2**, 9), which differs from *punctatus* in having a dark cloud at the tip of the wing, is very common and frequently found indoors; we have found it from time to time on tree-trunks, usually on one particular tree and not its neighbours, which is probably explainable by the larvae breeding there. A number of larval habitats have been recorded for *Sylvicola*, such as sap exuding from tree-wounds, decaying leaves in rain-filled tree holes, decaying roots of Burdock and Angelica, and sewage filter-beds. The larvae of *fenestralis*, in particular, have been reported as causing damage to honeycombs, cider and home-made wines. All the adults are usually found in damp situations.

Mycetobia pallipes Mg. used to be included in another family, the Mycetophilidae, to which, in the adult stage, it exhibits many similarities; the larva, however, has affinities with those of *Sylvicola* and lives in sap or fungi on decaying wood. For the identification of the British species of Anisopodidae, see Freeman.[7] The immature stages are dealt with by Keilin and Tate[12] and Brindle[8].

LITERATURE CITED

1. Alexander, C. P. Crane Flies of New York. II. Biology and Phylogeny. *Cornell Univ. Exp. Sta. Mem.* 38. 1920.
2. Miall, L. C. and Shelford, R. The structure and life-history of *Phalacrocera replicata*. *Trans. ent. Soc. Lond.* **1897**: 343–61. 1897.
3. Barnes, H. F. Some observations on the Mating Habits and Oviposition of the Limnobiidae (Dipt.). I. *Ent. mon. Mag.* **60**: 71–3. 1924.
4. Cuthbertson, A. The mating habits and oviposition of Crane Flies. *Ent. mon. Mag.* **65**: 141–5. 1929.
5. Miall, L. C. The Natural History of Aquatic Insects. McMillan, London. 1895.
6. Coe, R. L. Handbooks for the Identification of British Insects. Vol. IX. Pt. 2. Tipulidae. Royal Entomological Society of London. 1950.
7. Freeman, P. Handbooks for the Identification of British Insects. Vol. IX. Pt. 2. Trichoceridae, Ptychopteridae and Anisopodidae. Royal Entomological Society of London. 1950.
8. Brindle, A. Taxonomic notes on the Larvae of British Diptera. 11. Trichoceridae and Anisopodidae. *Entomologist*, **95**: 285–8. 1962.
9. Brindle, A. The Larvae and Pupae of British Tipulinae. *Trans. Soc. Brit. Ent.*, **14**: 63–114. 1960.

(Continued on p. 352)

CHAPTER II

CULICIDAE . . . . . Mosquitoes, Gnats.

9 genera. Over 50 species. Slender, long-legged flies; ocelli absent; ten veins or their branches reach wing-margin (Pl. **6**, 2–5); veins and hind-margin of wing fringed with scales, ♂ antennae plumed (except Dixinae); proboscis long and rigid, adapted for piercing in ♀ (Culicinae), short or concealed and not so adapted (Chaoborinae and Dixinae). (Pl. **3**: **6**).

SINCE the discovery that various diseases, including malaria, yellow fever and dengue fever, are transmitted by the agency of blood-feeding female mosquitoes, there has been a wealth of literature concerning them; consequently their story is widely known. As regards the British species, there is an excellent and easily-accessible monograph, with keys, by Marshall;[1] we may, therefore, content ourselves here with a cursory glimpse of the principal features of the sub-family Culicinae, to which the name "mosquitoes" is now considered to be properly applied. The term "gnat" has often been loosely used to denote various Nematocerous flies with long, narrow wings and body and long, delicate legs; today it is thought better to restrict it to the other two sub-families, the Chaoborinae and Dixinae. The mosquitoes are easily distinguished by the long proboscis, adapted for piercing in the female only and used for imbibing juices of flowers and fruits in both sexes (Pl. **3**, 4, 6); by the long, often ornate palps and beautifully-plumed antennae of the males (Pl. **3**, 6: Pl. **6**, 10, 11); and by the scales on the veins and hind margins of the wings in both sexes (Pl. **6**, 3–5). The Chaoborinae are similar in appearance, but have short, poorly-developed mouthparts (Pl. **3**, 5); while the Dixinae are characterised by small mouthparts, absence of scales and a distinctive wing-venation (Pl. **6**, 2).

CULICINAE.

The sub-family is divided into two clearly-definable sections, the Anophelines and the Culicines, of which there are four and twenty-six British species respectively. Adult Anophelines are distinguishable from Culicines by the knobbed palps of the males, which are rather like a golf-club in shape (Pl. **6,** 10); by the elongated palps, as long as the proboscis, in the females (Pl. **6,** 9); and by the fact that they have no scales, but only fine hairs, on the abdomen in either sex. The resting attitude is also characteristic, the proboscis being aligned with the abdomen and both inclined at an angle to the surface upon which the insect is resting. Malaria-carrying mosquitoes are to be found among the ranks of the Anophelines. Two members of the *Anopheles maculipennis* complex occur in Britain; the "ague" or indigenous malaria of this country, which was prevalent in some coastal districts up to the latter part of the nineteenth century, is thought to have been conveyed by *A. labranchiae atroparvus* VanThiel, which breeds in coastal saline waters. *A. messeae* Falleroni, however, is an inland species and affects only fresh water. Although extremely difficult to separate by their structural differences, they are clearly separable by their choice of habitat and differences in life-history. Another British Anopheline, *A. plumbeus* Stephens, breeds in water-filled tree holes, a characteristic shared by certain Culicine species.

Culicines may be known by the shortness of the female palps (Pl. **6,** 8); by the generic characters of the male palps, e.g. *Culex* (Pl. **3,** 6); by the scales on the abdomens of both sexes and the resting attitude. Here the proboscis forms an angle with the abdomen, which is held parallel with the surface upon which the fly is resting. There are five British genera, of which *Culex*, *Aēdes* and *Culiseta* (=*Theobaldia*) are the most familiar. The female of *Culex pipiens* L., often called the Common Gnat, may, for practical purposes, be recognised by the very short stalk of the anterior forked cell (Pl. **6,** 3), in which it differs from all othe British mosquitoes, except the closely-similar but uncommon *C. molestus* Forskål.* Females of *pipiens* are very frequently found hibernating in houses, even in towns, or in sheds and outhouses; they do not feed on the blood of mammals, but appear to confine their attentions to birds, or juices from flowers. The larvae are to be found in all kinds of stagnant water, such as wells, tanks, water-butts, puddles and ponds.

The 14 British species of the genus *Aëdes* differ a good deal in their choice of habitat in the immature stages. Some are restricted to salt-marshes on the coast, while others confine themselves to pools in such locations as birch or pine woods, or moorland sphagnum-patches, where there is rotting vegetation and the water is of an acid nature. Some regularly affect water which is shaded by vegetation, while others seem only to occur in open situations; one species is to be found only in water-filled tree holes. *A. punctor* Kirby (Pl. **3,** 4) is a common species of the acid water type; the bloodthirsty female will certainly attract the notice of the field-naturalist sooner or later by her ardent attentions.

Culiseta annulata Schrank is one of the largest and most distinctive of the British mosquitoes, with several dark spots on the wings formed by conglomerations of scales (Pl. **6,** 5), in this respect resembling to some extent *Anopheles maculipennis* Mg. (Pl. **6,** 4); *annulata* may be distinguished at once, however, by the presence of many whitish rings on the legs. Females of this species are often found hibernating indoors in company with those of *C. pipiens* L.; unlike *pipiens*, however, they do not hibernate completely and have been known, at times, to revive sufficiently from their winter torpor to regale themselves with a blood-meal from their human companions. The immature stages are passed in stagnant water in similar locations to those affected by *pipiens*.

As an example of special adaptation, the structure of mosquito eggs is very interesting. The eggs of *maculipennis*,

* *C. molestus* Forskål is now generally regarded as a subspecies of *pipiens*.

which are laid singly on the surface of the water, are boat-shaped and float horizontally; they are equipped with lateral floats which ensure that the "deck" will remain continuously uppermost. Those of *pipiens* and *annulata* are somewhat conical in shape and are gathered together, as they are laid, by the female, with the aid of her crossed hind legs. Since they have an adhesive envelope, they form a kind of raft, in which the eggs assume a vertical position, broad end downwards. The effect of this is to make the raft more or less boat-shaped, with a concave upper, and convex lower surface. After the requisite period of immersion, the larvae hatch and emerge from the lower part of the egg directly into the water. The eggs of *A. punctor*, which are laid separately on the ground, usually in places liable to be flooded by subsequent rain, do not exhibit such marked features. To counter the risk of all the larvae being destroyed by a premature drying-up of the pool, the eggs do not all hatch simultaneously, but in successive batches at fairly lengthy intervals; moreover, the females lay their eggs at different levels, so that those which are not reached by a temporary rise in water-level will be hatched by a subsequent deeper flooding. The eggs are well able to survive prolonged desiccation in such circumstances; according to Marshall,[1] eggs of another species, *Aëdes cantans* Mg., immersed in water seventeen months after being laid, hatched in 24 hours.

Mosquito larvae are very active in habit, progressing in the water by a jerky doubling and straightening of the body; they display some interesting external features (Fig. 10D). The specimen figured is a typical Culicine; breathing is accomplished with the aid of a process at the hind extremity, known as a siphon, by means of which the larva is able to suspend itself vertically, head downward, at the surface and take in atmospheric air. Brushes of hairs, with which the mouthparts are equipped, are used to sweep in tiny particles of food suspended in the water; some species feed by nibbling at dead leaves at the bottom, or algae and detritus of various

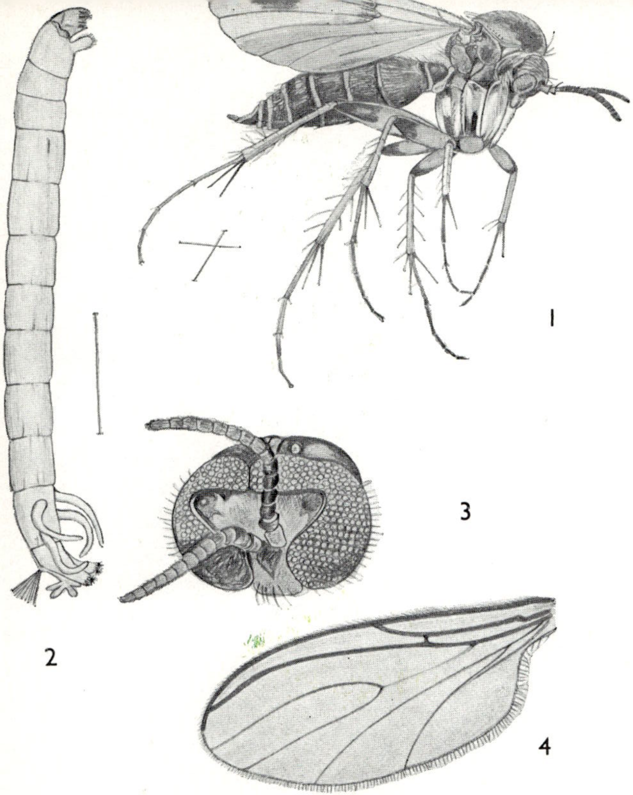

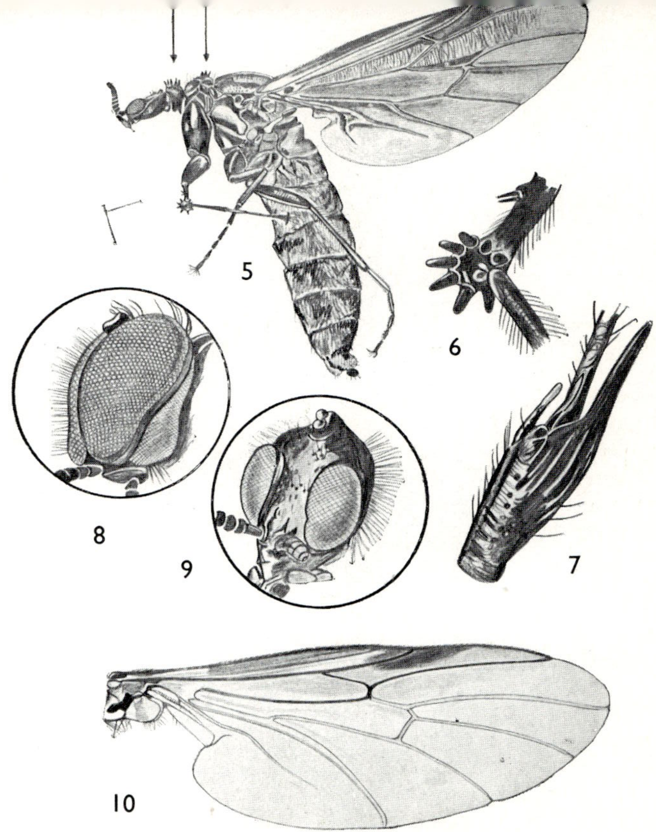

Plate 7

MYCETOPHILIDAE

1. *Dynatosoma fuscicorne* Mg., ♀, p. 72. 3. Sciarinae, "eye-bridge", p. 73. 4. *Bradysia*, wing, p. 73.

CHIRONOMIDAE-III

2. *Chironomus*, larva, p. 70.

BIBIONIDAE

5. *Dilophus febrilis* L., ♀, p. 74. 6. *Dilophus febrilis* L., apex of tibia. 7. *Bibio* sp., apex of tibia, p. 75. 8. *Bibio* sp., ♂ head, p. 74. 9. *Bibio* sp., ♀ head, p. 74. 10. *Bibio* sp., wing, p. 74.

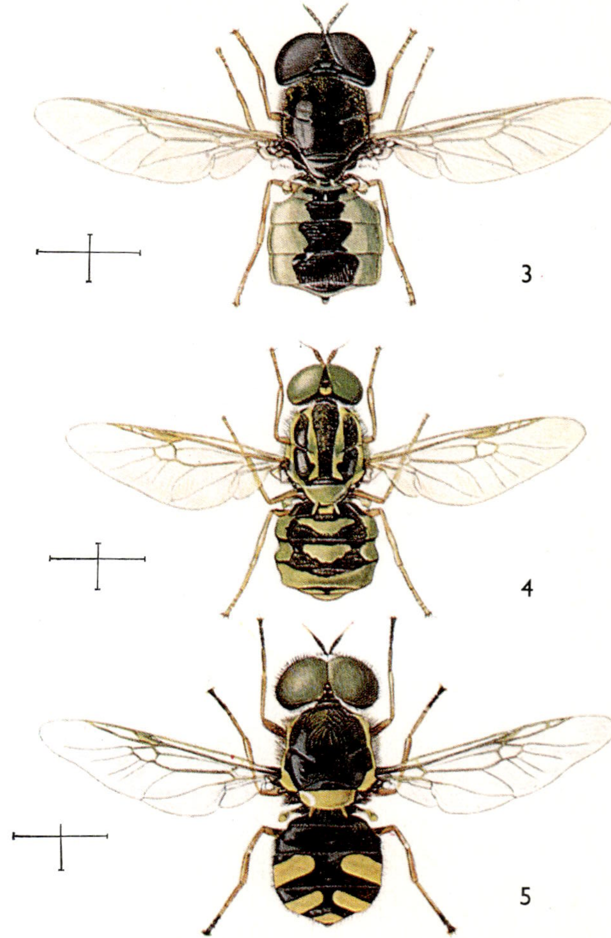

Plate 8

STRATIOMYIDAE-II

1. *Stratiomys potamida* Mg., ♂, p. 93. 2. *Stratiomys longicornis* Scop., ♂, p. 93. 3. *Odontomyia viridula* F., ♂, p. 94. 4. *Oxycera trilineata* F., ♂, p. 94.

kinds. Anopheline larvae have no siphon, but breathe through a pair of spiracles on the dorsal surface of the eighth abdominal segment, suspending themselves in a horizontal position at the surface, meanwhile, with the aid of special processes along the body; they feed in this position by sweeping with the mouth-brushes. As in Culicine larvae, there is an anal segment, set at an angle to the remainder of the abdomen, which bears brushes of hairs and gills. The function of these gills, although partly connected with respiration, is said to be mainly concerned with the elimination of waste from the alimentary tract by absorption of water which is subsequently expelled from the anal aperture.

In shape, mosquito pupae (Fig. 11D) are quite different from the larvae; they are also active, diving at the least sign of danger and progressing by lashing the abdomen, which is equipped with tail-paddles. Breathing is accomplished with the aid of prominent respiratory trumpets, a special adaptation of the anterior thoracic spiracles, which should be compared with the respiratory horns in Tipulid pupae (Fig. 11A). The pupa is enabled to float at the surface and maintain its respiratory trumpets in contact with the air by means of an air-space in the thorax; the abdomen, meanwhile, hanging downward.

When the adult fly is about to emerge, the pupal abdomen is extended horizontally; the dorsal surface of the thorax splits and the mosquito gradually makes its exit. Ultimately, it rests its front legs upon the water in order to support itself for the extraction of the other legs and the abdomen, after which it takes a short rest, either on adjacent vegetation or on the pupal skin, in order to mature sufficiently to fly away. It will readily be appreciated that calm conditions are required for this extremely delicate operation of emergence; the surface of a pool will often mutely testify to the hazards involved.

A particular feature of interest in both larvae and pupae of *Mansonia richiardii* Ficalbi is the fact that they do not breathe at the surface like the other species, but obtain their air by tapping submerged stems and roots of aquatic vegetation with their specially-adapted siphons and respiratory trumpets respectively.

Many of the British species of mosquitoes are not difficult to rear at home with very simple equipment and they provide an excellent field for study. Some species, however, will not mate in a confined space, so that full-cycle breeding is precluded in such cases.

In addition to Marshall,[1] keys to the British species of Culicinae are provided by Mattingly [2] and Edwards.[3]

CHAOBORINAE (now usually given family status).

The species of this sub-family resemble those of Culicinae in wing-venation and plumed male antennae but differ in having short mouthparts; in the males, the wings very noticeably do not extend as far as the tip of the abdomen (Pl. **3,** 5), a character which may, at first sight, cause them to be confused with some species of the family Chironomidae (Pl. **6,** 7). The larva of *Chaoborus crystallinus* Deg. (Pl. **6,** 6) is well-known to students of pond life as the " Ghost Larva " or " Phantom Larva ". An interesting account of the structure of this creature is given by Miall;[4] its extraordinary transparency makes it an unusually fascinating live subject for study with the microscope; in a suitable live-box, much of the internal structure can be made out. There are no spiracles, breathing being carried on by absorption, through the integument, of air dissolved in the surrounding water. While resting and waiting for its prey, which includes small Crustaceans such as Water Fleas, the larva maintains a motionless, horizontal position; when it moves, by a sudden lashing of the body, the motion is so rapid as to be almost imperceptible and the creature reappears, motionless, a little distance away. The pupa rests at the surface with the abdomen extended vertically downward, the effect being quite unlike the "comma" shape of a resting mosquito pupa, although, in the other Chaoborine genus, *Mochlonyx*, both larvae and pupae are much more like those of the Culicinae. Brindle[7] keys the larvae of Chaoborinae.

DIXINAE.

The sub-family Dixinae of the Check List has often been treated as a separate family, Dixidae; there are thirteen British species, all belonging to the genus *Dixa*. In both sexes, the antennae are slender and longer than the thorax and those of the males are not plumed; the mouthparts are small and not adapted for piercing; and the wing-venation is distinctive by virtue of the strongly-curved stalk of the anterior forked cell (Pl. **6**, 2). The general appearance of *Dixa* species is not unlike that of some of the smaller Tipulids, but the absence of a V-shaped suture on the dorsum will remove doubt at once. Miall [5] describes both larva and pupa; the larva is about 8 mm. long and bears a pair of pseudopods or false legs on the fourth and fifth segments; long spines equip the head and tail. In life, the larva adopts a characteristic attitude, resting in an inverted-U shape, apparently upon the surface of the water but, in fact, with a film of water extended over its body. It also progresses thus along the surface and sometimes wanders on to partially-submerged vegetation, still surrounded by a film. Feeding is accomplished with mouth-brushes; in this respect and in its general structure, it is said closely to resemble an *Anopheles* larva, which, it will be remembered, is also a surface-feeder. The pupae of *Dixa* are not unlike those of *Chaoborus*.

For the identification of British species of Chaoborinae and Dixinae, see Freeman.[6] Brindle[8] keys larvae of Dixinae.

The habit in the males of congregating in dancing swarms, which was noticed in the preceding chapter, is also prevalent in the family Culicidae.

LITERATURE CITED

1. Marshall, J. F. The British Mosquitoes. British Museum, London. 1938. (Reprint Johnson Reprint Co., London, 1966).
2. Mattingly, P. F. Handbooks for the Identification of British Insects. Vol. IX. Pt. 2. Culicidae, sub-fam. Culicinae. Royal Entomological Society of London. 1950.
3. Edwards, F. W. in Edwards, Oldroyd and Smart. British Bloodsucking Flies. pp. 1–24. British Museum, London. 1939.
4. Miall, L. C. The Natural History of Aquatic Insects. p. 113. *Corethra*. McMillan, London. 1895.
5. Miall, L. C. *Ibid.* p. 157. *Dixa.* 1895.
6. Freeman, P. Handbooks for the Identification of British Insects. Vol. IX. Pt. 2. Culicidae, sub-fams. Dixinae and Chaoborinae. Royal Entomological Society of London. 1950.
7. Brindle, A. Taxonomic notes on the larvae of British Diptera. 8. The subfamily Chaoborinae (Culicidae). *Entomologist*, **95**: 178–82. 1962.
8. Brindle, A. Taxonomic notes on the larvae of British Diptera. 15. The Dixinae (Culicidae). *Entomologist*, **96**: 237–43. 1963.

CHAPTER III

CHIRONOMIDAE .	.	Non-biting Midges.
MYCETOPHILIDAE	.	Fungus Gnats.
BIBIONIDAE	.	Fever Flies, St. Mark's Flies.

CHIRONOMIDAE.

26 genera. Over 380 species. Delicate gnat-like flies; head overhung by humped thorax; ♂ antennae plumed; * mouthparts poorly developed; ocelli absent; wing-venation reduced; usually 6, rarely 7 or 8, veins or branches reach margin; anterior veins somewhat more strongly marked than posterior; Vein 2 mostly faint or vestigial and posterior cross-vein mostly absent. (Pl. **6,** 1: **6,** 7: **7,** 2.)

THE habit of swarming is very highly developed in this family; large numbers of males, dancing up and down in sheltered places especially near water, usually over some prominence such as a tree or bush, are a common sight. Females are often to be found resting on vegetation near the swarm. Some species congregate an hour or so before sunset, some just before darkness sets in; others are more likely to be seen at midday or in the early afternoon, especially on mild days in winter or early spring, to which seasons their appearance seems to be confined. While the swarms of some species are always found at a considerable height, those of others will invariably be seen close to the ground or to the surface of water.

As we noticed in the preceding chapter, certain species of Chironomidae are of a form and size which may cause them to be confused, at first sight, with the Chaoborinae; the males, with certain exceptions, having plumed antennae and wings which do not extend as far as the tip of the abdomen (Pl. **6,** 7); there are, however, no scales on the veins or hind-margins of the wings and the venation is quite distinctive.

* Except Clunioninae, p. 71.

Chironomids are delicate in appearance, often of considerable beauty; seen from one side, both sexes have a characteristically humped thorax which overhangs the head, so much so that, from a dorsal aspect, the latter may be quite concealed. Wing-venation will be seen to exhibit a reduction in some of the veins causing the others, particularly the anterior ones, to assume greater emphasis (Pl. **6,** 1). Vein 2 is often ill-defined or vestigial and the posterior cross-vein only present in the sub-family Tanypodinae. Mouthparts are not well-developed and in some species no food is taken in adult life.

Owing to the delicacy of these flies and the nature of the characters used in the identification of species, often entailing the preparation of genitalia mounts, they are not easy for the field-naturalist to study, except in a general way. It should be possible for him, however, to rear many of them at home and to learn much of their biology. The immature stages are both terrestrial and aquatic, mainly the latter; habitats of the terrestrial forms include dung, moss, rotten wood and other forms of decaying vegetable matter.

In the aquatic types, the eggs are laid in masses or ribbons, within a protective envelope of glutinous nature, which swells on contact with water. The familiar red "bloodworms" of water-butts and stagnant pools, which loop themselves into figures-of-eight when they swim, are larvae of *Chironomus* species; their structure and life-history has been studied by Miall [1] and Miall and Hammond.[2] They build themselves little tubular homes from fragments of debris at the bottom, from which they extend the head to feed on small particles of decaying vegetable matter. The larval body (Pl. **7,** 2) is of an elongated, cylindrical shape and is adorned with various appendages including pseudopods, at both ends, which bear tiny hooks, probably used for anchoring to the burrow or for similar purposes. As in the "Ghost Larva" of *Chaoborus*, there is only a very poorly-developed tracheal system, with no spiracles. Species of Chironomid larvae, such as the bloodworms, which burrow in the mud or haunt the bottom, where the degree of aëration is much lower than at the surface, are equipped with haemaglobin in the blood for the storage of oxygen; hence their red colour. On the other hand, those which normally live near the surface do not need such a provision and are mainly greenish, or at least pale-coloured. Larvae of some foreign species of Chironomidae have been found at extraordinary depths.

Chironomid pupae vary a good deal in structure. Those of *Chironomus* resemble *Chaoborus* to some extent, except that, instead of the respiratory trumpets, there are bunches of filaments which, by absorption of air from the surrounding water, make it possible for the pupae to remain at the bottom. The tracheal system, contrary to that of the larva, is well-developed; when the time for emergence arrives, it is sufficiently full of air, obtained via the filaments, to allow the fly and the puparium containing it to float upwards to the surface. In the Tanypodinae, the pupa strongly resembles those of the Culicinae (cf. Fig. 11D).

Male antennae in Clunioninae differ from the remainder of the family in not being plumed; the sub-family contains some interesting seashore species, one of which, *Clunio marinus* Hal., occurs along the South Coast. It is a tiny species, about 1½ mm. in length, with wingless females; the larvae construct burrow-tubes of minute particles of sand and seaweed, in which they live beneath the water. Emergence of the adults takes place during the intervals between tides, when the larval habitat is exposed.

Parthogenesis, i.e. reproduction without fertilisation by a male, occurs in species of *Chironomus*, *Corynoneura*, and *Tanytarsus*; in the last-named, both pupae and newly-emerged adults are capable of producing larva-bearing eggs. Parthogenetic reproduction by immature stages is known as paedogenesis (cf. *Miastor*, p. 83).

According to Lloyd,[3] larvae of *Spaniotoma* and *Metriocnemus* species, in company with Psychodid larvae, play an important part in the efficient working of percolating filters in sewage systems.

Keys for the identification of British species of Chironomidae are provided by Coe.[4] Bryce[8] is useful for larvae.

MYCETOPHILIDAE.

c. 75 genera. About 500 species. Delicate gnat-like flies; head overhung by humped thorax; antennae in both sexes long, thread-like; coxae long and tibiae spurred; ocelli present; Vein 2 absent; anterior veins more strongly marked than posterior. (Pl. **7,** 1, 3–4.)

Mycetophilids may be found throughout the year, except in severe weather, in many kinds of location; they usually prefer damp or dark places, where there is rotten wood, rotting vegetation, excrement or fungi, the predilection of many species for the last-named accounting for their popular name, Fungus Gnats. Males sometimes congregate in dancing swarms like those of some of the families we have already surveyed; we have seen species of *Phronia* and *Mycetophila* behaving thus, close to the ground, in clearings and beneath bushes.

Many kinds of Fungus Gnats may readily be bred out at home; fungi in which larvae are discovered should be placed upon some sand or earth into which the larvae can migrate for pupation; otherwise the fungi are liable to decompose into a corrupt, liquescent mess, in which the larvae will perish. Some species have been bred from old birds' nests containing an accumulation of vegetable litter and excrement. Not infrequently, mushrooms purchased for the table will yield a profitable haul to the enterprising dipterist.

As a representative of the family, we may take *Dynatosoma fuscicorne* Mg. (Pl. **7,** 1), whose larvae are to be found in the familiar bracket-fungi on fallen tree-trunks and similar locations, where they often betray their presence by frass outside their burrows. *Macrocera* species are remarkable for their enormously long, thin antennae, which are sometimes as much as three times the length of the whole body; they may be found not uncommonly, in summer, on umbels such as Hedge Parsley and Hogweed.

The mostly minute species of the sub-family Sciarinae*, especially of the genus *Bradysia*, which seldom exceed 4 mm. in length, abound in many places, notably, together with Spring-tails, in accumulations of dead leaves, or moss; or, in the larval state, in manure, roots of seedling plants in greenhouses, various wild fungi and cultivated mushrooms, where they are said sometimes to cause much damage. As a group, they are fairly easily recognised by their characteristic wing-venation (Pl. **7**, 4) and by the fact that their compound eyes curve to meet one another above the antennae (Pl. **7**, 3), forming, as it were, an "eye-bridge" (cf. Lestremiinae, p. 84). Larvae of some foreign *Bradysia* species have a habit of migrating, gregariously, in columns, which has gained for them the name "army worms".

Mycetophilid larvae, generally, are long, worm-like grubs, soft, whitish and partly transparent, with a small darkened head. Certain species have a habit of building a kind of tubular web, within which they are able to move rapidly backwards and forwards. The pupae have no respiratory horns or trumpets.

Edwards[5] may be consulted for the identification of the British species of this family. Madwar[9] is useful for larvae. Buxton[11] may be consulted for all Diptera breeding in fungi.

BIBIONIDAE.

2 genera. 18 species. Mostly black, hairy flies (some females reddish); antennae short, inserted below eyes near mouth-opening, segments stout and closely set; ocelli present; tibiae with characteristic "beaks" or circlets of thorns at apex; Vein 2 absent; anterior veins more strongly marked than posterior. (Pl. **7**, 5–10.)

* Now usually given family status.

Although possessing certain features in common with the two preceding families, Bibionidae present, on the whole, a sharp contrast. They are mainly black, hairy flies, of a stouter appearance, with shorter, broader wings and more robust-looking legs ; the females of *Bibio hortulanus* L. and *B. anglicus* Verrall have the dorsum and abdomen reddish and traces of this colour appear in the legs of the females of some other species. Prominent apical processes are present on the tibiae, but these exhibit characteristic modifications (Pl. **7,** 6–7) which preclude any risk of confusion with Mycetophilidae. The tendency of the posterior veins to atrophy, with consequent emphasis of the anterior veins is again to be seen ; Vein 2 is also absent (Pl. **7,** 10). In both sexes, the heads are highly distinctive. Those of the males are holoptic, appearing to consist of little save the compound eyes, below which the antennae are inserted, close to the oral opening (Pl. **7,** 8); the eyes are divided into two clearly-defined areas, one with large facets, the other with smaller ones, a condition which occurs also in male Simuliidae (p. 81) and other insects, such as the well-known Whirligig Beetles. Female heads are narrower, with the eyes small and widely-separated (Pl. **7,** 9). There are only two genera, *Dilophus* with four British species and *Bibio* with fourteen.

Dilophus febrilis L. (Pl. **7,** 5), which used to be called (without justification) the Fever Fly, is abundant almost everywhere, particularly in early spring, and continues to be in evidence until as late as October ; we have occasionally seen enormous numbers of females clustered on stems of grasses ; the males are usually found hovering, rather sluggishly, in the air. As will be seen from the plate, the wings of the female are smoky, but in the male they are almost clear, with a black stigma. *Dilophus* species are easily distinguished from those of *Bibio* by the circlet of spurs, which may aptly be described as an inverted " crown of thorns ", at the apex of the fore tibia (Pl. **7,** 6); and by the double collar of spines on the front part of the thorax (Pl. **7**, 5). The adult flies are said to be of considerable significance in the pollination of fruit trees; on the other hand, the larvae have been recorded as causing damage to the roots of hops and other plants. They are known to breed frequently in manure and farmyard litter, or leaf-mould, which no doubt accounts for their introduction, at various times, into agricultural land, with consequent depredations upon crops.

Bibio species differ from *Dilophus* in lacking the double collar of spines on the pronotum and in possessing, instead of the apical circlet of spurs on the fore tibia, large projections one of which is usually very prominently claw-like or beak-like (Pl. **7**, 7).* The nature student will probably be most familiar with *B. marci* L., long known as "St. Mark's Fly", possibly owing to the fact that it generally commences to emerge near St. Mark's Day. At 10–12 mm., it is by far the largest of the British species and a heavy-looking, hairy black fly, sluggish in flight. Two smaller species, about 5–7 mm. long, *B. reticulatus* Loew and *B. johannis* L., are usually abundant in spring. The males will generally be found in large swarms over short grass or downland turf, close to the ground; the females, however, are not so readily encountered and should be sought on neighbouring flowers, or on the ground, where they will be more easily noticed when in copula. All the species occurring in Britain are fairly common; most of them fly in spring, one or two up to about the middle of June; two species, *B. clavipes* Mg. and *B. lepidus* Loew are on the wing in September and October. We have seen males of *lepidus*, on a damp, still day in early October, lazily dancing up and down in gnat-like fashion, about seven or eight feet from the ground, in a sheltered spot between tall hedges bordering a narrow cart-track.

Structurally, *Bibio* larvae are considered to be the most primitive in the Order; they have a comparatively large head

* This illustration is inverted, i.e. the tibia is lowermost.

with prominent biting mouthparts and open spiracles on 10 of the 12 segments. Here is an interesting case of larval development not having marched in step with that of the adult in the evolution of the species. Morris[6] and Brindle[10] have studied the immature stages; the eggs are deposited and pupation takes place in little subterranean cells; the larvae are often gregarious. No trace of respiratory horns is to be seen in the pupa.

For the identification of the British species of Bibionidae, see Edwards.[7]

LITERATURE CITED

1. Miall, L. C. The Natural History of Aquatic Insects. p. 222. *Chironomus*. McMillan, London. 1895.
2. Miall, L. C. and Hammond, A. R. The Structure and Life-History of the Harlequin Fly (Chironomus). Oxford. 1900.
3. Lloyd, Ll. The Biology of Sewage Disposal. New Biology No. 2. Penguin Books, London. 1947.
4. Coe, R. L. Handbooks for the Identification of British Insects. Vol. IX. Pt. 2. Chironomidae. Royal Entomological Society of London. 1950.
5. Edwards, F. W. British Fungus Gnats. *Trans. ent. Soc. Lond.* **57**: 505–70. 1924.
6. Morris, H. M. The larval and pupal stages of the Bibionidae. *Bull. Ent. Res.* XII (3): 221–32. 1921. *Ibid.* XIII (2): 189–95. 1922.
7. Edwards, F. W. A Synopsis of British Bibionidae and Scatopsidae. *Ann. appl. Biol.* **12**: 263–75. 1925.
8. Bryce, D. Studies on the Larvae of the British Chironomidae (Diptera) with keys to the Chironominae: and Tanypodinae. *Trans. Soc. Brit. Ent.* **14**: 19–62. 1960.
9. Madwar, S. Biology and Morphology of the immature stages of Mycetophilidae. *Phil. Trans. Roy. Soc. Lond.* (B). **227**: 1–110. 1937.
10. Brindle, A. Taxonomic notes on the larvae of British Diptera. 6. The Family Bibionidae. *Entomologist*, **95**: 22–6. 1962.
11. Buxton, P. A. British Diptera associated with fungi. III. Flies of all families reared from about 150 species of fungi. *Ent. mon Mag.*, **96**: 61–94. 1961.

CHAPTER IV

CERATOPOGONIDAE . . .	Biting Midges.
SIMULIIDAE . . .	Black Flies.
CECIDOMYIIDAE . .	Gall Midges.
PSYCHODIDAE . . .	Owl Midges, Moth Flies.
SCATOPSIDAE.	
THAUMALEIDAE (= ORPHNEPHILIDAE).	

CERATOPOGONIDAE.

17 genera. Over 150 species. Minute flies; ♂ antennae plumed; mouthparts short, adapted for piercing in ♀; ocelli absent; wing-venation reduced, similar to Chironomidae; six veins or branches reach margin, but Vein 4 forked; posterior cross-vein absent; anterior veins stronger than posterior. (Pl. **10**, 2–3.)

THE mouthparts in this family are short; in the females they are adapted for cutting wounds in the skin of their hosts from which to suck blood. Many species confine themselves to feeding either on the blood of insects or on nectar from flowers; some prey on caterpillars, some upon adult butterflies and dragonflies, where they pierce the wing-veins; others attack beetles and bugs and one foreign species is said to suck the blood of mosquitoes. Blair [1] records *Atrichopogon meloesugans* Kieffer as feeding on Oil Beetles and Varley [2] has found *Palpomyia semifumosa* Goetghebuer piercing the thorax of Mayflies.

In Britain, only the genus *Culicoides* contains species which bite mammals; *C. pulicaris* L. (Pl. **10**, 3), a very common insect, will serve as a representative. The dark spots or clouds on the wings are characteristic of most species of the genus and assist in distinguishing it from nearly all the other genera. It has been proved that *Culicoides* may convey disease organisms in the form of certain blood-parasite worms; a British species, *C. nubeculosus* Mg., has been shown to carry such an organism which causes a disease of horses. The irritation caused to human beings by the bites of these flies constitutes, at times, a serious nuisance in some parts of this country, notably Scotland and various coastal areas. Their attentions, particularly in the late evening, near sunset, are often such as to make an outdoor stroll something of an ordeal. Attacks are generally more pronounced when the weather is sultry and the skin therefore perspiring; the forehead, neck, and unprotected parts of the scalp are especially vulnerable places. At some seaside resorts during the holiday season, when, in accordance with popular custom, the human anatomy is much exposed, the Biting Midges seize the opportunity of taking a heavy toll; it is not unusual to hear the blame for their misdeeds being attributed to much larger insects, "mosquitoes" in particular. In America, they are rather humorously referred to, by some, as "no-see-ums".

Life-histories of Ceratopogonidae may be broadly divided into two groups. In the first, the larvae are narrow, elongated worm-like creatures, which are aquatic and swim in an undulating fashion; they live in the mud or at the surface of ponds or ditches, and are associated with confervae or algae; or in the liquid products of farmyard litter-heaps and similar situations. The adults of this group are, as a rule, relatively bare-winged, as in *Culicoides*. Larvae of the second group are terrestrial and said to live in such locations as under the bark of trees, or in sap flowing from tree-wounds; in this case, the adults are generally hairy-winged, as in *Forcipomyia bipunctata* L. (Pl. **10**, 2), a common species whose larvae occur in rotting vegatation and old manure heaps. The immature stages of Ceratopogonidae are described by Miall[3], Saunders[4] and Kettle and Lawson[22]; the larvae are not equipped with spiracles, respiration being effected by absorption through the integument. The adults when at rest, unlike Chironomidae in which family they were for long included, carry the wings folded one above the other.

For the identification of British species, see Edwards;[5] a separate account of the genus *Culicoides* is given by the same author in Edwards, Oldroyd and Smart,[6] by Kettle and Lawson[22] and by Campbell and Pelham-Clinton[28].

SIMULIIDAE.

3 genera. 41 species. Small flies; body short and stout; short, thick legs; antennae short and bare, segments not distinctly separated; ocelli absent; mouthparts short, adapted for piercing in ♀; wings broad, anterior veins strengthened, posterior faint and of characteristic arrangement, posterior cross-vein absent. (Pl. **10**, 1.)

Mouthparts of female Simuliidae correspond closely to those of Ceratopogonidae in their method of operation; they are used to obtain mammalian blood, including that of human beings. As in that family, attacks seem to be more concentrated when the weather is sultry and the skin perspiring. The females have a characteristic habit of flying, quite noiselessly, in a cloud around the head of their victim; this can be almost as irritating as the minor inflammations from the bites, which may be felt for some time afterwards. We have known these flies, on occasions, to congregate in such numbers as to make it almost impossible to open the eyes, as a result of which it has been difficult not to yield to a wild impulse to break into a run in an effort to escape. In such cases tobacco smoke affords some relief from their attentions. The males, also, have a similar habit of assembling in clouds around the head; their mouthparts, however, are weak and not adapted for piercing.

The type of locality in which the adults are most likely to be found will be that which is most suitable for the development of the immature stages, which are aquatic; consequently they are well known to anglers; sometimes, however, adults may be found at considerable distances from water, where, no doubt, they are windborne. Simuliid larvae and pupae demand a degree of aëration of the water not normally found in stagnant ponds or pools; they are, therefore, only to be found in brooks, streams and rivers, where there is constant movement of the water. The degree of flow favoured by various species varies a good deal; some common species manage quite well where the flow is not much more than a trickle; others prefer to attach themselves to weeds in rivers with a powerful, steady current; yet others affect only hill streams, attaching themselves to rocks, over which water cascades with considerable force.

Miall [7] gives an account of the immature stages of *Simulium*. The eggs are laid in a glutinous mass attached to aquatic plants; in some species, the females are said to crawl right into the water for this purpose, often descending to a considerable depth. In its structure, the larva displays a degree of adaptation appropriate to the special type of habitat affected (Fig. 10C). At the hind end of the body is a kind of sucker-pad, by means of which the larva is able to attach itself to the surface of stones or water weeds, thus preventing itself from being washed away by the current. It is able, also, to spin a silken web upon the spot which it chooses for its station and augments its hold by clinging to this with the aid of minute hooks on the sucker. Food, in the form of tiny particles of debris carried along by the current, is trapped by the mouth-brushes. Not surprisingly, the larvae are usually found in "colonies" in particular spots where optimum conditions obtain.

Special adaptation to environment is also evident in the pupa (Fig. 11C) which is equipped with two bunches of respiratory filaments and enclosed within a silken cocoon. As development proceeds, the puparium becomes distended with an accumulation of air; up to this point, some similarity in the process with that of *Chironomus* (p. 71) may be discerned. The method of escape adopted by the adult fly, however, is very different; the puparium, still submerged, and now with the anterior part protruding from the cocoon, splits dorsally along the thorax. Surrounded by gases retained by the hairy

PLATE 9

RHAGIONIDAE-I

1. *Rhagio scolopacea* L., ♂, p. 98, showing venation-teratology. *Inset.* Normal venation.

CYRTIDAE-I 2. *Oncodes pallipes* Latr., ♂, p. 113.

TABANIDAE-I

3. *Chrysops caecutiens* L., ♀, p. 103. 4. *Chrysops quadratus* Mg., ♀ abdomen, p. 103. 5. *Chrysops sepulchralis* F., ♀ abdomen, p. 103. 6. *Chrysops relictus* Mg., ♀, p. 103.

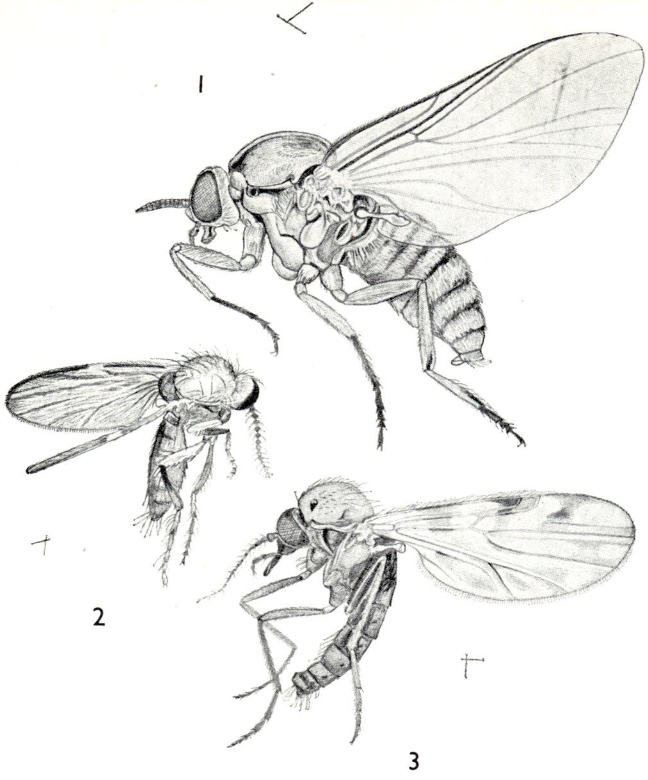

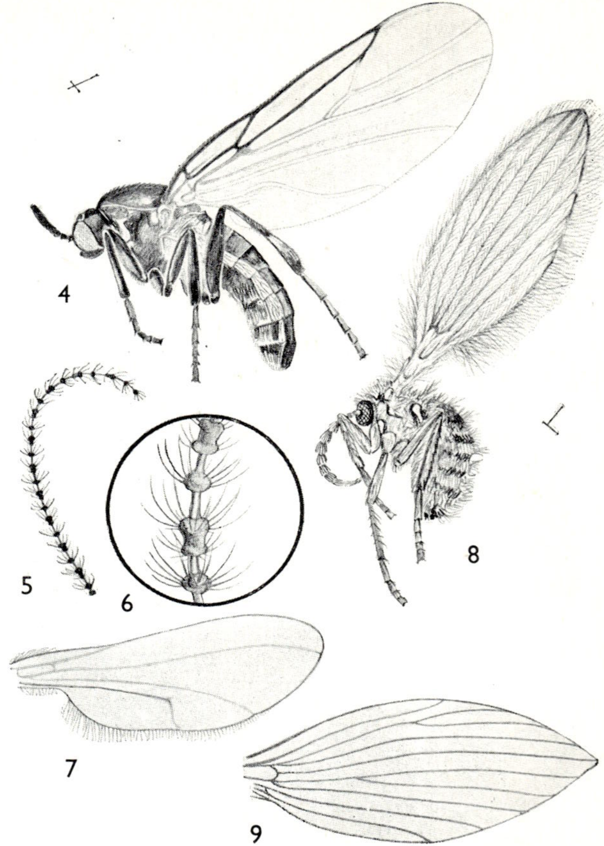

PLATE 10

SIMULIIDAE

1. *Simulium equinum* L., ♀, p. 81.

CERATOPOGONIDAE

2. *Forcipomyia bipunctata* L., ♀, p. 78. 3. *Culicoides pulicaris* L., ♀,

SCATOPSIDAE

4. *Scatopse notata* L., ♀, p. 85.

CECIDOMYIIDAE

5. Cecidomyiid antenna, p. 83. 6. Cecidomyiid antenna (enlarged), p. 83.
7. Cecidomyiid wing, p. 83.

PSYCHODIDAE

surface of the body, the fly emerges, ultimately rising to the surface in a bubble; there the bubble bursts and the fly immediately takes to its wings, unwetted. Mating takes place soon after emergence; vast swarms of males may sometimes be seen over water, extending from a foot or so above the surface up to seven or eight feet; hovering, almost motionless, facing the wind; it is probable that they assemble thus to await the emergence of females from the water.

Adult Simuliidae are easy to recognise by their very characteristic form and the particular features summarised earlier; in the United States, they are commonly known as "Buffalo Gnats", the name apparently arising from a fanciful resemblance of the humped thorax to the profile outline of an American Bison, or as "Turkey Gnats", due to the propensity of certain species for attacking poultry. Males in this family are holoptic; moreover, the compound eyes are divided, as in Bibionidae, into two areas, one with larger facets, and one with smaller (cf. p. 74).

Whilst the British species are not of any great economic or medical significance, in other parts of the world the attacks of Simuliidae are said to cause intense suffering and frequently the death of domestic animals and beasts of burden; some species are known to convey disease organisms, including parasitic worms, to man and beast. *S. ornatum* Mg. is known to convey a worm to cattle in this country, but no serious harm is caused in this case. *S. equinum* L. (Pl. **10,** 1), an abundant species, sometimes causes much distress to horses in England by its habit of feeding, in large numbers, inside the ears.

For the identification of British species, reference should be made to Smart,[8] who also describes the rearing of *S. ornatum* Mg. under artificial conditions in the laboratory[9]; the family is also dealt with by Edwards in Edwards, Oldroyd and Smart[10] and Davies[23].

CECIDOMYIIDAE.

152 genera. Over 600 species. Small or minute, delicate flies; wings broad, often hairy and fringed; legs long and thin; antennae long, with bead-like segments characteristically adorned with circlets or whorls of hairs; genitalia in both sexes often prominent, ovipositor often very long; venation greatly reduced, sometimes as few as three long veins reach margin, posterior cross-vein absent; ocelli present or absent. (Pl. **10,** 5–7.)

The attention of the observer in the field will most probably be more readily attracted to the larvae of Cecidomyiidae than to the adults; those of the majority of species live in plants where they give rise to the formation of galls. These galls are often characteristic of the species which cause them and thus serve as an aid to identification; a good deal of information concerning Cecidomyiid galls in Britain may be obtained from the works of Connold,[11] Swanton [12] and Niblett.[13] Some of these flies are of economic significance. *Mayetiola destructor* Say is the well-known Hessian Fly, which is said to have earned its name by its alleged introduction into America by Hessian troops; it causes in that country and in Southern Europe serious damage to wheat crops, although, for various reasons, it does not appear to be a very important pest in England. Two other "wheat midges", however, *Contarinia tritici* Kirby and *Sitodiplosis mosellana* Géhin, are said to be destructive in some parts of this country, especially the eastern and southern counties. Species of *Rhabdophaga*, *Helicomyia* and *Dasyneura* attack willows and osiers; there is a series of species which affect cultivated plants and vegetables, clovers and grasses. Accounts of British Cecidomyiidae of economic importance are given by Barnes [14] and Smith.[15]

Cecidomyiid larvae are rather short, narrowed at both ends and of various colours, including white, yellow, orange, red and brown. A peculiar process on the ventral surface of the anterior part of the body is known variously as a sternal spatula, anchor-process, or "breast-bone"; its purpose is somewhat obscure. It is said to assist the larvae in effecting the remarkable leaps which they are capable of making; these are apparently accomplished by linking the anchor-process to certain hook-like structures near the tail, thus looping the body; the hold is then suddenly released (cf. Piophilidae, p. 196, and Clusiidae, p. 216). The shape of the " breast-bone " is often quite distinctive and provides assistance in the identification of species.

Paedogenesis, to which we referred earlier (p. 71) is known to occur in the genus *Miastor*, which is represented in Britain. The female deposits four or five very large eggs, which give rise to very large larvae. Each of these develops within itself up to thirty more larvae which feed on the tissues of their parent-host, ultimately boring their way out to give rise to further larvae themselves. After a number of such generations, male and female adult flies develop in the normal way and start the whole process all over again. It seems that different types of larvae are produced; those which are paedogenetic have no " breast-bones ", while those which will pupate and produce normal flies possess this organ. An interesting feature of this phenomenon is that flies arising from a particular larva are invariably of the same sex.

Apart from the gall-causers, the larvae of a few species of Cecidomyiidae are predatory upon other insects, mostly plant bugs, mites and immature stages of other Diptera, including other species of Cecidomyiidae; a further few species develop in the excrement of leatherjackets or caterpillars and in rotting vegetation.

The adults in this family are small to minute flies, seldom more than 4–5 mm. in length; they are especially characterised by long, slender antennae, with bead-like segments, each bearing a circlet of small spines or hairs (Pl. **10,** 5–6). Wings are relatively large and broad, often with long hairs and fringes on the hind margin; long veins are few; cross-veins, when present, are near the base (Pl. **10,** 7). In the sub-family

Lestremiinae, as in Sciarinae (Mycetophilidae, p. 73), the eyes are approximated or joined above the antennae ; the former may be distinguished by the absence of apical spurs on the tibiae.

There is no single work dealing with the identification of all the British species of Cecidomyiidae ; Edwards [16] may be consulted in regard to the Lestremiinae, while much information may be gained from the works already cited, the bibliograhies contained therein, and Harris.[24]

PSYCHODIDAE.

7 genera. About 75 species. Minute flies; body, legs and wings adorned with long hairs ; wings often folded rootlike over body ; sometimes pointed at apex ; antennae with bead-like segments ; ocelli absent ; wing-venation characteristic, many long veins or branches reach margin, cross-veins inconspicuous and near base. (Pl. **10,** 8, 9.)

The Owl Midges or Hairy Moth-Flies are easily recognisable ; they may often be seen on windows, where they either run actively with a curious, jerky gait or remain perfectly still ; or they may be seen outdoors, particularly early or late in the year, in large numbers on tree-trunks after rain, when the atmosphere is humid. Another typical habitat is the undersurface of leaves in very damp places. *Psychoda phalaenoides* L. is essential in the pollination of the curious flowers of *Arum maculatum* L. (vide Grensted[27]).

Whilst the British representatives of the family have no medical significance, species of the sub-family Phlebotominae are known to be carriers of various diseases, including the well-known " 3-day fever " ; they are the Sand Flies so familiar to members of the Forces abroad.

Larvae of *Psychoda alternata* Say and other species sometimes occur in astronomical numbers in sewage filters (cf. p. 72) ; it has been recorded that the dust produced by the disintegration of dead adults of the above-mentioned species at a South African sewage works was responsible for attacks of chronic bronchial asthma in native workers.

The tiny larvae of Psychodidae are whitish, long-cylindrical and tapered at both ends; there is a distinct head, while the hind segment is elongated and bears outgrowths which surround the rear spiracles. Mainly aquatic habitats are selected, but development may take place in cow-dung, other excrement, or wet, decaying matter generally.

Freeman [17] may be consulted in regard to the identification of the adults; papers by Satchell[18] and Smith and Grensted[25] on the immature stages will be found interesting and useful.

SCATOPSIDAE.

10 genera. 35 species. Small to minute blackish flies; ocelli present; antennae short, segments not very distinct; wing-venation reduced, anterior veins very strong, posterior very faint; posterior cross-vein absent. (Pl. **10**, 4.)

Scatopsidae have been regarded by some as a sub-family of the Bibionidae; in contrast with the members of that family, however, the body is not hairy; the tibial spurs are wanting; and the eyes in both sexes are approximated above the antennae in a manner which resembles, to some extent, the Sciarinae (Mycetophilidae, p. 73) and the Lestremiinae (Cecidomyiidae, p. 84).

Scatopse notata L. (Pl. **10**, 4) is a very common one; a ♀ is depicted in the plate; the male may be distinguished from the other British species by the extremely short first segment of the hind tarsi. Holoplagia lucifuga Loew (=S. *transversalis* Loew), a much smaller species, about 2 mm. in length, is associated with the black ant, *Lasius fuliginosus* Latreille; we have observed it for many years in succession, running actively about among the ants on an old, rotten elm stump which houses a colony.

Ectaetia clavipes Loew, another tiny species about 2 mm. long, occasionally turns up on windows. We have noticed that, when killed, the females not infrequently exude a long ribbon of eggs, joined end to end, several times the length of the abdomen. The laying of eggs after death is a phenomenon which will be quite often encountered in Diptera; it will not be unfamiliar to the breeder of Lepidoptera. Females kept in captivity will often, when death approaches, deposit their eggs in places which appear to be most unsuitable and quite unlike the normal type of location selected; the muscles concerned with the ejection of the eggs apparently continue to function involuntarily for some little time after death.

Anapausis soluta Loew (=*Psectrosciara tenuicauda* Duda) has been recorded as swarming in large numbers in the hot noonday sun on wooden fences and similar places.

Scatopsidae develop in many media, including dung and various forms of decaying organic matter. The larva of *Scatopse* is not unlike a miniature version of *Bibio*, with nine pairs of open spiracles. For the identification of British species, see Edwards[19] and Collin.[26]

THAUMALEIDAE.

1 genus. 3 species. Small, bare, drab-brownish or yellowish flies; head small, ocelli absent; eyes holoptic in both sexes; palps longer than antennae; the two basal segments of antennae stout and remainder long and thin, giving impression of long, thin style; wing-venation simple, *sc* short, Veins 2 and 3 curved toward one another apically; ♂ genitalia large.

The species of this family, which are all about 3–4 mm. in length, are usually found on herbage on the banks of hill streams, mainly in Scotland, Wales and the northern counties; the larvae bear a strong resemblance to those of *Chironomus*; a pseudopod is present near the head (cf. Pl. **7**, 2). Immature stages are aquatic, the larvae favouring situations where a very thin trickle of water flows over rocks or stones, while the pupae are mostly found at the bottom of the stream. For further reading on the family, reference may be made to Edwards [20] and Séguy.[21]

LITERATURE CITED

1. Blair, K. G. Midges attacking *Melöe* Beetles. *Ent. mon. Mag.* **73**: 143. 1937.
2. Varley, G. C. A Ceratopogonid Fly, *Palpomyia semifumosa* Goet., feeding on Mayflies. *Ent. mon. Mag.* **84**: 36–8. 1948.
3. Miall, L. C. The Natural History of Aquatic Insects. p. 155. *Ceratopogon*. McMillan, London. 1895.
4. Saunders, L. G. On the life-history and anatomy of the early stages of *Forcipomyia*. *Parasitology*. **16**: 164–212. 1924.
   Saunders, L. G. On the life-history, morphology and systematic position of *Apelina* Kieff. and *Thyridomyia* n.g. *Parasitology*. **17**: 252–77. 1925.
5. Edwards, F. W. On the British Biting Midges. *Trans. ent. Soc. Lond.* **74**: 389–426. 1926.
6. Edwards, F. W. in Edwards, Oldroyd and Smart. British Bloodsucking Flies. 25–50, 129–48. British Museum, London. 1939.
7. Miall, L. C. The Natural History of Aquatic Insects. p. 175. *Simulium*. McMillan, London. 1895.
8. Smart, J. The British Simuliidae. *Freshwater Biol. Assn. Brit. Emp. Sci. Pub. No.* 9. Wray Castle, Ambleside. 1944.
9. Smart, J. On the Biology of the Black Fly, *Simulium ornatum* Mg. (Diptera, Simuliidae). *Proc. roy. Phys. Soc. Edin.* **22**: 217–38. 1934.
10. Edwards, F. W. in Edwards, Oldroyd and Smart. British Bloodsucking Flies. 50–66. British Museum, London. 1939.
11. Connold, E. T. British Vegetable Galls. Hutchinson, London. 1901.
    Connold, E. T. Plant Galls of Great Britain. Adlard & Son, London. 1909.
12. Swanton, E. W. British Plant Galls. Methuen, London. 1912.
13. Niblett, M. Notes on Some Gall-causing Cecidomyiidae: 1. *Entomologist.* **74**: 221–4. 1941.
    Niblett, M. *Ibid.*: 2. *Entomologist.* **75**: 14–16. 1942.
    Niblett, M. *Ibid.*: 3. *Entomologist.* **75**: 42–5. 1942.
    Niblett, M. *Ibid.*: 4. *Entomologist.* **75**: 109–12. 1942.
    Niblett, M. *Ibid.*: 5. *Entomologist.* **81**: 142–4. 1948.
    Niblett, M. *Ibid.*: 6. *Entomologist.* **82**: 159–62. 1949.
14. Barnes, H. F. Gall Midges of Economic Importance. (7 vols.) Crosby, Lockwood, London. 1946–8. (Vol. 8 completing the series is in preparation.)

15. Smith, K. M. A Text Book of Agricultural Entomology. University Press, Cambridge. 1931.

16. Edwards, F. W. On the British Lestremiinae, with Notes on Exotic Species. *Proc. R. ent. Soc. Lond.* (B) **7.** pt. 1: 18–24. pt. 2: 25–32. pt. 5: 102–8. pt. 9: 173–82. pt. 10: 199–210. pt. 11: 229–43. pt. 12: 253–65. 1938.

17. Freeman, P. Handbooks for the Identification of British Insects. Vol. IX. Pt. 2. Psychodidae. Royal Entomological Society of London. 1950.

18. Satchell, G. H. The larvae of the British species of *Psychoda* (Diptera: Psychodidae). *Parasitology.* **38**: 51–69. 1947.

Satchell, G. H. The ecology of the British species of *Psychoda* (Diptera: Psychodidae). *Ann. appl. Biol.* **34**: 611–21. 1948.

Satchell, G. H. The respiratory horns of *Psychoda* pupae (Diptera: Psychodidae). *Parasitology.* **39**: 43–52. 1948.

Satchell, G. H. The early stages of the British species of *Pericoma*, Walker. (Diptera: Psychodidae). *Trans. R. ent. Soc. Lond.* **100**: 411–47. 1949.

19. Edwards, F. W. A Synopsis of British Bibionidae and Scatopsidae. *Ann. appl. Biol.* **12**: 263–75. 1925.

20. Edwards, F. W. Revision of Thaumaleidae. *Zool. Anz.* **82**: 121–42. 1929.

21. Séguy, E. Faune de France. **12.** Orphnephilidae. Lechevalier, Paris. 1925.

22. Kettle, D. S. and Lawson, J. W. H. The early stages of British biting midges *Culicoides* Latreille (Diptera, Ceratapogonidae) and allied genera. *Bull. ent. Res.* **43**: 421–67. 1952.

23. Davies, L. Taxonomy of British Black-flies (Dipt., Simuliidae). *Trans. R. ent. Soc. Lond.* **118**: 413–508. 1966.

24. Harris, K. M. Gall midge genera of economic importance (Dipt., Cecidomyiidae), Part I: Introduction and subfamily Cecidomyiinae supertribe Cecidomiidi. *Trans. R. ent. Soc. Lond.* **118**: 313–58. 1966.

25. Smith, K. G. V. and Grensted, L. W. The larva of *Psychoda humeralis* Mg. (Dipt., Psychodidae). *Parasitology,* **53**: 155–6. 1963.

26. Collin, J. E. Notes on some British Scatopsidae (Diptera). *J. Soc. Brit. Ent.* **5**: 72–5. 1954.

27. Grensted, L. W. Diptera in the spathes of *Arum maculatum L. Ent. mon. Mag.*, **83**: 1–3. 1947.

28. Campbell, J. Allan and Pelham-Clinton, E. C. A taxonomic review of the British species of *Culicoides* Latreille (Diptera Ceratopogonidae). *Proc. R. Soc. Edinburgh,* **67**: B, 14–302. 1960.

KEY TO THE BRITISH FAMILIES OF NEMATOCERA (SECTION I)

(Based on Edwards, *Trans. Soc. Brit. Ent.* **5**: 2. 1938)

1. **V**-shaped suture on dorsum; Veins 6 and 7 both reach margin..2
   Suture on dorsum absent or not **V**-shaped; Veins 6 and 7 not both reaching margin 3
2. Vein 7 long (Fig. 7A); ocelli absent
   TIPULIDAE (p. 45)
   Vein 7 short, abruptly curved (Pl. **2,** 13); ocelli present
   TRICHOCERIDAE (p. 57)
3. Ocelli present 4
   Ocelli absent 8
4. Tibiae with apical spurs 5
   Tibiae without apical spurs; wing-venation reduced 7
5. Fork of Veins 2 and 3 at or before anterior cross-vein (Pl. **2,** 9)
   ANISOPODIDAE (p. 59)
   Fork of Veins 2 and 3, if present, well beyond anterior cross-vein 6
6. Antennae inserted below compound eyes, near mouth-opening; fore tibiae with characteristically modified apical spurs (Pl. **7,** 6–9)....BIBONIDAE (p. 73)
   Antennae inserted near middle of compound eyes. (Pl. **7,** 1, 3).....................MYCETOPHILIDAE (p. 72)
7. Antennae short, segments not very distinct; wings not hairy; small to minute blackish flies (Pl. **10,** 4)
   SCATOPSIDAE (p. 85)
   Antennae long, distinct bead-like segments with circlets of hairs; wings usually hairy, with fringe on hind margin (Pl. **10,** 5, 6)..................CECIDOMYIIDAE
   (Lestremiinae, p. 84)
8. Ten or eleven veins or branches reach margin 9
   Eight or less veins or branches reach margin 11
9. Principal cross-veins near or beyond middle of wing; medium-sized to large flies 10
   Principal cross-veins near base of wing; small to minute moth-like flies. (Pl. **10,** 8, 9) ......PSYCHODIDAE (p. 84)
10. Vein 3 forked; tibiae spurred; **U**-shaped suture on dorsum; resembling Crane Flies (Pl. **3,** 2)
    PTYCHOPTERIDAE (p. 58)
    Vein 3 not forked; tibiae without spurs; no suture on dorsum; gnat-like flies (Pl. **3,** 4) ....CULICIDAE (p. 61)

11. Four or less veins reach margin (Pl. **10,** 7)
CECIDOMYIIDAE (p. 82)
Six or more veins reach margin 12
12. Cross-veins near base of wing .... THAUMALEIDAE (p. 86)
Cross-veins, if present, near middle of wing 13
13. Wings broad; antennae short and bare (Pl. **10,** 1)
SIMULIIDAE (p. 79)
Wings not very broad; antennae hairy 14
14. Vein 4 forked (Pl. **10,** 2, 3); wings folded, one above the other, at rest .... CERATOPOGONIDAE (p. 77)
Vein 4 not forked (Pl. **6,** 1); wings usually separated, when at rest .... CHIRONOMIDAE (p. 69)

CHAPTER V

SECTION II. BRACHYCERA. SHORT-HORNS

STRATIOMYIDAE . . . . . Soldier Flies.
RHAGIONIDAE (= LEPTIDAE). . Snipe Flies.

STRATIOMYIDAE.

13 genera. Over 50 species. Small to fairly large, flattened flies, without bristles; often with metallic lustre, or strikingly coloured; 3rd antennal segment annulated; ♂ eyes usually holoptic (except *Solva*); scutellum often bearing spines; tibiae without spurs (except *Solva*); empodium pad-like; squamae small; costa not surrounding wing; veins crowded anteriorly; discal cell small: small veins issuing apically therefrom usually faint. (Pl. **5**: **8**: **11**.)

THE name "Soldier Flies" is not inappropriate for many of the species of this family owing to their striking colours and distinctive markings; here indeed are to be found some insects to delight the artist's eye. Ranging in size from about 3 to 20 mm., they are somewhat sluggish in their habits and will more frequently be seen resting than in flight. Many species will be found on waterside vegetation, especially in coastal areas. At rest, the wings are folded over the abdomen, one above the other, with the anterior edges or costal margins more or less parallel. In order to assess the range of the family it will be necessary to consider representatives from each of the six sub-families.

SARGINAE.

One of the commonest, as well as a very striking species, is *Chloromyia formosa* Scop. (Pl. **5**, 1), both sexes of which have a beautiful metallic sheen; in the female, the abdomen is deep blue-green with violet reflections; the scutellum, it will be noted, as usual in this subfamily, is not armed with spines. This fly is in evidence on vegetation in damp places from May to July.

A number of rather elongated flies, all with a handsome metallic lustre, make up the genus *Sargus*. *S. bipunctatus* Scop. (Pl. **5,** 4) is not uncommon; like *formosa*, it displays a difference in colouring between the sexes. The male differs from the female figured in the plate in having the abdomen all shining coppery-green and lacking the orange-red markings at the base. Edwards [1] records a female of this species as immersing itself completely in cow-dung, probably for the purpose of oviposition. Also not uncommon, *G. iridatus* Scop. is nearly as long, with a brilliant green thorax and deep violet abdomen with bronze reflections; both sexes are alike in colour, but the female has the abdomen more club-shaped, i.e. much broader posteriorly. *G. cuprarius* L. is very similar, but here there is a conspicuous dark cloud about the middle of the wing, as compared with a general darkening in *iridatus*.

Michrochrysa polita L. (Pl. **9,** 2), a pretty little shining-green fly, is common in many kinds of location from May till early autumn, particularly on vegetation near water; often it will be found in considerable numbers on willows. In this species, the antennae and legs are black, which serves to distinguish it from *M. flavicornis* Mg.; here the antennae and legs are reddish or yellowish and the fly is rather smaller. *M. cyaneiventris* Zett. is even smaller still and differs in having the abdomen shining-black in both sexes.

BERIDINAE.

In this sub-family, the scutellum is adorned with four or six spine-like processes on the hind margin, which are pubescent like the surface of the scutellum itself; the males have the hind metatarsus swollen or dilated. Two species of the genus *Beris*, the common *B. vallata* Forster (Pl. **5,** 5) and the less common *B. clavipes* L. will be known by their black thorax and orange abdomen, which gives an appearance not unlike that of certain small, common species of Tenthredinidae (Hymenoptera), such as the Poplar and Gooseberry Sawflies. Males of *vallata* have black abdominal pubescence compared with pale in *clavipes*; in the females, *vallata* has the wings fairly clear with a dark stigma, while in *clavipes*, they are smoky all over. Other species of the genus are deep black with a more or less metallic green thorax and are distinguished by minor differences. Verral [2] mentions males of *B. geniculata* Curtis dancing in shade with a wild kind of flight like *Rhamphomyia* (Empididae, p. 127); also a male *vallata* "hovering", i.e. remaining in a more or less fixed position in the air, with the wings "fanning" very rapidly, a phenomenon found elsewhere in Diptera, particularly in the Syrphidae (p. 154) but not normally in such sluggish fliers as the Stratiomyidae. Oldroyd[14] records *B. morrisii* Dale as swarming at a height of seven feet over bare ground in a muddy lane; only males were present in the swarm, and the direction of flight was horizontal, not up and down as in the type of swarm discussed under Nematocera. Similar swarms are recorded by Parmenter[15] and Edwards.[3]

In *Chorisops tibialis* Mg. (Pl. **5,** 3), the wing-venation is characteristic, likewise the development of the hind legs; the scutellum bears four yellow spines. The species is not at all uncommon in the southern counties, and sometimes occurs in fair numbers; males are said to assemble in dancing swarms.

STRATIOMYINAE.

The largest species in the family are to be found in the genus *Stratiomys*; there are four British representatives, of which we illustrate *S. potamida* Mg. (Pl. **8,** 1) and *S. longicornis* Scop. (Pl. **8,** 2); the twin scutellar spines should be noted, also the structure of the antennae (Pl. **11,** 4) with their compound 3rd segments. *Stratiomys* species are usually to be found on waterside vegetation, particularly umbels such as Water Hemlock.

S. potamida flies rather heavily with a wasp-like hum. In the female, the transverse yellow abdominal bands are less in evidence. *S. chamaeleon* L. is about the same size but less common ; here the bands are broken, leaving a row of yellow spots on the sides of the abdomen. Both of the preceding species have bare eyes in both sexes ; *S. furcata* F., which is, like *potamida*, moderately common in the south and south-east of England, has hairy eyes in the male ; while *longicornis* has the eyes hairy in both sexes. *S. furcata* is somewhat smaller than *potamida* and *chamaeleon* ; it may easily be distinguished by its very narrow linear spots on the sides of the abdomen. *S. longicornis*, as may be seen from the plate, differs from all the others in having the abdominal spots either very faint or more often completely absent ; the unusually dense pubescence on the body gives it a marked bee-like appearance, which may account for the fact that records of its occurrence are not numerous. It is likely to be found in much the same sort of locality as the others ; we have, in fact, found it, together with *potamida* and *furcata*, by a marshy pond in a London suburb.

In general build, *Odontomyia tigrina* F. is not unlike the species of *Stratiomys*, but is much smaller (about 9 mm.), shining black and with very transparent, filmy wings and hardly-noticeable veins ; it is quite common in many places near water from early June to August. The same genus contains another common species, *O. viridula* F. (Pl. **8**, 3), in which the abdomen varies, in both sexes, from whitish to yellow and orange, or apple-green, with a dark median band.

Clitellariinae.

In this subfamily the scutellum is armed or unarmed. Some very strikingly-coloured species, less common than the two foregoing, are contained in the genus *Oxycera*. *O. trilineata* F. (Pl. **8**, 4) gets its name from the three prominent longitudinal bands on the dorsum. In *O. pulchella* Mg.

(Pl. **8,** 5), the rather startling colour-scheme, with its daub-like patches of yellow, gives a veritable " toy-soldier " appearance ; like the preceding species, *pulchella* appears to be fairly widely distributed in the southern half of the country.

Nemotelus species may be recognised by their characteristic " snout " or forward prolongation of the lower part of the head (Pl. **11,** 2) ; here again the annulations in the third antennal segment are well seen. These are small black and white flies, about 5 mm. in length, in which the males usually have the abdomen mainly white with black markings and the females, mainly black with white or pale yellow spots ; sometimes they are quite numerous in coastal or estuarine localities, where the larvae develop in the brackish water of pools and dykes and in salt-marshes. Males are said occasionally to indulge in dancing flights over the marshes.

PACHYGASTRINAE.

The scutellum in the British species of this sub-family is not armed. A little black, rather rotund species, *Pachygaster atra* Pz. (Pl. **11,** 1) may sometimes be found in considerable numbers near water on the foliage of such trees as Willow and Alder. Males of another species, *P. leachii* Curtis, have been recorded as hovering, by Parmenter.[4]

SOLVINAE (= XYLOMYINAE).

This sub-family, which is represented by one genus, with three species, presents a sharp contrast to the remainder of the family, especially by virtue of the fact that the mid- and hind-tarsi are spurred. None of the three species are well known in the adult stage, at least in the wild state. *Solva* (= *Xylomya*) *maculata* Mg. and *S. marginata* Mg. (Pl. **11,** 5) have, according to Verrall,[2] been bred on a number of occasions ; we have, ourselves, bred *marginata* from larvae taken, in company with several other kinds of Dipterous larvae in February, beneath the bark of an old, rotten oak stump in Epping Forest. The adults, looking, whilst alive, somewhat like Ichneumon Flies (Hymenoptera) with their black and yellow bodies and yellowish legs, emerged in late May and early June. Here again, the scutellum is unarmed; superficially, these flies bear less resemblance to the rest of the Stratiomyidae than to the species of Xylophaginae, fam. Rhagionidae (cf. Pl. **14,** 2); they are, however, of somewhat more robust build and may positively be distinguished by the presence of a closed cell immediately below the discal cell. *Solva* larvae are dark, reddish-brown, flattened, leathery creatures, narrowed anteriorly (Pl. **11,** 6) and of very sluggish habits; on the other hand, the puparium is of delicate texture and pale, transparent yellow colour, with rows of backwardly-directed spines on the abdominal segments (Pl. **11,** 6, 7).

Austen [5] has discussed the method of pupation and escape of the adult fly, together with its bearing on the systematic position of the genus; it is stated that pupation takes place within the hard larval skin, which retains its shape, but instead of the fly escaping direct from the puparium contained therein, the pupa emerges for about three-quarters of its own length from a longitudinal fissure in the dorsal surface of the anterior segments. After the emergence, the empty puparium remains wedged by its hind extremity in the fissure. In the case of some of our own *marginata* specimens, the puparium remained wedged (Pl. **11,** 6) but in others, it was found some distance away from the larval skin.

Immature stages of *Stratiomys*, *Nemotelus*, *Oxycera* and *Odontomyia* are aquatic while those of *Pachygaster*, *Microchrysa*, *Chloromyia*, *Geosargus*, *Beris*, *Chorisops* and *Solva* lead a terrestrial existence in earth, fungi, bark, rotten wood or debris in tree-holes. Terrestrial forms are more or less flattened-cylindrical in shape with a long conical head-segment, as we have seen in *Solva*. In the aquatic forms, the eggs are laid on plants near or in water, or sometimes actually on the surface. An account of the structure and biology of the immature stages of

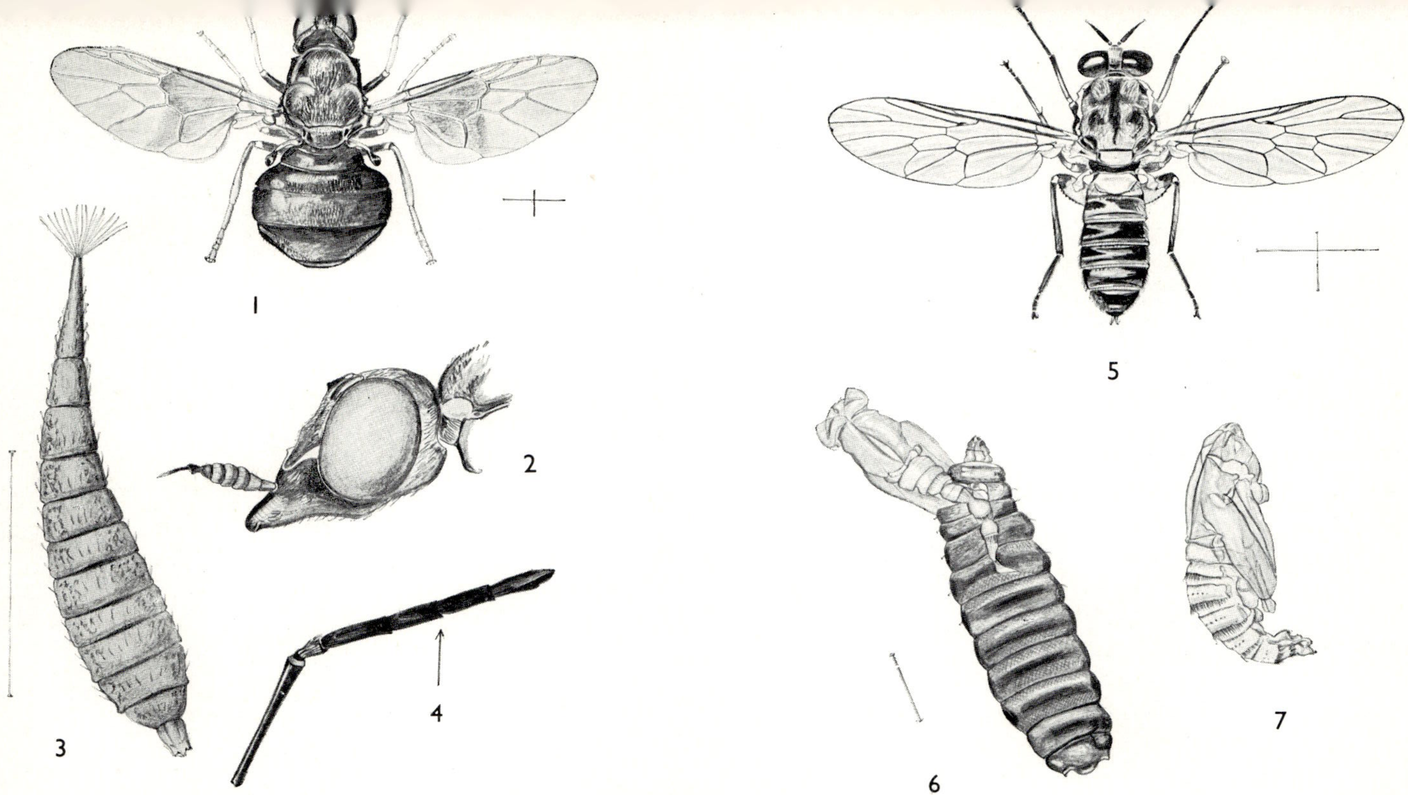

PLATE II

STRATIOMYIDAE-III

1. *Pachygaster atra* Pz., ♀, p. 95. 2. *Nemotelus*, head, p. 95. 3. *Stratiomys*, larva, p. 97. 4. *Stratiomys*, antenna, p. 93.

5. *Solva marginata* Mg., ♀, p. 95. 6. *Solva marginata* Mg., larval skin with puparium extruding, p. 96. 7. *Solva marginata* Mg., puparium, p. 96.

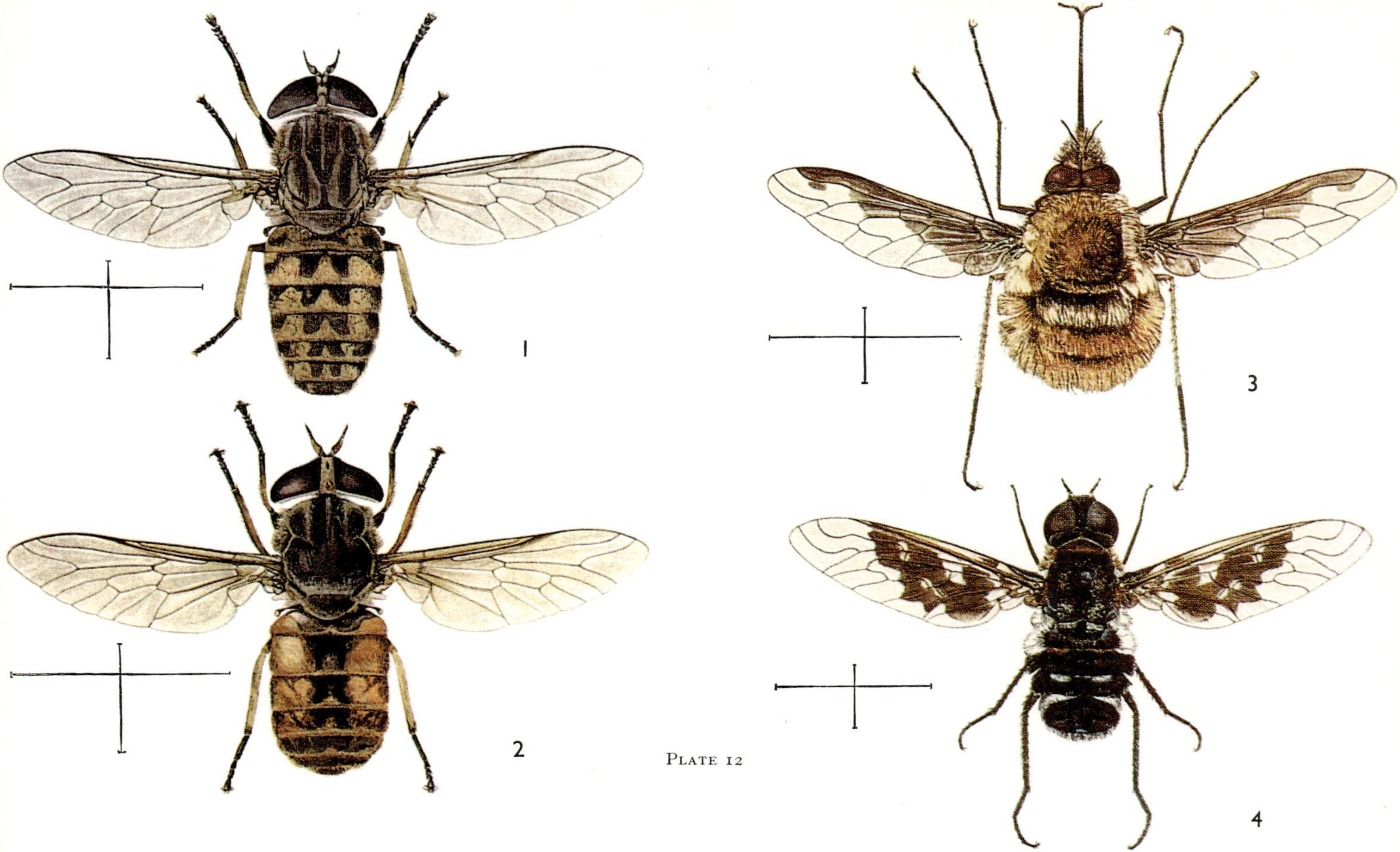

PLATE 12

TABANIDAE-II

1. *Tabanus bromius* L., ♀, p. 104.

BOMBYLIIDAE

3. *Bombylius major* L., ♂, p. 110.

Stratiomys is given by Miall ;[6] the larvae are equipped with a long, narrow extensible tail-segment, at the end of which there is a rosette of hairs surrounding the spiracle (Pl. **11,** 3). These hairs serve the purpose of buoying-up the hind extremity of the larva, when floating at the surface, and keeping the rear spiracles exposed to the air ; similarly when resting at the bottom, in very shallow water. When the larva descends, which it does by lashing its body, the rosette of hairs closes inwards, trapping a supply of air and preventing the entry of water into the spiracles. The fully-grown larva pupates in its last skin (cf. terrestrial form, *Solva*), sometimes emerging from the water for this purpose or sometimes floating at the surface. Only a part of the skin is occupied by the pupa, the remainder being filled with air. Some immunity to the attack of natural foes is probably acquired through its resemblance to a piece of floating debris, and some through the fact that the larval skin provides a tough armour ; this, being partly empty, no doubt deceives marauding beetle larvae and the like when they probe it with investigating mandibles. Aquatic larval forms are said to be mainly carnivorous.

Keys to the British species of Stratiomyidae are to be found in Verrall.[2] Papers by Brindle[11] deal with immature stages.

RHAGIONIDAE.

7 genera. 18 species. Small to fairly large, slender, long-legged flies, without bristles ; antennae not prominent ; 3rd segment not annulated (except *Xylophagus*) and usually with style ; ♂ eyes usually holoptic (except *Xylophagus*) ; some or all tibiae spurred ; empodium pad-like ; squamae small. (Pl. **9,** 1 : **14,** 1, 2.)

XYLOPHAGINAE.

Very little resemblance to Stratiomyidae can be seen in most species of Rhagionidae ; in this sub-family, however, the third antennal segment is annulated. Xylophaginae comprises the single British genus *Xylophagus*, with two species, both of which are uncommon and bear an even greater superficial resemblance to Ichneumon Flies (Pl. **14,** 2) than the Solvinae; from which, apart from the fact that they are of a longer and narrower build, they may be distinguished by the open cell immediately below the discal cell. In this case also, the larvae live in decaying tree-trunks in old forest districts such as the New Forest and parts of the Scottish Highlands; they are said to feed on other insect larvae occurring in the same habitats. The immature stages have been described by Buchanan White [7] and Perris.[8]

RHAGIONINAE.

The genus *Rhagio* (= *Leptis*) contains the largest species in the family. *R. scolopacea* L. (Pl. **9,** 1) is very common from May to July; the specimen illustrated exhibits what is known as venation-teratology, i.e. abnormality of the vein-arrangement, in the right wing and to a less extent in the left. The normal arrangement is shown in the inset.

R. scolopacea will usually be seen sitting, head downward, on tree-trunks, walls and fences and looking rather formidable. Moreover, it has a peculiar habit of suddenly taking to flight and making a dart toward the observer. Verrall [2] doubted the authenticity of certain records by Continental authors, which purported to show that this species is guilty of sucking the blood of mammals, but Séguy [9] takes the opposite view. We have not encountered an instance of such habit in any species of the family, although there appear to be indubitable records involving foreign species of *Symphoromyia*, *Atherix* and other genera. Furthermore, we have not seen any evidence of Rhagionidae preying on other insects, a mode of living generally imputed to the family; although it is to be admitted that the behaviour of *scolopacea* is sometimes such as to suggest "stalking"; the structure of the mouthparts, on the other hand, is not strongly indicative of predatory habits.

The larvae, however, appear to have a definite tendency to

carnivorous habits. Marchal[10] quotes the larva of *R. tringaria* Mg., the adult of which is closely similar to *scolopacea* in size, build and colour, but with clear instead of spotted wings, as preying upon small earthworms in leaf-mould or decaying wood. *Rhagio* larvae are long-cylindrical in shape, with a small, scarcely-retractable head, bearing strong mouthparts; the ventral surfaces of the body-segments being roughened to assist in locomotion and the hind segment bearing longitudinal grooves and terminal protuberances surrounding the rear spiracles. The pupae, unlike those of Stratiomyidae, are free and bear upon their middle segments transverse rows of strong, short spines with a battery of stronger ones on the hind segment.

In the genus *Atherix*, the three British species are not common. Females of *A. ibis* F. are stated to congregate, in large clusters, to oviposit on vegetation overhanging water and to die as they do so. The result is a mass of dead flies and eggs, usually pear-shaped, containing literally thousands of bodies. Larvae hatching from the eggs fall into the water, where they develop in pieces of wood-debris or the submerged parts of wooden piles and similar locations. They are long-cylindrical, considerably narrowed anteriorly, with a small head-capsule containing well-developed mouthparts; each abdominal segment bears ventral pairs of fleshy pseudopods, crowned with tiny spines; the last segment has a much larger single median pseudopod ventrally and a posterior forked appendage, the branches of which are long-haired.

CHRYSOPILINAE.

This sub-family is represented by four genera containing eight species from which we will select, as an example, the common *Chrysopilus cristatus* F. (Pl. **14,** 1), a delicate fly which may be found from June to August in damp, shady places such as bracken on the borders of woods, the undergrowth of close thickets, or vegetation bordering streams. The body, in fresh specimens, is clothed with golden pubescence, which soon becomes worn and lost. A female is depicted in the plate; in the male, which is deep black in ground-colour instead of light grey, the head with its holoptic eyes is wider than the thorax. *Chrysopilus* larvae live in decomposing vegetable matter or rotten wood, like those of *Rhagio*, and are said to be carnivorous.

For the identification of British species of *Rhagionidae*, see Verrall.[2] For early stages see Brindle[12], Hobby and Smith.[13]

LITERATURE CITED

1. Edwards, J. Oviposition of *Geosargus bipunctatus* Scop. *Ent. mon. Mag.* **83**: 64. 1947.
2. Verrall, G. H. British Flies. Vol. V. Gurney & Jackson, London. 1909.
3. Edwards, J. Swarming of *Beris morrisii* Dale. *Entomologist.* **80**: 27. 1947.
4. Parmenter, L. The Hovering of a Stratiomyid (Dipt.). *Ent. Record.* **58**: 114. 1946.
5. Austen, E. E. On the Preliminary Stages and mode of escape of the Imago in the Dipterous genus *Xylomyia* Rond. (*Subula* Mg. et Auct.) with especial reference to *Xylomyia maculata* F. and on the systematic position of the genus. *Ann. Mag. Nat. Hist.* **3**: 181–90. 1899.
6. Miall, L. C. The Natural History of Aquatic Insects. p. 189. *Stratiomys*. McMillan, London. 1895.
7. Buchanan White, F. Metamorphoses of *Xylophagus cinctus* F., and *X. ater* F. *Ent. mon. Mag.* **13**: 160–2; 216. 1876–7.
8. Perris, E. Histoire des insectes du pin maritime. *Ann. Soc. ent. France.* (4) X: 202–7. 1870.
9. Séguy, E. Faune de France. **13.** Rhagionidae. Lechevalier, Paris. 1926.
10. Marchal, P. Sur la biologie larvaire de *Leptis tringaria* Meigen. *Bull. Soc. ent. France.* 233–7. 1903.
11. Brindle, A. Taxonomic notes on the larvae of British Diptera. 3. The genus *Solva*, Walker (Stratiomyidae). *Entomologist,* **94**: 202–5. 1961. 7. The genus *Pachygaster* Meigen (Stratiomyidae). *Entomologist,* **95**: 77–82. 1962. 17. The Stratiomyinae (Stratiomyidae). *Entomologist,* **97**: 91–6. 1964. 17. The Clitellarinae (Stratiomyidae). *Entomologist,* **97**: 134–9. 1964. 23. The Geosarginae (Stratiomyidae). *Entomologist,* **98**: 208–16. 1965.

(Continued on p. 352)

CHAPTER VI

TABANIDAE . Horse Flies, Gad Flies, Stouts, Clegs.

3 genera. 28 species. Robust, more or less flattened flies, without bristles; head large, as wide as, or wider than thorax; occiput concave; 3rd antennal segment annulated; eyes holoptic in ♂, separated in ♀; mouthparts stout, adapted for piercing in ♀; palpi prominent, porrect in ♂, pendulous in ♀; squamae large;* tibiae spurred; empodium pad-like; costa extends round entire wing; Vein 3 with wide apical fork embracing wing-tip. (Pl. **9**: **12**: **14,** 3, 4.)

SOME of the largest and most rapid-flying of our British flies occur in this family. The females are equipped with biting mouthparts in the form of dagger-like organs (Pl. **14,** 4) and are fierce bloodsuckers. Apart from the distress which they cause to their normal hosts, which are mainly cattle, horses and deer, they often inflict painful bites upon human beings. An account of the feeding process is given by Cragg.[1] The effect of the bite varies with individual victims; sometimes very little discomfort is felt after the initial puncture, but in other cases severe irritation is experienced for several days. Although the family is not of serious medical significance in this country, Tabanids are known to transmit certain diseases in other parts of the world. It is noteworthy that these strong, fierce insects are included among the recorded prey of other Diptera, such as Asilidae (p. 117) and Dung Flies (Scatophagidae, p. 250), also of certain wasps and dragonflies.

Bloodsucking is not indulged in by the males, which subsist chiefly on nectar; they are, as a rule, much less in evidence

* Especially thoracic squamae; these are not very evident in the plates, since they are inclined to the vertical in the dried specimens (but see Fig. 2).

than the females and are usually to be found on trunks of trees, foliage or flowers; sometimes they may be encountered drinking at the margin of ponds or streams, or hovering in glades.

In life, the eyes of many species are beautifully coloured, with wavy, iridescent bands reminiscent of watered silk; these colours fade and disappear after death. Various names have been bestowed upon Tabanidae, including burrel-flies, gad flies, whame-flies, breeze-flies, horse flies, dun flies, stouts and clegs. Of these, the term "cleg" appears to be generally used today in respect of *Haematopota* species, while the name "stout" is used by the local folk of the New Forest area to denote the larger species of *Tabanus*. These stouts were, at one time, thought to be responsible for the phenomenon of "gadding" in cattle, i.e. rushing about in a state of panic, with tails erect, but according to Austen in Oldroyd,[2] this is properly attributable to the attentions of Warble Flies (*Hypoderma*, p. 293).

Haematopota species are able to approach, undetected, and to settle upon exposed parts of the body without being felt, until the proboscis pierces the skin. Some, but not all, of the species of *Tabanus* announce their approach by a humming noise, which varies in pitch according to species. A faint hum can usually be detected from *Chrysops* females as they draw close to their favourite point of attack, i.e. the back of the head or neck. Almost invariably, a rapid turn on the part of the prospective victim fails to reveal their presence, since they have already settled on the clothing and are stealthily making their way to a place where they may bite.

HAEMATOPOTINAE.

Two of the four British representatives of the sole genus *Haematopota* are common from May to September; in fact too common on occasions for the peace of mind of the field-naturalist. *H. pluvialis* L. is generally more abundant in the south and the closely-similar *H. crassicornis* Whlbg. (Pl. **14,** 3) more common in Scotland and the north. In *pluvialis*, the swollen basal segment of the antennae, in fresh specimens, is dull and the third segment bears clear yellowish markings, especially at the base; while in *crassicornis*, the basal segment is nearly all shining-black, the third segment all black and the wings generally more darkly mottled. *H. italica* Mg. is very local and mainly confined to coastal and estuarine neighbourhoods; in this species, the basal segment of the female antennae is characteristically thin; while the male may be distinguished from *crassicornis* by the reddish third antennal segment; and, by a tuft of short, tawny hairs behind the vertex, from *pluvialis*, in which the hairs in the corresponding position are long and black.

At rest, *Haematopota* species carry the wings folded, roof-like, over the abdomen; the females, in our experience, seem more persistent in their attentions in sultry weather, especially when thunder is imminent.

SILVIINAE.

The species of the sole genus, *Chrysops*, are easily distinguishable from one another, in the female, by the abdominal markings and coloration (Pl. **9**); these, with the pictured wings and iridescent eyes, make them a fine sight to see when at rest, with the sun shining on them. *C. caecutiens* L. (Pl. **9,** 3), locally known in Herefordshire as the "thunder fly", has black middle tibiae in both sexes. *C. quadratus* Mg. (Pl. **9,** 4) and *C. relictus* Mg. (Pl. **9,** 6) have the middle tibiae brownish-yellow; in the former, the male palps are pointed: in the latter, short and rounded. The sexes display little difference in colour in *C. sepulchralis* F. (Pl. **9, 5**), the abdomen and legs being all black or brownish-black; the species is decidedly local in the extreme south and scarcely recorded elsewhere.

Chrysops species are on the wing from June to August.

TABANINAE.

A single genus, *Tabanus*, represents the sub-family; here are included flies of bigger and more robust build, with characteristic scimitar-shaped antennae (Pl. **14,** 4). *T. sudeticus* Zeller, *T. verralli* Oldroyd, and *T. bovinus* L., which are usually abroad in July and August, are the largest British Tabanids and among the largest of all the British flies. Females of *sudeticus* may measure up to 25 mm. in length and nearly 50 mm. in wing-span; the others a little less. These very large stouts are local and mostly confined to old forest areas, such as the New Forest; the typical form of *sudeticus*, which is a rather variable species, seems to have been mostly recorded from Scotland. Males are much less frequently seen than the females, which are usually encountered in the vicinity of forest ponies or cattle in suitable localities; their very deep hum will betray their presence while flying.

Some of the smaller species of the genus are to be found quite commonly in varied types of locality from June to August, occasionally as early as May or as late as September. *T. bromius* L. (Pl. **12,** 1) and *T. bisignatus* Jaennike, which is roughly the same length, but somewhat stouter and darker in colour, are two particularly common and drab-looking species; they will often be found on the flanks or bellies of cattle in meadows. Their attacks seem to be borne with great fortitude by these patient creatures, which seldom display alarm or panic; the bites are often severe, since not infrequently blood may be seen still oozing from recent incisions. We are reminded of one particularly docile-looking Jersey cow, which at our approach, instead of moving away with the usual slow, bovine gait, came very deliberately toward us; turning her flank, she revealed three *bromius* females busily engaged in feeding at a spot where she was apparently unable to dislodge them with her tail. Her action seemed so very much like an appeal for assistance that we felt constrained to attempt to remove them with a wide-necked killing-bottle, a successful operation to which she submitted with great calmness. Subsequently, she even allowed us to make passes with a net around her ears, eyes, horns and hind legs to remove some of the numerous other flies swarming about her—a somewhat fruitless occupation, since any relief which she may have felt was speedily nullified by the arrival of hordes of reinforcements. However, this very friendly animal seemed to appreciate our attentions so much that she followed us around the hedges of the field where we were collecting; ultimately, she effectively put an end to our operations by continually pushing a moist nose against the nets we held in our hands and against our arms and backs. It was interesting to find that she behaved in a similar, effusively affectionate manner a day or two later, when we again visited her field.

T. distinguendus Verrall (Pl. **12,** 2) contrasts strongly with the two previously mentioned drab species by virtue of its orange-red lateral and dark median markings on the abdomen; it is common and widely distributed throughout the country.

Tabanid larvae are more or less long-cylindrical in shape, a little narrowed at both ends, with a pale or flesh-coloured integument, which is usually marked with fine longitudinal striations; transverse rows of protuberances are borne on the segments, presenting an easily-recognisable appearance (Fig. 10E). The retractile head is armed with two hook-shaped mandibles; with these, the larva makes an incision, for the purpose of sucking the body-contents, in the integument of its prey, which consists principally of insect larvae and worms. Our specimen illustrated in the figure was found in early February under the bark of a rotten oak stump in Epping Forest, in company with other Dipterous larvae, including leatherjackets, with which it was removed to a breeding-jar. After it had disposed of all its cohabitants, it was successfully reared on a diet of small earthworms from the garden. This creature's appetite was truly remarkable; several worms, each two or three times its own length, were sucked dry in the course of each week. Only once, however, did we discover it in the act of feeding, this occasion being in the middle of the night; in the day-time, it concealed itself beneath the fragments of bark and other litter in the jar. Throughout its larval state in captivity, care was taken to ensure moist conditions for it, similar to those in which it was discovered. Late in April it became quiescent, ignored the prey provided for it, and eventually pupated. About ten days later, a fine male *bisignatus* emerged.

The pupa (Fig. 11E), it will be seen, bears transverse rows of backwardly-directed spines, similar to those we saw in the Stratiomyidae (cf. Pl. **11**, 7). Escape of the adult fly is effected through a dorsal split in the anterior segments, assisted by subsequent lateral splitting.

Accounts of the immature stages of Tabanidae are given by Marchand,[3] Cameron [4] and Isaac.[5] Generally, the larvae appear to require moist conditions in which to live, such as damp soil or sand adjoining ponds or streams, or wet, rotten wood. Some, indeed, seem to have adapted themselves to a completely aquatic existence; eggs, in such cases, are laid on leaves and stems of plants overhanging water, into which the newly-hatched larvae fall. *T. bromius* and certain other species seem to prefer drier locations, such as the soil beneath meadow-turf. Choice of habitat does not appear to be uniform in any particular group, and may vary in an individual species. In all cases, pupation takes place out of the water.

For the identification of the British species of Tabanidae see Verrall,[6] Oldroyd[2] and Goffe.[7] Lyneborg[9] should also be consulted. Brindle[8] keys the immature stages.

LITERATURE CITED

1. Cragg, F. W. The Structure of *Haematopota pluvialis* L. *Sci. Mem. Med. and Sanit. Depts. India, N.S.* **55**: 1–36. 1912.
2. Oldroyd, H. in Edwards, Oldroyd and Smart. British Blood-

sucking Flies: 67–113: 149–152. British Museum, London. 1939.

3. Marchand, W. The Early Stages of Tabanidae (Horse-flies). *Monog. Rockefeller Inst. Med. Res.* No. 13: 1–203. 1920.
Marchand, W. Notes on the Early Stages of *Chrysops*. *J.N.Y. Ent. Soc.* **25**: 149–63. 1917.

4. Cameron, A. E. Bionomics of the Tabanidae of the Canadian Prairie. *Bull. Ent. Res.* **17**: 1–42. 1926.
Cameron, A. E. The Life History and Structure of *Haematopota pluvialis* L. *Trans. Roy. Soc. Edinb.* **58**(1): 211–50. 1934.

5. Isaac, P. V. A Practical and Simple Method of Rearing Tabanid Larvae. *Mem. Dept. Agric. India, Ent. Ser. VIII.* No. 5: 53–9. 1924.

6. Verrall, G. H. British Flies. Vol. V. Gurney & Jackson, London. 1909.

7. Goffe, E. R. British Tabanidae. *Trans. ent. Soc. S. England.* **6**: 43–114. 1931.

8. Brindle, A. Taxonomic notes on the larvae of British Diptera. 1. The genus *Haematopota* Meigen (Tabanidae). *Entomologist*, **94**: 121–4. 1961.

9. Lyneborg, L. A revision of the Danish species of *Hybomitra* End. (Dipt., Tabanidae). *Ent. Medd.*, **29**: 78–150. 1959.
Lyneborg, L. On *Tabanus tropicus* and other Linnean species of Palaearctic Tabanidae (Diptera). *Ent. Medd.*, **31**: 97–103. 1961.

CHAPTER VII

BOMBYLIIDAE . . .	Bee Flies.
CYRTIDAE	Spider-parasite Flies.
THEREVIDAE . . .	Stiletto Flies.
SCENOPINIDAE . . .	Window Flies.

BOMBYLIIDAE.

5 genera. 12 species. Small to fairly large, furry, bee-like flies, with long, thin legs; 3rd antennal segment not annulated; style small or vestigial, segmented; proboscis either very long and thin (*Bombylius, Phthiria*) or short, with prominent labella (*Anthrax, Thyridanthrax, Villa*); ♂ eyes holoptic or approximated (*Bombylius, Phthiria*) or separated (*Thyridanthrax*, etc.); empodium vestigial, pulvilli sometimes minute; wings often pictured, opened at rest; venation complex, costa surrounds wing, Vein 3 forked. (Pl. **12**.)

THE name, " Bee Flies ", accorded to members of this family has a double significance; firstly, on account of the resemblance of certain species to some of the smaller Humble-Bees and, secondly, on account of the fact that the immature stages of many species of Bombyliidae are parasitic upon various Solitary Bees. Fabre [1] tells the interesting story of a Continental species, *Anthrax trifasciata* Mg. Eggs are laid on the ground near the nest of the host, a species of Mason-Bee, which seals the cell containing its grubs, just before they pupate, with a kind of mortar. The newly-hatched *Anthrax* larvae are minute, active, hair-like creatures, which wander off and search for a tiny orifice in the mortar, through which to penetrate to the cell. In this condition, they are able to subsist for long periods without food, as well they need, since gaining an entry is no easy task, even for so microscopic an organism. Once ensconced in the bee-cell, the larva does not necessarily begin feeding on its host immediately; the attack is made at a time when the bee-grub becomes quiescent, in preparation for the pupa-stage. At this point, the fly-larva undergoes a transformation into a smooth, fat grub with a small sucker-mouth. It obtains its sustenance from the bee-pupa by applying this sucker-orifice to its skin and, without causing a lesion, withdraws, slowly but surely, the whole of the body-substance of its host. This process is so regulated that the early death and subsequent putrefaction of the host, with its attendant disaster for the parasite, is avoided ; when the fly-larva is small, the withdrawal from the host is almost imperceptible ; as it grows, the process speeds up ; the host is ultimately sucked dry when the fly-larva is ready to pupate. This it does, perforce, inside the cell, being enclosed by an apparently impenetrable wall. The pupal stage has two phases ; the first, or pre-pupal phase, is normal, the transformation from larva to fly taking place within a normal puparium ; but now the problem of escape from the cell arises. In the case of the rightful occupant, exit would be effected with the aid of the powerful mouthparts, but the parasite-fly has no such armament ; so, another remarkable transformation takes place ; the second-phase pupa develops strong spines on the head, in the form of a coronet and, with this as a ram, commences an onslaught upon the imprisoning wall. Assistance in this process is gained from the transverse rows of strong spines on the abdominal segments. Eventually the wall is breached, the pupa protrudes its anterior part and the adult fly emerges to freedom.

Such a life-history, with minor variations, is fairly representative of many species of Bombyliidae, particularly of the genus *Bombylius*. In spring, one may often see a tawny-coloured, furry fly, very much like a small Humble-Bee, hovering above a primrose, with its long, thin proboscis extended into the flower, a performance akin to that of the Humming-bird Hawk and Bee Hawk Moths. Close observation will reveal that, in fact, the fly rests its two front legs, or sometimes the two hind ones, upon the flower, while feeding, and keeps its body suspended in the air by a vibration of the wings so rapid as to make them virtually invisible. When at rest, dark foremargins become apparent on the wings; in this state, the wings are held partly open, more or less at a right angle to one another. This extremely agile insect is the common *Bombylius major* L. (Pl. **12,** 3); its average length is about 11 mm., but very much smaller and rather larger specimens may be encountered; we possess a male which barely exceeds 7 mm. *B. major* is on the wing very early in the year, from about the middle of March until early May, and is generally distributed over the southern half of the country. Males of this species may not infrequently be seen hovering in clearings, in the spring sunshine, at five or six feet from the ground.

B. discolor Mikan, which is about the same size, and recognisable by the brown dots or cloudy patches at the junctions of veins and cross-veins and at the ends of the veins, is distributed over much the same area, but is less common. *B. canescens* Mikan and *B. minor* L. both have clear wings and are only about 8 mm. long. *B. canescens* may be known by the long black hairs amid the short, yellowish pubescence behind the eyes and by its mainly black femora, while *minor* lacks the black hairs and has mainly orange femora. Both are uncommon, *minor* being apparently confined to the south and south-west, while *canescens* is recorded from Scotland and Ireland in addition.

Chapman [2] gives notes on the life-history of *major*; eggs are laid broadcast by the female, when in flight; the host recorded by this author was a Mining Bee of the genus *Andrena*. *B. canescens*, also, is said to be associated with Mining Bees, particularly of the genus *Halictus*. Nielsen [3] has observed the immature stages of *minor*, the host in this case being a Solitary Bee of the genus *Colletes*; these bees construct burrows in the ground, within which they store pollen and honey for the sustenance of their larvae. First-stage larvae of ***minor*** are of thread-like form and feed on this pollen and honey store. Eventually, they moult, assume a maggot-form and attack the bee-grubs.

The life-history of some species of *Anthrax*, *Thyridanthrax* and *Villa* differs in that the immature stages are spent as internal parasites of other insects, notably Lepidopterous caterpillars, or even as hyperparasites in the parasitic larvae of Ichneumons (Hymenoptera) or Tachinidae (p. 259) within them; certain species affect the egg-pods of locusts. Consequently, there are variations in the form of the larvae, but the pupae are armed in a very similar manner to those which we have already discussed; no doubt the armature is necessary for the purpose of effecting an escape from the hard pupal shell of the host. As for our British representatives, the striking *Thyridanthrax fenestratus* Fall. (Pl. **12**, 4) appears to be more or less confined to sandy commons such as occur in Surrey, Hampshire and Dorset, where it is fond of basking on paths and bare patches in hot sunshine. Like other species of the family, it is remarkably quick in its movements. The closely allied species of *Villa*, which have little dark marking on the wings, save along the costal margin, are similar in build and habits; the same type of locality is favoured, except in the case of *V. paniscus* Rossi, which seems to prefer coastal sandhills. These flies are on the wing from June till September, but are mainly in evidence in July and August.

Phthiria pulicaria Mikan is the smallest British Bombyliid, at a little over 3 mm., and is yellowish-grey, only lightly pubescent, with a long proboscis; it is said sometimes to occur in numbers on the flowers of Mouse-ear Hawkweed and Cat's-ear, on or near coastal sand-dunes, from May to July.

Keys to the British species of Bombyliidae, with descriptions and records, are furnished by Verrall.[4]

CYRTIDAE.

2 genera. 3 species. Small, bristleless flies, with remarkably bulbous bodies and small heads; head almost entirely occupied by enormous compound eyes, holoptic in both sexes; antennae with only two apparent segments (basal segment concealed), inserted at top of head, with very long arista (*Acrocera*), or at bottom, with long, thin style (*Oncodes*); proboscis absent; thorax humped; thoracic squamae greatly developed; empodium pad-like; wing-venation much reduced, costa not extending beyond apex of wing; veins rather weak and indistinct (*Oncodes*). (Pl. **9**, 2: **14**, 5.)

Life-histories in this family, so far as they are known, match for interest those of the Bombyliidae. The eggs, which are said to be black, have been found on dead branches and twigs, and on Horse-tails; those of the genus *Oncodes*, which is represented in Britain, are deposited in masses. Since the older larval stages are known to be internal parasites of spiders, Séguy [5] argues that the first-stage larvae must be active wanderers, like those of Bombyliidae, in order to reach their hosts. Now, the first-stage larva of a foreign species, *Pterodontia flavipes* Gray, has been studied by King [6] and described as possessing bands of strong spines on both dorsal and ventral surfaces, with a sucker and two long bristles at the hind extremity. It is, apparently, capable of progressing in similar fashion to that of the well-known " Looper " caterpillars, but also has the power of leaping. This latter feat is said to be accomplished by standing upright upon the sucker, with the two tail-bristles bent, suddenly releasing the hold of the sucker and simultaneously straightening the bristles. As much as 5–6 mm. may be traversed in a single leap. Entry to the host's body is gained through the leg-joints. Early larvae of *Oncodes*, also, are said to be capable of leaping.

A foreign species, *Astomella lindeni* Erichs., has been stated to live, in the larval stage, in the body of a spider, with its rear spiracles connected to the lung-books of the host; this device for breathing, while totally enclosed in the body of a host,

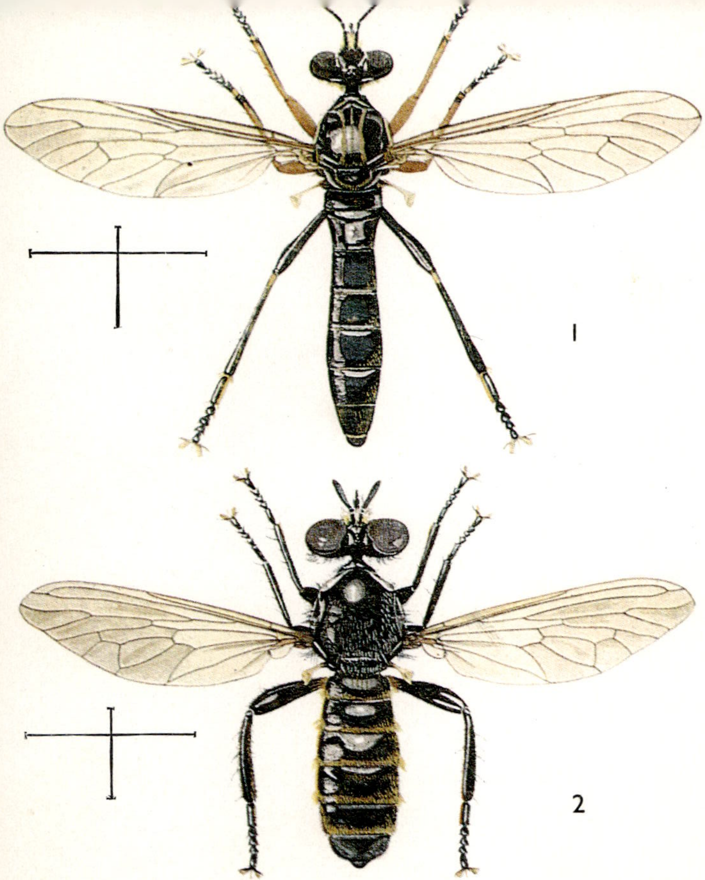

PLATE 13

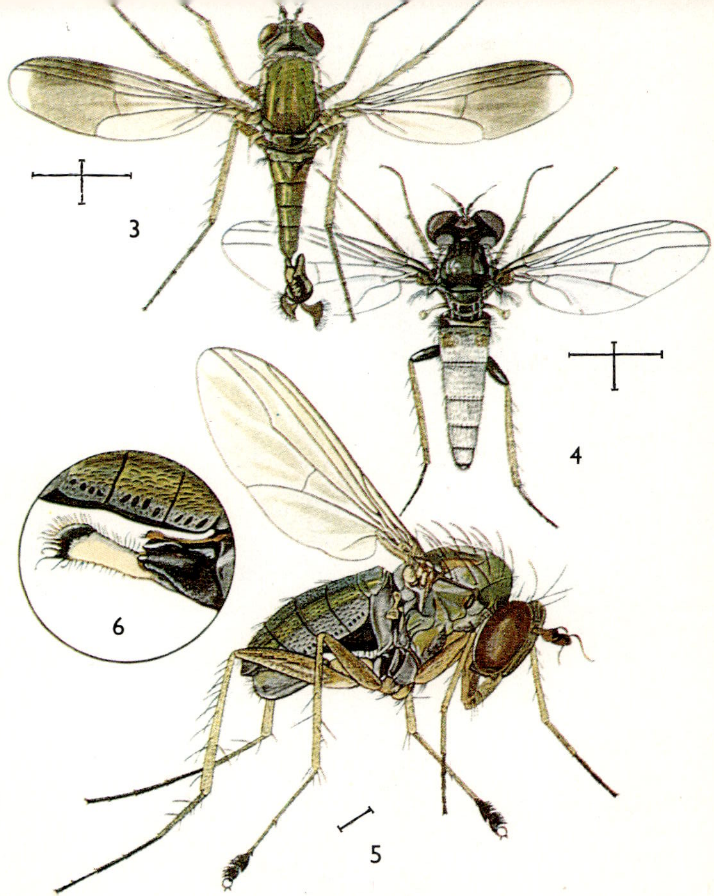

ASILIDAE-II

1. *Dioctria rufipes* Deg., ♀, p. 122.
2. *Laphria marginata* L., ♀, p. 121.

DOLICHOPODIDAE-I

3. *Poecilobothrus nobilitatus* L., ♂, p. 134 (genitalia everted). 4. *Argyra diaphana* F., ♀, p. 137. 5. *Dolichopus popularis* Wied., ♂, p. 133. 6. *Dolichopus ungulatus* L., p. 134, ♂ genitalia.

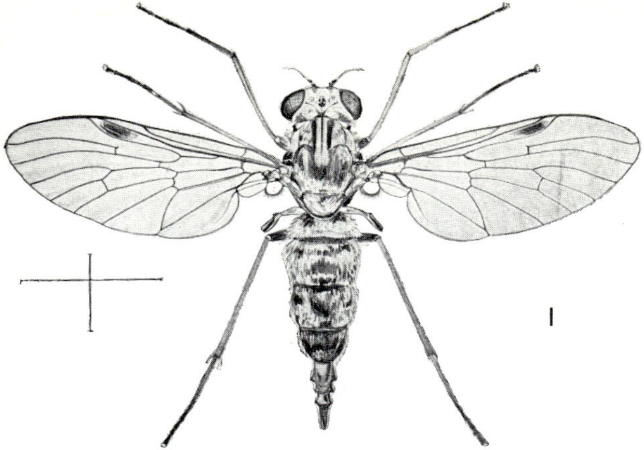

1

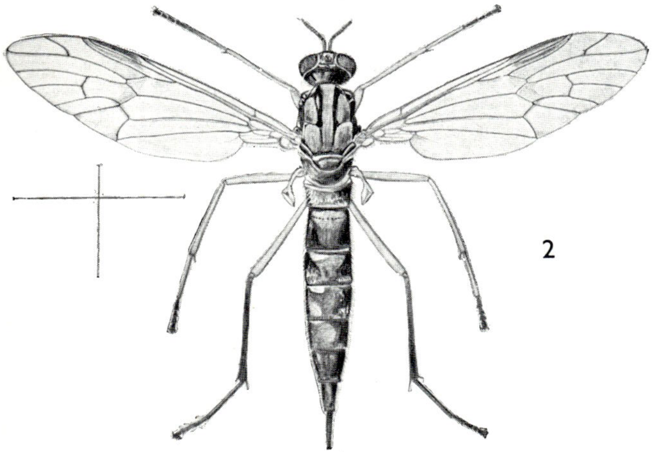

2

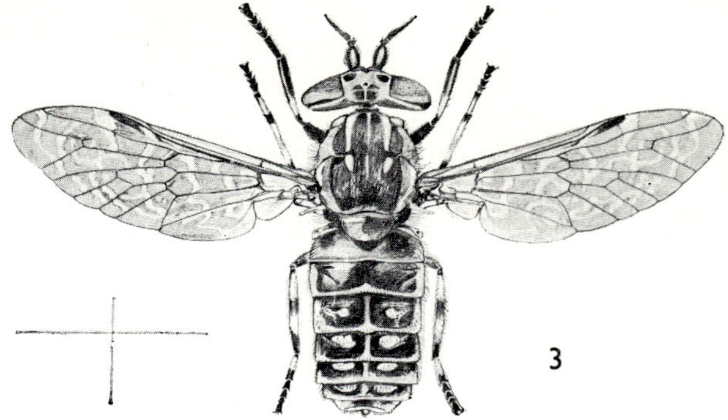

3

4

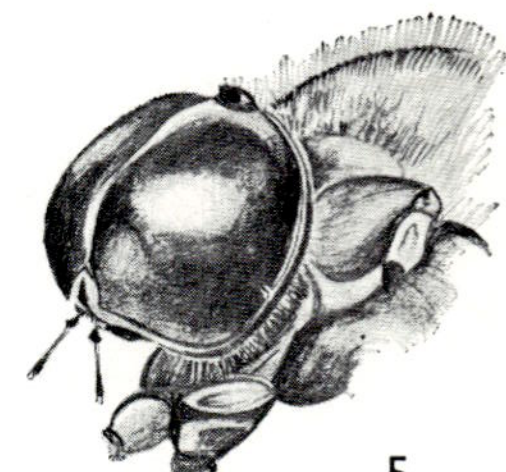

5

PLATE 14

RHAGIONIDAE-II

Chrysopilus cristatus F., ♀, p. 99

TABANIDAE-III

3. *Haematopota crassicornis* Whlbg., ♀, p. 103. 4. *Tabanus*, ♀ head, p. 101.

CYRTIDAE-II

f.p. 112

we shall encounter elsewhere (cf. Conopidae, p. 177, Tachinidae, p. 259). Lundbeck[7] also refers to a Cyrtid species in this connection. Emerton[8] relates his discovery of larvae and pupae of a species of *Acrocera,* the other genus represented in this country, resting on spiders' webs, each being adjacent to the shrivelled, empty skin of a full-grown spider. The larvae were fat and fleshy, with the hind part greatly swollen, the whole body being roughly the shape of an electric bulb. They were hanging on the web by their mouthparts and supported by strands of web beneath them.

Obviously, a good deal more remains to be discovered concerning the life-history of these flies; as, for instance, what, exactly, happens when the young migratory larva encounters its prospective host. Unfortunately, Cyrtidae are rare in this country and opportunity for observation thus limited; it is not unlikely that more knowledge will ultimately be forthcoming from students of spiders rather than from dipterists.

Our example, *Oncodes pallipes* Latr. (Pl. **9,** 2), like the other two British species of the family, seems to be local and confined to the south; we have found it by sweeping under trees, and once as the prey of the Asilid, *Leptogaster cylindrica* Deg. (p. 119). The highly characteristic head (Pl. **14,** 5) should be noted. *O. gibbosus* L. has entirely white squamae and darkened femora and may thus be distinguished from *pallipes,* in which the squamae are smoky, with darkened margins, and the legs entirely pale. There are records of considerable numbers of *gibbosus* occurring at times in the New Forest.

Acrocera globulus Pz. has also occasionally been recorded as abundant in the same locality; it may be readily distinguished from the two species of *Oncodes* by its distinct wing-veins and the insertion-point of the antennae at the top of the head.

The flight of Cyrtidae is said to be slow and feeble; hovering is accompanied by very slight backward and forward movements, like the swinging of a very slow pendulum, or of a small caterpillar on a gossamer thread; the places chosen for hovering thus are sunlit clearings in woods, or high up in the air.

For detailed descriptions of the British species of Cyrtidae, with notes on localities, etc., see Verrall.[4] Schlinger[12] discusses the biology of *Oncodes*.

THEREVIDAE.

> 3 genera. 12 species. Medium-sized, somewhat elongated, bristly and pubescent flies, with slender legs; basal segment of antenna long, 2nd very short, 3rd elongate-conical with short, sharp apical style; face and frons densely hairy (*Thereva*) or face bare (*Psilocephala, Tabuda*); ♂ eyes holoptic; proboscis short with prominent labella; empodium bristle-like or absent; wing-venation complex, costa surrounds wing, Vein 3 forked; ovipositor with a circlet of spines. (Pl. **15,** 2.)

Adult Therevidae have been stated by various authors to be predatory upon other insects, but there appears to be no very definite supporting evidence; we, personally, have never witnessed an instance of such habit. They are often to be seen resting on foliage; some species favour sandy banks or coastal sand-dunes; males are said, on occasions, to execute a kind of wild dance in hot sunshine in such locations.

Thereva nobilitata F. (Pl. **15,** 2) is one of the commonest British Therevidae; its life-history and habits are discussed by Collinge.[9]

The larvae are long, thin and worm-like; they appear to consist of 20 segments, but this is due to the fact that the abdominal segments are marked by a strong, transverse constriction. A distinct, darkened head is present; the thoracic segments each bear a pair of bristles and the tenth abdominal segment, three pairs; while the hind extremity bears a pair of small, thin protuberances. Therevid larvae are rapid and snake-like in their movements. Humid conditions are favoured; recorded habitats include rotten wood, leaf-mould and fungi; not infrequently they are turned up in garden soil. Normally, they are said to be predatory, but damage to cabbages and potatoes has been imputed to them. The pupa is very short, squat and particularly mummy-like in appearance; there are spur-like projections enclosing the antennae of the developing fly and a very stout, curved spine projecting backward from the base of each wing-sheath; the abdominal segments also bear backwardly-directed spines.

No doubt the name, "Stiletto Flies", refers to the characteristic shape of the body of the adults, the thorax constituting the "handle" and the long, pointed abdomen, the "blade".

For the identification of the British species of the family, see Verrall[4], Collin[10] and Taylor[11].

SCENOPINIDAE.

1 genus. 3 species. Small, black flies, without bristles; antennae pendulous, the two basal segments short and the 3rd long, without style; ♂ eyes holoptic (*fenestralis*) or separated; proboscis very short; squamae small; empodium bristle-like; wing-venation simple and distinctive, costa not extended beyond apex of wing, Vein 3 forked. (Pl. **15**, 1.)

Of the three British species of the sole genus, *Scenopinus*, the most likely to be seen is *S. fenestralis* L. (Pl. **15**, 1), a dull black fly, with reddish-brown legs, which may bear blackish markings; the antennae are situated low, in a kind of pit, while the third segment is elongated and tapers slightly toward the apex, where there is a curious oval depression. On the second abdominal tergite, in both sexes, there is a pair of peculiar oval, pitted areas.

The species is usually encountered on the windows of old houses, stables and outhouses, which no doubt accounts for the popular name. In such situations, the fly is sluggish and somewhat Stratiomyid-like in its movements; it is more often seen resting, motionless, with the wings folded, parallel, over the abdomen. At one time, according to Verrall,[4] *fenestralis* was known as the "Carpet Fly", due to a belief that its larvae were destructive to carpets, stable-rugs and similar articles;

it is now thought that they are predatory upon the larvae of clothes moths and fleas living in such situations. Dry, dusty habitats are favoured by the larvae, unlike those of Therevidae, to which, although smaller, they bear a great resemblance in form ; their movements are also rapid and serpentine. The head is somewhat elongated and pointed, yellow in colour and furnished with strong mouthparts ; a pair of small bristles adorns each of the thoracic segments, also the hind segment, which bears a short, fleshy, pseudopod-like protuberance in place of the two processes in Therevid larvae. Transverse rows of backwardly-directed spines equip the abdominal segments of the short, stout pupa.

Verrall [4] deals with the British species of Scenopinidae in detail.

LITERATURE CITED

1. Fabre, J. H. The Life of the Fly. Hodder & Stoughton, London. 1927.
2. Chapman, T. A. On the economy, etc. of *Bombylius*. *Ent. mon. Mag.* **14**: 196–200. 1878.
3. Nielsen, J. C. Ueber die Entwickelung von *Bombylius pumilus* Mg., einer Fliege, welche bei *Colletes daviesana* Smith., schmarotzt. *Zool. Jahrb. Abt. Syst.* **18**: 647–58. 1903.
4. Verrall, G. H. British Flies. Vol. V. Gurney & Jackson, London. 1909.
5. Séguy, E. Faune de France. **13**. Oncodidae. Lechevalier, Paris. 1926.
6. King, J. L. Observations on the Life-history of *Pterodontia flavipes* Gray. *Ann. ent. Soc. Am.* **9**: 309–21. 1916.
7. Lundbeck, W. Diptera Danica. Vol. 1. Wm. Wesley & Son, London. 1907.
8. Emerton, J. H. An internal Dipterous parasite of Spiders. *Psyche.* **5**: 404. 1890.
9. Collinge, W. E. Observations on the life-history and habits of *Thereva nobilitata* Fabr. and other species. *J. Econ. biol.* **4**: 14–18. 1909.
10. Collin, J. E. British Therevidae. *Proc. Roy. Phys. Soc.*, **23**: 95–102. 1948.
11. Taylor, E. *Thereva lunulata* Zett. in Northumberland. *Ent. mon. Mag.* **89**: 251. 1953.

(Continued on p. 352)

CHAPTER VIII

ASILIDAE . . . Robber Flies, Assassin Flies.
EMPIDIDAE.

ASILIDAE.

15 genera. 27 species. Medium-sized to large, robust, bristly and hairy flies, usually of narrow build and with strong bristly legs ; 3rd antennal segment not annulated ; style present (except *Laphria*) ; eyes separated in both sexes and vertex sunken ; face with prominent " beard " ; mouthparts horny and adapted for piercing in both sexes ; empodium bristle-like, pulvilli large, or absent (*Leptogaster*) ; wings folded over abdomen at rest ; venation complex, Vein 3 forked ; genitalia usually prominent in both sexes. (Pl. **1** : **13**: **15**, 3–7.)

WHEREAS in the Tabanidae we saw an adaptation of the female mouthparts to a piercing function for the purpose of sucking the blood of mammals, in the Asilidae and the succeeding family, Empididae, the mouthparts of both sexes are in the form of rigid or semi-rigid piercing organs, but sustenance is obtained by sucking the body-substances of insects.

Methods adopted by these flies to effect the capture of their prey fall into two categories. Species of the sub-family Asilinae and of the genus *Laphria* are in the habit of remaining motionless on foliage, sunlit paths, bare earth or sandy patches, tree-trunks and similar locations, waiting for their quarry. While so doing, they often affect a kind of crouching position, resting more or less on their sides, with the legs drawn up in such a way as to remind the observer, appropriately even if somewhat fancifully, of a leopard crouching to spring. When a potential victim arrives at a suitable proximity, the Asilid makes a sudden flying hop or " capture-dart ", during which the legs are allowed to hang slackly until contact is made.

Leptogaster and *Dioctria* species usually take their prey on the wing, and fly with the front legs extended like a scoop.

Often, the prey is bigger and stronger than the hunter, and sometimes formidably armed, but the sudden nature of the attack and the immediate quietus administered while the victim is still surprised, enables very considerable odds to be overcome. There is some evidence for the view that the instantaneous collapse of victims of Asilid attack is due to the injection of poison, since the presence of glands opening into the proboscis has been established by Whitfield.[1]

Asilids, themselves, are subject to attack by other insects, such as Tiger Beetles, Hunting Wasps and Biting Midges (Ceratopogonidae, p. 77), by entomophagous fungi, mites, spiders and birds. According to one authority, they are sometimes accompanied by tiny flies of the genus *Desmometopa* (Milichiidae, p. 238), which attach themselves to the thorax and participate in the feast when prey is wounded and juices exuding.

Male Asilids may become the victims of their voracious partners at the opening stage of the mating process, an unhappy fate not unknown elsewhere among Arthropods; the approach by the suitor is therefore usually made with caution. There is little evidence to suggest that females preoccupied with prey are especially selected by males for courtship and none to suggest that offerings are made by the male to the female as we shall see is the case in the Empididae (p. 126).

Various types of location, including sand, earth, rotten wood and decomposing vegetable matter generally are chosen for depositing the eggs. Those of *Asilus crabroniformis* L. are laid in dung. The resulting larvae are very similar in structure to those of the Tabanidae. Unlike them, however, Asilid larvae are not carnivorous and seem to prefer a diet of decomposing vegetable matter; moreover, they affect dry or moderately dry situations under dead leaves, the bark of fallen trees or rotten wood generally. They do not appear to be adapted to life in wet or saturated conditions. The pupae are not unlike those of the Bombyliidae, being armed with spines at head and tail and along the dorsal surface.

There are four sub-families, namely, Leptogastrinae, Asilinae Laphriinae and Dasypogoninae.

LEPTOGASTRINAE.

Our example of the first, which contains the single genus *Leptogaster*, with two species, is the common L. *cylindrica* Deg. (Pl. **15,** 3). This extraordinarily slender insect conveys to the observer, particularly when seen in flight, an almost Tipulid-like impression. Closer examination, however, reveals the stouter legs, the short antennae with the third segment bearing a long, stout apical style, and the characteristic Asilid mouthparts. The absence of an alula and axillary lobe should be noted; either or both may be present in the other sub-families. An interesting feature of *Leptogaster* is the absence of discernible pulvilli, with the empodium developed like a claw; the normal claws are greatly elongated and only slightly curved. Mouthparts are comparatively much more weakly developed than those of other species of Asilidae, which no doubt accounts for the normal choice of prey; this is stated by Hobby [2] to include small, soft-bodied Nematocerous flies, small Plant Bugs and the like. We have taken a ♂ *cylindrica* with a ♂ *Oncodes pallipes* Latr. as prey; the *Oncodes* was dead, but critical examination revealed no sign of a lesion.

L. cylindrica, which is generally more in evidence than the second species, *L. guttiventris* Zett., may be distinguished by dark longitudinal stripes on the hind femora and an unbroken dark median stripe on the dorsal surface of the abdomen. In *guttiventris,* there is a darkish band just before the apex of the hind femora and the abdomen has faint transverse pale bands. Both species are to be found from late May or early June until August, usually on long grasses in meadows or near woods.

ASILINAE.

Here are included 13 British species in 9 genera. *Asilus crabroniformis* L. (Pl. **1**) is one of our largest and most impressive British flies. The closed, stalked cell bounded by Veins 1 and 2 should be noted ; the third antennal segment bears a thin style. On first acquaintance, *crabroniformis* is often regarded as a rather fearsome creature ; it is, however, quite harmless to Man. This handsome fly occurs from July to September in dry localities, especially heaths, chalk-downs and clearings in pine-woods. Its flights are usually short, mainly consisting of " capture-darts " ; the distance covered in these darts may be as much as 20 inches. Hunting is done from the ground or small eminences, such as pieces of stick, stones, or dung. Among the recorded prey are various large insects, including Grasshoppers, Beetles, Bees, Wasps and, among the Diptera, *Eristalis* and *Volucella* species (Syrphidae, pp. 161–4), *Mesembrina meridiana* L. (Muscidae, p. 301) and *Sarcophaga carnaria* L. (Calliphoridae, pp. 276–8).

The remainder of the sub-family consists mainly of rather drab grey or brownish-grey flies ; one of the commonest species is *Machimus atricapillus* Fall., which is about 12–15 mm. in length, and may be positively recognised, in the male, by the curious flap-like, two-horned process on the hind-margin of the eighth abdominal sternite (Pl. **15**, 5). In both sexes, the legs are reddish and the tarsi black. *M. atricapillus* occurs from June to September in, or on the outskirts of woods, where it may be found on foliage or tree-trunks. It seems not to be easily alarmed ; on more than one occasion, when we have been standing very still, observing insect life in such surroundings, specimens have alighted on our clothing and remained there for some considerable time. " Capture-darts " of a foot or so are made quite frequently, but often without success. We have noticed that, on occasions, a dart is made at some small object which moves suddenly in a manner resembling the movement of an insect, such as a leaf-fragment in a gust of wind, or an airborne seed. It might be inferred from this that vision is not so good in this species as in others. The prey of *atricapillus* forms the subject of a special note by Parmenter.[3]

Equally common, at roughly the same times and in similar locations, is *Epitriptus cingulatus* F., which, at 11–12 mm., is the smallest of the British Asilinae. The legs are very similar in colour to those of *atricapillus*, but the fly is generally of a sandier hue and may be differentiated at once, by the presence of long, pale hairs on the frons, whereas in *atricapillus* they are all black. Hunting is carried out from the bare ground, or from heather, bracken or other raised objects. In both species, the recorded prey includes a wide range of insects, but mainly consists of soft-bodied Diptera.

A larger species, usually about 15 mm. long, and found frequently on sand-dunes in coastal localities is *Philonicus albiceps* Mg.; it also occurs inland occasionally in sandy places. This is of a lighter grey than the previous two species and has considerable whitish pubescence and bristles. Recorded prey, not unnaturally, includes a large proportion of Diptera peculiar to the seashore; it is interesting to note that cannibalism is sometimes practised by this species.

Laphriinae.

In this sub-family there is the single genus *Laphria*, two of the three British species of which are rare. The third, *L. marginata* Mg. (Pl. **13,** 2) is comparatively common in deciduous woodland in the southern half of England, where it normally hunts in a similar manner to the Asilinae. Freshly-emerged specimens are rather fine-looking insects, with shining blue-black thorax and abdomen, bearing golden hairs of varying density. Wing-venation is similar to that of the Asilinae, but the strap-shaped third antennal segment without a style (Pl. **15,** 4) is in contrast; a wide range of prey is taken, also similar to that of the Asilinae. Parmenter,[4] in an interesting note on the rare *L. gilva* L., gives an up-to-date key to the British Laphriinae.

DASYPOGONINAE.

Nine British species in four genera occur here, and may be known by the wing-venation, in which the cell bounded by Veins 1 and 2 is open (Pl. **15,** 7), in contrast to the closed, stalked cell in the Asilinae and Laphriinae. The wing illustrated refers to *Lasiopogon cinctus* F., a small undistinguished-looking species, about 9 mm. long, which is fond of lying in wait for its prey on bare, sunlit patches or footpaths. Recorded prey includes various small Tipulids and even a moderately large *Nephrotoma.* On account of its cryptic coloration and its flight when disturbed, which consists of an extremely rapid dart to another vantage-point, an alert eye is required to notice this fly ; it is in evidence in early May, but the flight period is short and over by the end of June. *Isopogon brevirostris* Mg., a blackish fly about 9 mm. in length, is noteworthy on account of the extraordinarily attenuated hind metarsus in the male, which contrasts sharply with the dilated hind tibia (Pl. **15,** 6).

One or two very common species occur in the genus *Dioctria. D. rufipes* Deg. (Pl. **13,** 1) flies from May to July, usually in meadows, where it may hunt from the grasses or from higher points in the hedgerows. The shape of the antennae and the pubescent prominence from which they spring should be noted ; the third segment bears a short, thick style. *D. rufipes* may be distinguished from the similar but smaller *D. baumhaueri* Mg. by the larger size of the frontal prominence coupled with its orange front and mid-femora ; in *baumhaueri,* the prominence is much smaller and there is a dark stripe on the dorsal surface of the anterior femora. This species occurs mainly in the south from June to August. *D. atricapilla* Mg. is a very dark insect, about 10–12 mm. long, of stout build, with entirely black legs and darkened wings, which are blackish in the male and brownish in the female ; it is fairly common in meadows and wastes of certain southern localities, in June and July ; hunting is carried out from grasses by darts. *D. oelandica* L., at 14–15 mm., is the largest species of the genus, and, like *atricapilla*, occurs in certain localities in the southern half of England, but seems to prefer woodland situations to the open ; it hunts from higher points such as bushes and young trees. *D. oelandica* may be recognised by its size, the yellow or orange legs and the darkened wings, which, when viewed obliquely, have a peculiar violet sheen.

Records of prey of *Dioctria* species show a high proportion of Hymenoptera, especially Ichneumons.

For the identification of the British species of Asilidae, see Verrall [5] and Hobby ; [6] additional reading may be found in Lundbeck.[7] Brindle[12] deals with the larvae.

EMPIDIDAE.

35 genera. Over 350 species. Small to medium-sized, mainly bristly flies ; head more or less spherical ; 3rd antennal segment not annulated ; style or arista, if present, terminal ; proboscis rigid, of varying length, projecting downward or forward, adapted for piercing in both sexes ; ♂ eyes holoptic or separated ; thorax often humped ; legs frequently with structural modifications, e.g. thickened femora, tarsi, or metatarsi, or with secondary sexual characters ; empodium usually thin or bristle-like, occasionally pad-like (e.g. *Clinocera*) ; wings folded over abdomen at rest ; venation often greatly reduced ; anal vein sometimes abbreviated or absent, therefore anal cell sometimes absent ; alula and/or axillary lobe mostly small, sometimes absent ; darkened stigma often present, membrane sometimes clouded or banded. (Pl. **15**, 8: **18**.)

Like the previous family, Empididae are also predatory upon insects, but whereas the Asilidae are catholic in their choice of prey, the species of this family seem to be chiefly addicted to Diptera. The complex genus *Empis*, in the sub-family Empidinae, contains the largest and most noticeable species, in which the proboscis is long and downwardly-projecting

(Pl. **15,** 8) ; while the remainder of the family consists, on the whole, of small to very small flies in five sub-families, viz., Tachydromiinae, Hybotinae, Ocydromiinae, Hemerodromiinae and Clinoceratinae, in which the proboscis is either downwardly-directed and short (Pl. **18,** 10, 11), forwardly directed (Pl. **18,** 9) or almost imperceptible (Pl. **18,** 8). In some cases the wing-venation is considerably reduced (Pl. **18,** 4), even to the extent of the sub-costa being quite vestigial and Veins 6 and 7 being completely absent (Pl. **18,** 5). In these cases, the discal cell, as such, also disappears since the cross-vein dividing it from the second basal cell is missing. A constant feature in all Empididae is the presence, on the inner eye-margins near the base of the antennae, of an incision or nick, which, though sometimes not very prominent, is, nevertheless, always perceptible (Pl. **18,** 2).

EMPIDINAE.

The largest British Empid is the common *Empis tessellata* F. (Pl. **15,** 8), a sombre, grey fly, which sometimes occurs in considerable numbers on hawthorn blossom or umbelliferous flowers. It seldom makes long or rapid flights but usually progresses in a series of leisurely hops ; not infrequently it will be seen to probe flowers with its long proboscis, no doubt for the purpose of sipping the juices ; at other times, it may be seen in slow, heavy flight, bearing an insect victim, which may be quite large, as for example, a Flesh Fly (*Sarcophaga*, p. 275) or *Bibio marci* L. (p. 75). In the specimen illustrated, the shape of the antennae, the prominent male genitalia and the wing-venation should be noted. An anal cell and well-developed axillary lobe are present ; the forked cell at the apex of Vein 3 is characteristic in shape for the genus, i.e. the upper branch of the fork is short and at a steep angle. Another common species of a drab, brownish-grey colour, about 8–9 mm. long, is *E. livida* L., which may immediately be known by the fact that the two vein-branches emanating from the outer anterior corner of the discal cell stop short of the wing-margin. *E. stercorea* L., which is about 7 mm. long, is common in spring and easily recognised by its yellow or orange colouring and sharply-defined black median longitudinal stripe on the dorsum. *E. femorata* F., about 6 mm. long and dark grey in colour, may also be easily recognised by the very dilated hind femora.

Certain smaller species of the genus which includes nearly 40 British species have, in the females, curiously adorned and flattened legs, particularly the hind ones; the hairs thereon being dilated or flattened so as to resemble scales. These are known as "pennate" legs; an example is *E. aestiva* Loew, a small, shining black species, about 3 mm. long, which is not at all uncommon, from June to August, on flowers of thistles and other Compositae, in open spaces or on the borders of woods.

Most Nature-observers will have noticed, at various times in summer, over ponds and ditches, or even small puddles after heavy rain, congregations of small flies performing rapid aerial evolutions, weaving back and forth in intricate flight patterns, or skimming the surface of the water at great speed. If an attempt be made to net some of them, it will soon be realised that nice judgment is required to bring the net sufficiently near the surface to catch the elusive quarry, without making a prodigious splash, wetting the net, rendering useless the specimens, and being out of action for at least some minutes until the net is once more dry. When one of the flies is eventually secured in good condition and examined under a lens, it will quite probably be found to have an appearance similar to the fly depicted on Pl. **18,** 1. This is a male of the common species *Hilara maura* F., in which it will be seen that the proboscis is of moderate length and the wing-venation, in most respects similar to that of *Empis*, except that the forked cell at the apex of Vein 3 is elongated, i.e. the upper branch is long and at an acute angle. Particularly striking are the enormously-dilated fore metatarsi, which character, in greater or lesser degree, is diagnostic for the males of this genus, the British species of which number over 60.

Not infrequently, when *Hilara* species are taken, and incidentally, they occur in situations other than over water, there will be found, adhering to the side of the net, a small insect, tangled with, or enclosed in, what appears to be a kind of silken web. It may also strike the collector that, often when this happens, both a male and female *Hilara* are captured; the explanation of this seeming coincidence makes an interesting biological story. *Hilara* species mate whilst in flight and, almost invariably when a mating pair is caught, they are in possession of prey. All the evidence adduced by various authors points to the conclusion that the male catches a small victim, such as a midge, envelops it in a tangle of silken threads, which, according to Eltringham,[8] is secreted by glands opening on the fore tarsi, and offers it to the female. She accepts this, sucks the juices during copulation and discards it after the mating is completed. It is not established whether the juices are necessary for fecundation or whether the offering is merely a blandishment. A similar process occurs in *Empis* and in the next genus for our consideration, *Rhamphomyia*, but the prey is not enveloped in a web; in *Empis*, after pairing has commenced, the male will perform the prodigious feat of suspending himself from some such object as a twig or leaf-stem by his front legs and supporting the female, with her offering, by means of his mid- and hind-legs. The male of an American species of *Empis* encloses the offering in a sort of frothy balloon, which, it is said, appears to excite the female. It is noteworthy that, according to Hamm,[9] the species of the sub-genus *Xanthempis*, which have not developed the practice of making offerings, do not copulate in the air.

Rhamphomyia species are very small to medium-sized grey or black flies with a wing-venation similar to *Empis* and *Hilara*, except that Vein 3 is not forked at all (Pl. **18**, 3). In some species of the genus, Vein 6 does not reach the margin. The proboscis is directed downwards and may be fairly long, but not so long as in *Empis* and generally not so short as in *Hilara*. Males sometimes indulge in a sort of wild, erratic, dancing flight, while on other occasions, they may hover or fly over the surface of the water in much the same manner as *Hilara* species, in the company of which they often occur. The common, early spring species, *R. sulcata* Mg. may not infrequently be taken with Alder Flies (Megaloptera) as prey.

TACHYDROMIINAE.

Included in the large genus *Platypalpus* is a series of very small flies, ranging from about 1½ mm. to 4 mm., which will usually be found running in a rapid, characteristic fashion on the leaves of bushes and low plants ; they turn up quite often in the sweep-net, where their activity attracts early attention. When they are examined, it will be found that the mid-femora are greatly dilated and armed with formidable spines, while the mid-tibiae are equipped with a prominent, curved apical spur ; the tibiae fold back upon the femora in pocket-knife fashion (Pl. **18,** 7) and according to Lundbeck,[10] the prey, which has been recorded as including small plant-bugs, midges and even *Microchrysa polita* L. (p. 92) and *Urophora solstitialis* L. (Tephritidae, p. 186), is held fast in this trap. A similar device is to be seen in Hemerodromiinae (p. 129) and *Ochthera mantis* Deg. (Ephydridae, p. 289), where the fore-legs are modified for this purpose. Wing-venation is reduced in *Platypalpus*, the anal cell being bounded by a very weak anal vein (Vein 6), which is hardly more than a fold (Pl. **18,** 4). This vein is quite absent, consequently so is the anal cell, in *Tachypeza nubila* Mg. (Pl. **18,** 5), an elongated, shining black or brownish-black fly, about 3½ mm. long, with brownish-tinged wings and long legs ; it is usually to be found, from May to September, running in a swift, jerky fashion on tree-trunks or fences, in company with species of *Medeterus* (Dolichopodidae, p. 135).

When danger threatens, this fly does not take to its wings, which are held folded over the abdomen, but removes itself from sight by running with extraordinary rapidity to the opposite side of the tree-trunk, a procedure also adopted by *Medeterus*. A further reduction is seen in the wing-venation of *Tachydromia arrogans* L., for both Vein 6 and the cross-vein between Veins 5 and 6 have disappeared ; this species is about the same size and closely similar in build to *Tachypeza*, but has two broad dark bands across the wings ; locations and habits are also very similar.

HYBOTINAE.

Hybos femoratus Mueller (Pl. **18,** 9) will exemplify the few British species of this sub-family. *H. femoratus* is a not uncommon, shining black fly about 4–5 mm. long, with a characteristic humped thorax, curved abdomen and dilated hind femora ; the proboscis is directed forward and the palpi are prominent and porrect ; a discal cell is present ; a weak anal vein reaches the wing-margin, and the anal cell is longer than the basal cells. The flight of *Hybos* species is comparatively slow and ponderous ; they are usually to be found on the low vegetation of damp, wooded places, in late summer and autumn. Prey includes small Diptera, such as midges and *Bradysia* species.

OCYDROMIINAE.

Twenty-six species in seven genera make up the British list ; considerable diversity of form is exhibited, particularly as regards the shape of the antennae, which, for example, may have the third segment long and strap-shaped, with a very short style (Pl. **18,** 6) or short-ovate, with a long arista-like extension as in *Ocydromia glabricula* Fall. (Pl. **18,** 11), in which it will also be noted that the proboscis is extremely short, barely protruding from the mouth-opening. In species of this sub-family where the proboscis is longer, it may be

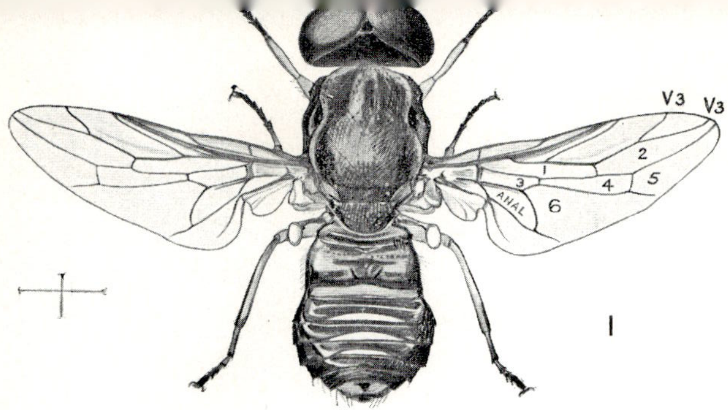

1

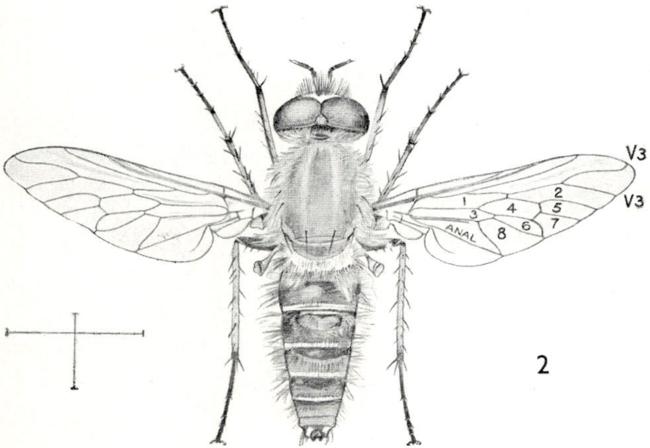

2

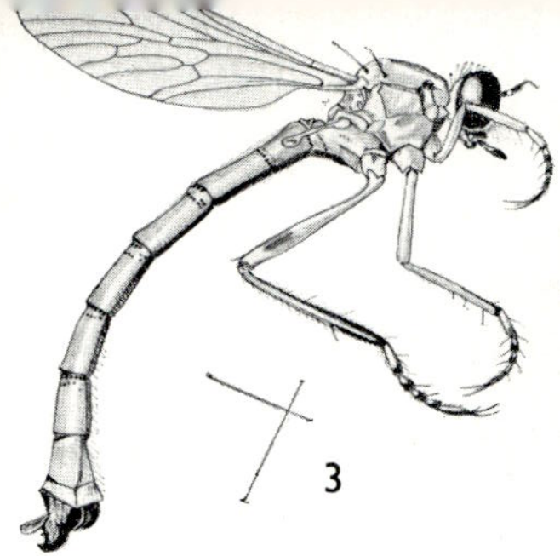
3

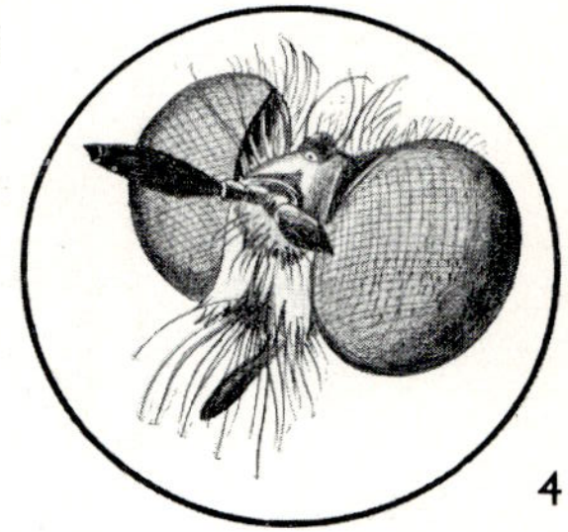
4

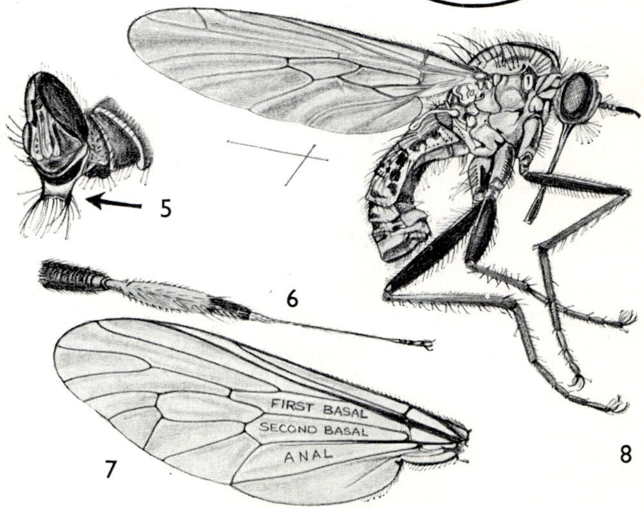

5

6

7

8

PLATE 15

SCENOPINIDAE

1. *Scenopinus fenestralis* L., ♂, p. 115.

THEREVIDAE

2. *Thereva nobilitata* F., ♂, p. 114.

ASILIDAE-III

3. *Leptogaster cylindrica* Deg., ♂, p. 119. 4. *Laphria marginata* L., head, p. 121. 5. *Machimus atricapillus* Fall., apex of ♂ abdomen, p. 120. 6. *Isopogon brevirostris* Mg., ♂ hind leg, p. 122. 7. *Lasiopogon cinctus* F., wing, p. 122.

EMPIDIDAE-I

8. *Empis tessellata* F., ♂, p. 124.

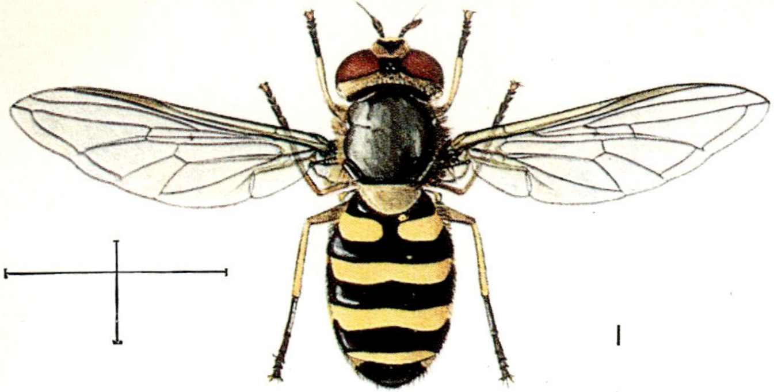

1

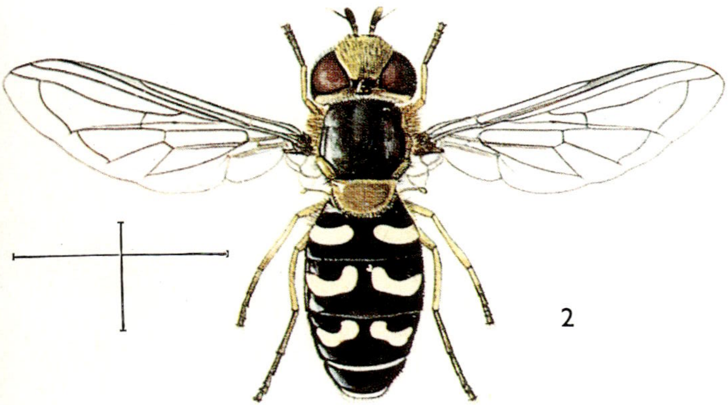

2

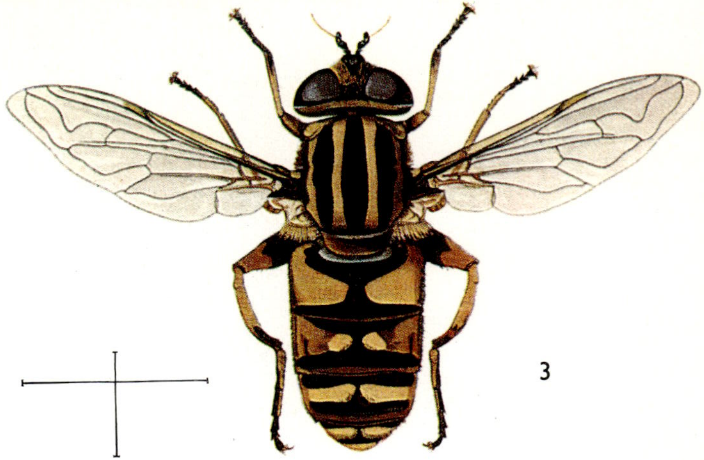

3

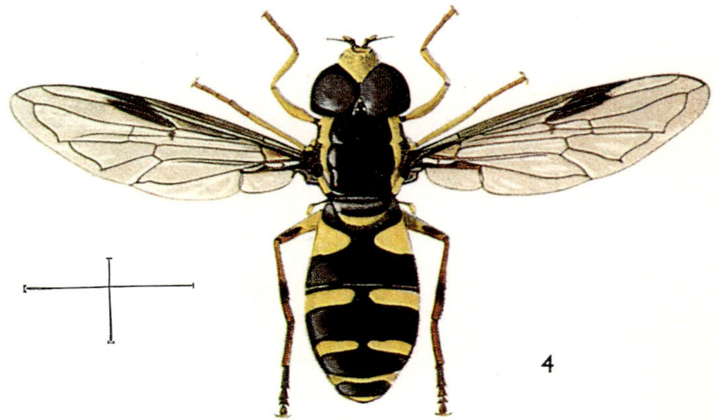

4

Plate 16

SYRPHIDAE-I

1. *Syrphus ribesii* L., ♀, p. 156.

3. *Helophilus pendulus* L., ♀, p. 164.

directed forward or obliquely downward. Wing-venation is generally complex, with the anal cell shorter than the second basal cell; the thorax as a rule is strongly humped.

Our example, *glabricula*, may be found in damp places, hovering or settled on waterside vegetation, from early summer to autumn. It has a shining metallic black thorax and abdomen in the male, with pale yellowish triangular basal markings on the sides of the abdominal segments. The female is yellowish or reddish, with black markings.

HEMERODROMIINAE.

This sub-family is generally characterised by greatly elongated front coxae, swollen front femora, mostly armed with spines, and the absence of an alula or axillary lobe. The front legs are adapted for holding prey in similar fashion to the mid-legs of *Platypalpus* (p. 127), the tibiae folding back upon the femora like the blade of a pocket-knife upon its handle. From the twenty-three species in six genera, we give, as an example, *Phyllodromia melanocephala* F. (Pl. **18,** 10), a very small fly, which has a quite striking appearance under a lens; the thorax is yellow or reddish, the abdomen darker, the head black as the specific name indicates, the legs and basal antennal segments yellow, while the short pear-shaped third segment and long, terminal arista are black. Wing-venation is reduced and the proboscis is very short and downwardly directed. *P. melanocephala* is not at all uncommon, from June to August, in damp shady places on the borders of ponds and streams.

CLINOCERATINAE.

Some fourteen British species in four genera make up the list; our example is *Clinocera stagnalis* Hal. (Pl. **18,** 8), in which the elongated front coxae, undilated femora, almost invisible proboscis, clouded cross-veins and short third antennal segment with downwardly-directed terminal arista, are noteworthy. *C. stagnalis* is a bluish or slaty-grey fly, with the dorsal surface of the head and body dull brownish or greenish-brown ; it may be encountered, from May to August, at the borders of ponds or marshy places, where it sometimes runs on the surface of the water itself. Lundbeck [10] found this species running on the surface of shallow pools as far north as the Faroe Islands and Greenland. The illustration should be compared with that of *Hydrophorus* (Pl. **19,** 9) and *Campsicnemus* (Pl. **19,** 10), two species of Dolichopodidae, which also have the habit of running on the surface of water ; certain species of Ephydridae (p. 218) likewise are capable of the feat.

Not a great deal appears to be known of the habits of the immature stages of Empididae, except that some of the larvae live in decaying wood or vegetable matter, or in the earth, especially that containing a high proportion of humus. In the case of *Hemerodromia,* they are aquatic, or mud-dwellers. Instances of carnivorous habit have been recorded. Generally, the larval structure is narrow-cylindrical, with twelve segments including a small, retractable head ; the ventral surface of the hinder segments bear transverse swellings or protuberances. In the pupae, the thorax is armed with bristles, and the abdominal segments with transverse rows of small spines, between which lie bristles.

The family is monographed by Collin[13]. Adult courtship behaviour and immature stages are dealt with by Hobby and Smith[14], and Brindle[15].

LITERATURE CITED

1. Whitfield, F. G. S. The Relation between the Feeding-habits and the Structure of the Mouth-parts in the Asilidae. *Proc. Zool. Soc. Lond.* II : 599–638. 1925.
2. Hobby, B. M. The British Species of Asilidae (Diptera) and their Prey. *Trans. ent. Soc. S. England.* **6** : 1–42. 1930.
3. Parmenter, L. The Prey of a Population of *Machimus atricapillus* Fallén (Diptera). *Proc. R. ent. Soc. Lond.* (A) **17** : 71–2. 1942.
4. Parmenter, L. *Laphria gilva* L. (Dipt., Asilidae), in Surrey,

with a key to the British Laphriinae. *Ent. mon. Mag.* **82**: 304–5. 1946.

5. Verrall, G. H. British Flies. Vol. V. Gurney & Jackson, London. 1909.
6. Hobby, B. M. A Key to the British Species of Asilidae. *Trans. ent. Soc. S. England.* **8**: 45–9. 1932.
7. Lundbeck, W. Diptera Danica. Vol. II. Wm. Wesley & Son, London. 1908.
8. Eltringham, H. On the Production of Silk by Species of the Genus *Hilara* Meig. *Proc. Roy. Soc.* (B): 102. 1928.
9. Hamm, A. H. Further observations on the Empinae. *Ent. mon. Mag.* **45**: 157–62. 1909.
10. Lundbeck, W. Diptera Danica. Vol. III. Wm. Wesley & Son, London. 1910.
11. Collin, J. E. Notes on the Empididae with Additions and Corrections to the British List. *Ent. mon. Mag.* **62**: 146–59, 185–90, 213–19, 231–7. 1926. **63**: 20–9, 61–7, 93–8. 1927.
12. Brindle, A. Taxonomic notes on the larvae of British Diptera. 10. The Asilidae. *Entomologist,* **95**: 241–7. 1962.
13. Collin, J. E. British Flies. VI. Empididae. Cambridge. 1961.
14. Hobby, B. M. and Smith, K. G. V. The bionomics of *Empis tessellata* F. (Dipt., Empididae). *Ent. mon. Mag.*, **97**: 2–10. 1961.

    Hobby, B. M. and Smith, K. G. V. The immature stages of *Rhamphomyia* (*Megacytarus*) *anomalipennis* Mg. (Dipt., Empididae). *Ent. mon. Mag.*, **97**: 138–9. 1961.

    Hobby, B. M. and Smith, K. G. V. The bionomics of *Empis opaca* Mg. (Dipt., Empididae). *Ent. mon. Mag.* **97**: 204–8. 1961.

    Hobby, B. M. and Smith, K. G. V. The larva of the viviparous fly *Ocydromia glabricula* (Fln.) (Dipt., Empididae). *Ent. mon. Mag.*, **98**: 49–50. 1962.
15. Brindle, A. Taxonomic notes on the larvae of British Diptera. 18. The Hemerodrominae (Empididae). *Entomologist,* **97**: 162–5. 1964.

CHAPTER IX

DOLICHOPODIDAE

47 genera. Over 250 species. Mostly small, bristly, metallic bluish or greenish flies ; head, in profile, usually long-oval, higher than wide ; antennae inserted about middle of head or above ; 3rd segment mostly short and more or less oval, but sometimes long and pointed or strap-like ; short or long terminal styles or dorsal aristas may be present ; eyes separated in both sexes ; proboscis short and stout, retractable ; palpi flat, often greatly developed ; ♂ tarsi often ornamented ; empodium usually bristle-like (but pad-like in *Hydrophorus*, *Medeterus*, etc.) ; wings sometimes with white or dark markings ; venation simple, *sc* weak or absent, second basal and discal cells confluent, anal cell short, Vein 6 short or absent, Vein 4 sometimes bayonet-shaped (*Dolichopus*) or forked (*Sciopus*); ♂ genitalia often very prominent and beautifully adorned. (Pl. **13**: **19**.)

As in the two preceding families, predatory habits are characteristic of the Dolichopodidae, but the method of feeding differs. The retractable proboscis is short and furnished with peculiar plate-like palpi ; the well-developed labella are used to envelop the prey, which normally consists of small, soft-bodied insects or mud-dwelling larvae. Obviously for the purpose of releasing the body-juices, the prey is crushed or milled between the labella.

Dolichopodids, in most cases, frequent damp situations near water ; some species actually run about on the surface of water, or on the mud at the margin and may be found in such locations, in the case of one or two species, even on mild days in winter ; the greater number of species, however, are in evidence only in the summer months or early autumn. A few are to be found on tree-trunks and foliage remote from water. The characteristic resting attitude of the family, with the head and thorax raised, gives these flies what might be appropriately described as an expectant or peering look.

In the Empididae, especially as regards the smaller species, a tendency toward departure from the usual Brachycerous complex wing-venation and build of antennae was becoming perceptible ; in the Dolichopodidae, wing-venation has become very simple (Pl. **13**: **19**) and, while a few instances of long and pointed, or strap-shaped third antennal segments, with long or short styles are to be found (Pl. **19,** 2,* 4, 5), generally this segment is shorter and more rounded at the apex, while the style is replaced by an arista, which may be apical, subapical or dorsal (Pl. **19,** 6, 7 : **13,** 5). Sexual dimorphism may occur in the antennae, as, for example, in *Syntormon,* where the third segment is long and pointed in the male, but much shorter and stouter in the female (Pl. **19,** 6). A curious prescutellar depression frequently occurs on the dorsum. Males are often equipped with enormous, strikingly-ornamented genitalia (Pl. **13,** 6), with plumed front or mid-tarsi, or with other structural modifications of the tarsi (Pl. **13,** 5: **19,** 10); owing to these very noticeable characteristics, they are less difficult to identify than the females.

Eight sub-families present themselves for consideration.

DOLICHOPODINAE.

Dolichopus popularis Wied. (Pl. **13,** 5) provides an example of the largest genus in the family, with nearly 50 species, in some of which the males are highly ornamented. Distinguishing features of this fly are found in the plumed third and fourth segments of the mid-tarsi, double row of acrostichals, three preapical bristles on hind femora and pale lower post-ocular cilia. Vein 4 is somewhat abruptly bent, bayonet-fashion, at about two-thirds of its length ; the general bristly nature of the legs should be noted, particularly as regards

* Illustrated from dry specimen ; antennae porrect in life.

the basal segment of the hind tarsi, which bears dorsal bristles, a characteristic of the genus. *D. ungulatus* L. is an exceedingly common species, occurring, as a rule in large numbers, in most localities near water. It may be recognised by the black lower postocular cilia and yellow hind femora, which bear three or more preapical bristles; the male tarsi are not plumed, but the genitalia (Pl. **13,** 6) are enormous, extending almost the length of the abdomen, beneath which they are turned. The external lamellae are prominent, of a whitish colour, with blackish edges bearing spines, which are bent rectangularly in the middle, the apical ones being very stout.

Poecilobothrus nobilitatus L. (Pl. **13,** 3) is easily recognised, in the male, by the handsome wing-markings; a fringe of prominent, curled spines adorns the external lamellae of the genitalia, which are shown everted in the plate. In both sexes, the species may be known by the convergent, wavy Veins 3 and 4, absence of dorsal bristles on the basal segment of hind tarsi, single preapical bristle on hind femora, single erect bristle on hind coxae, and double row of acrostichals. *P. nobilitatus* is generally abundant; we have seen pools and ditches, especially those covered with duckweed, literally teeming with vast numbers settled on the weed, rising in a cloud, with an audible noise, when disturbed. The fascinating courtship of *P. nobilitatus* is described by Smith and Empson[8].

HYDROPHORINAE.

A very common example of this sub-family is the little coastal species, *Hydrophorus oceanus* Macq. (Pl. **19,** 9), which is not metallic-coloured, but of a dusty, grey hue and has a single row of acrostichals. When alive and in their natural haunts, this species and some of its allies, which have a habit of running on the surface of the water, are not unlike *Clinocera* (Empididae, p. 130) and some of the species of Ephydridae (p. 218) which may occur in company with them. Under a lens, however, the Empids will be differentiated at once by the quite different wing-venation (cf. Pl. **18,** 8) and the Ephydrids by the costal breaks and the extreme shortness of Vein 1 (cf. Pl. **35, 9**); other characters, such as an arista, plumed on the upper surface, or the typical wide Ephydrid mouth-opening (cf. Pl. **35,** 12) may also assist. *H. oceanus* may be known from the other British species of the genus by its unspotted wings and the fact that the scutellum bears only two bristles, which are apical and crossed.

Another and larger frequenter of salt mudflats, which may be encountered from time to time is *Thinophilus flavipalpis* Zett., a drab, grey fly, about 6 mm. long; here the acrostichals are entirely absent, the palpi remarkably developed as large, scale-like plates (Pl. **19,** 1), the fore metatarsi bent and the mid-tarsi with the two apical segments dilated.

APHROSYLINAE.

In the four British species of the sole genus, *Aphrosylus*, acrostichal bristles are absent. As an example, we may consider *A. raptor* Walker; here, the palpi are also developed as large, scale-like organs, like those of *Thinophilus*; the antennae are quite characteristic, however (Pl. **19,** 4) and a series of long, crinkly hairs spring from the male genitalia (Pl. **19,** 3). *A. raptor,* which is about 5 mm. long and dark brownish-black in colour, is common, in southern coastal districts, on rocks periodically inundated by the tides; the larvae are said to live in cast-up seaweed.

MEDETERINAE.

Included in this sub-family is the interesting genus *Medeterus*, with some 26 British species; these are mostly very small flies, commonly to be found on tree-trunks, walls and fences, often in company with *Tachypeza* (Empididae, p. 127) and allied species. *Medeterus* species are usually of a dusty or frosted greyish or brownish colour, although, in some cases, traces of metallic blue or green may be detected, beneath. Their resting attitude, in which the body is held in an odd, erect pose, generally characteristic of the family, is especially noticeable. When alarmed, they run with extraordinary rapidity and have a remarkable faculty of moving forward, sideways, or backward at will. Usually, like *Tachypeza*, they disappear to the other side of a tree-trunk with a lightning, half-forward and half-sideways dart, but, on occasions, will make a short, jumping flight. The example illustrated, *M. flavipes* Mg. (Pl. **19**, 11) is common, in fact abundant in some places, and is of a greenish-blue to blackish metallic base-colour, thickly obscured by a greyish-white to yellowish-grey bloom, which gives iridescent reflections in certain lights. A sub-apical arista is present on the short, rounded, third antennal segment and acrostichals are in a double row. Keys to the British species of *Medeterus*, with notes by Collin,[1] will be found very useful.

RHAPHIINAE.

Syntormon pallipes F. is a common coastal species, occurring mostly on salt mudflats; two rows of acrostichals are present. The antennae (Pl. **19,** 6) are interesting on account of the peculiar development of the second segment, which is extended into a thumb-like projection; this is inserted into an excavation in the third segment, which is moderately short, with a fairly long jointed arista, in the female as illustrated, but longer and more pointed, with a shorter arista, in the male. *S. pallipes* is about 3½–4 mm. long and of a deep, greenish-black or bluish-black metallic colour, dusted with grey; the legs are yellow and the wings tinged with brown. Sometimes the abdomen is yellow at the base, but this character is very variable; the basal segment of the male hind tarsi bears a curved, two-pronged thorn.

Another example is *Xiphandrium appendiculatum* Zett., in which the third segment of the antennae is enormously lengthened and strap-like, with a tiny, jointed apical style (Pl. **19,** 2).*

* See note on p. 133.

The species is about $3\frac{1}{2}$ mm. long, of a deep greenish-black metallic hue, faintly obscured by a reddish-brown bloom, which gives coppery-purple reflections in certain lights. It is common on low plants near water, or in marshy places generally.

DIAPHORINAE.

Species of the genus *Chrysotus* are small, active, sparkling metallic green flies, about 4 mm. in length, or less ; on sunlit foliage, their iridescent wings are particularly noticeable. There are two rows of acrostichals, one erect bristle on the hind coxae, no preapical bristle on hind femora and the 3rd antennal segment is more or less short and rounded, with an apical or sub-apical arista (Pl. **19**, 7).

Some of the most handsome flies in the family are the silvery species of the genus *Argyra* ; our example is the large and common *A. diaphana* F. (Pl. **13**, 4), which cannot fail to attract the attention of the field-naturalist, sooner or later, as it gleams and flashes while moving in sunlight. On close examination, the colouring, in the male, will be found to be a beautiful, deep greenish-blue on the thorax and brownish-black on the abdomen, with semi-transparent yellow patches on the basal segments. The whole, more particularly the apical part of the abdomen, is covered with a lovely, silvery sheen, which, if regarded vertically, is, in places, almost invisible, but, viewed obliquely, especially from in front, glistens and gleams so as completely to obscure the ground colour. Several rows of long, thin, rather weak acrostichals are present, barely distinguishable from the numerous similar hairs all over the dorsum ; the elongated, triangular third antennal segment bears a dorsal arista ; the eyes occupy nearly the whole of the head save for a moderately narrow, velvety-black, sunken frons and narrower face ; hairs are present on the disc of the scutellum, and the genitalia are comparatively small. In the remaining species of the genus, the acrostichals are in a double row, either regular or irregular, and the individual bristles are clearly differentiated from the surrounding hairs. *Argyra* males have a habit of flying back and forth very rapidly, close to the ground; when seen against a dark background of, say, dead leaves or mud, the effect is very striking, little flashes and gleams of silver alternately appearing and disappearing with great rapidity.

A paper by Collin [2] on the British species will be found most useful.

CAMPSICNEMINAE.

The genus *Campsicnemus* contains some interesting little flies which may sometimes be found running on the surface of pools or sluggish ditches, even in winter (cf. *Hydrophorus*, p. 134). In some species of the genus, the acrostichals are in a single row, or two irregular rows towards the rear, but in *C. scambus* Fall. (Pl. **19,** 10), one of the commonest species, they are entirely absent, or present only in the form of a tiny hair or two near the front. A long dorsal arista springs from the triangular third antennal segment; the wide inverted-triangular frons exhibits beautiful violet and blue metallic reflections, in certain lights; the reddish-yellow face is narrow just beneath the antennae and widens out towards the oral margin, a feature which is characteristic of the genus. In *Campsicnemus* males, the legs are curiously contorted or adorned; as regards *scambus* in particular, the middle tibia is extraordinarily dilated, spiny and bent in the middle; the middle metatarsus is short, oddly deformed and ornamented, while the remaining tarsal segments are normal.

CHRYSOSOMATINAE.

Species of the sole genus *Sciopus* may be recognised at once by the apical fork of Vein 4, the lower branch of the fork, i.e. the apparent continuation of Vein 4, not infrequently being weakened, but nevertheless easily discernible. *S. platypterus* F. (Pl. **19,** 8) is commonly to be found in the summer months on tree-trunks, the foliage of hedgerows and similar locations. A female is illustrated; the male has the third, and basal half of the fourth segment of the mid-tarsi, white, while the apical half of the fourth and whole of the fifth segment is blackish, giving a conspicuous effect. The third and fourth segments are somewhat dilated. In both sexes, the basal segments of the hind tarsi are very noticeably longer than the succeeding segments. Other noteworthy features are the long, thin curved abdomen, short wide wings and long legs; a double row of acrostichals is present; the third antennal segment is very short with a dorsal arista and male genitalia are very prominent.

We have seen a female of another species, *S. maritimus* Becker, on a window, catch and "masticate" a small Psychodid barely more than 1 mm. long, the prey being held by the hind end of the abdomen. The process continued for fully fifteen minutes, the unfortunate Psychodid being vigorously moved between the labella at first, with only an occasional movement towards the end. It was still retained for a time after the captor had been imprisoned in a tube.

Dolichopodid larvae are long-cylindrical in shape and somewhat narrowed anteriorly, with a small retractable head; the hind end is broader and more or less truncate, with four short tubercles bordering the rear spiracles. Small, fleshy pseudopod-like protuberances are borne on the ventral surface of the middle segments; these are armed with spines, doubtless to aid locomotion. The larvae live mainly in wet or damp situations, such as sand or mud at the margins of pools and marshy places, or seaweed, or rotting vegetation and wood generally; in some cases, they actually live in the water. In all probability, most species are carnivorous in habit. *Medeterus* larvae have been thought by some authors to prey on the larvae and pupae of wood-boring beetles, beneath the bark of trees, in whose galleries they are often found, but whether they are true predators or merely scavengers seems not to be definitely established.

In *Dolichopus*, pupation is said to take place in a little, earthen cell, lined with some form of secretion which is thought to be impermeable ; this probably accounts for the fact that the pupae are equipped with a pair of long, respiratory horns, which protrude through the covering. The pupae are short and squat in shape. This habit of constructing some sort of cocoon occurs in other species, small pieces of debris being woven together with a kind of silk.

For the identification of the British species of Dolichopodidae, reference may be made to the papers already mentioned and to keys by Verrall ;[3] the latter are somewhat out-of-date and will not suffice for the females throughout the family. Their use may be augmented, however, by the works of Lundbeck[4], Parent[5] and Thuneborg[6]. Dyte[7] comments on the immature stages.

LITERATURE CITED

1. Collin, J. E. The British Species of the Dolichopodid Genus *Medeterus* Fisch. *Ent. mon. Mag.* **77**: 141–53. 1941.
2. Collin, J. E. A Revised Table of the British Species of *Argyra* Macq. *Ent. mon. Mag.* **79**: 114–17. 1943.
3. Verrall, G. H. List of British Dolichopodidae with Tables and Notes. *Ent. mon. Mag.* **40**: 164–73, 194–9, 223–8, 241–5. 1904. **41**: 51–7, 81–3, 108–12, 167–72, 189–96, 248–52. 1905.
4. Lundbeck, W. Diptera Danica. Vol. IV. Wm. Wesley & Son, London. 1912.
5. Parent, O. Faune de France. **35.** Dolichopodidae. Lechevalier, Paris. 1938.
6. Thuneborg, E. A revision of the Palaearctic species of the genus *Medetera* Fischer (Dipt., Dolichopodidae). *Ann. ent. Fenn.*, **21**: 130–7. 1955.
7. Dyte, C. E. Some interesting habitats of larval Dolichopodidae. *Ent. mon. Mag.* **95**: 139–43. 1959.
8. Smith, K. G. V. and Empson, D. Note on the courtship and predaceous behaviour of *Poecilobothrus nobilitatus* L. (Dipt., Dolichopodidae). *Brit. J. Animal Behaviour*, **3**: 32–4. 1955.

CHAPTER X

LONCHOPTERIDAE . . . Pointed-wing Flies.
PHORIDAE.

In the Check List, the last two families of the Brachycera, Lonchopteridae and Phoridae, carry the tendency to diverge from the general Brachycerous type, which was becoming evident in the Empididae and Dolichopodidae, a stage further; in fact, they begin to exhibit not a few of the characters generally associated with Cyclorrhapha, especially in the immature stages. Their systematic position has formed the subject of some controversy and they are included by some authors in Cyclorrhapha Aschiza (Section III). Reasons for this will become apparent as we consider some of their characteristics, and, subsequently, those of the first two families of Aschiza, Platypezidae (= Clythiidae) and Pipunculidae (= Dorilaidae). What we shall see is a gradual transition from Brachycera into Cyclorrhapha; the problem of where one ends and the other commences is one which need not concern us greatly. In any case, the characters by means of which Brachycera and Aschiza are differentiated (cf. Introductory Chapter) are not, in these borderline families, of a nature to be readily appreciated by the field-naturalist, especially with the aid of a pocket-lens; it is fortunate, therefore, that the families in question are highly characteristic in general appearance, so that we may conveniently consider the three families of the Aschiza together with the Brachycera. For this reason the key on p. 175 covers both Sections II and III.

LONCHOPTERIDAE.

1 genus. 7 species. Small brownish or yellowish, bristly flies, with characteristically pointed wings; 3rd antennal segment more or less spherical, with long terminal arista;

eyes separated in both sexes; legs fairly long, bristly; pulvilli small; empodium absent; wing-venation differs in sexes (Vein 6), Vein 1 very short, Vein 4 forked, cross-veins basal; ♂ genitalia prominent. (Pl. **22,** 4, 5.)

Species of Lonchopteridae are easily recognisable by their characteristic, pointed wings and simple wing-venation, in which there are only basal cross-veins (Pl. **22,** 4, 5). Whilst somewhat similar pointed wings occur in some species of Psychodidae (cf. Pl. **10,** 9), there can be little risk of confusion.

In the sweep-net, one or two Lonchopterid species turn up very frequently, when working shady situations near brooks or leaf-carpets in woods; they will soon become familiar to the collector by virtue of the rapid, jerky manner in which they run. This agitated gait is not unlike that of certain other families, but, with a little experience, there should be no danger of confusing them, even with the naked eye, since the pointed wings are clearly evident, even when folded over the abdomen, as is normally the case.

Lonchoptera is the sole British genus, with seven species. *L. lutea* Pz. (Pl. **22,** 4) is very common and widely distributed; in colour, it varies from pale yellow to very dark brown, almost blackish. It is interesting to note that the wing-venation differs as between the sexes, Vein 6, in the male (Pl. **22,** 4), curving downwards to end in the wing-margin, whilst, in the female, it curves forward to end in Vein 5 (Pl. **22,** 5), a characteristic which is constant in all the species. The head is very wide from the dorsal aspect, wider in fact than the thorax, and like the body and legs, armed with strong bristles. Male genitalia are very prominent and turned under the abdomen, extending for about half its length; the lamellae bear long hairs, and the general effect is reminiscent of the male genitalia of some of the species of Dolichopodidae (cf. Pl. **13**).

Not a great deal appears to be known of the life-history, but the larvae are said to live under dead leaves and in decaying vegetable matter, generally. According to de Meijere,[1] the larva is oval and very much flattened, with long, curved bristles at either end and lateral fringes of cilia, which give it an odd appearance ; the head is poorly developed. Pupation is said to take place in two stages, the first, or pre-pupal stage within the larval skin. Other biological notes are to be found in Lubbock.[2]

Keys to the adults of the British species, with notes, are furnished by Collin.[3]

PHORIDAE.

20 genera. Over 250 species. Minute to small, very active, greyish-black, brownish or yellowish flies of humped-back appearance, with highly characteristic wing-venation ; frons wide, generally with stout, upward-curved bristles ; 3rd antennal segment large, concealing two basal segments ; arista long, dorsal or apical ; eyes separated in both sexes ; proboscis fleshy, palpi large, usually spatulate and bristled ; empodium bristle-like ; wings folded over abdomen at rest, sometimes vestigial or absent in female ; venation reduced, sub-costa barely discernible, anterior veins (1–3) very strong and crowded basally, remainder (4–7) weakly developed and characteristically arranged ; ♂ genitalia usually prominent, with a peculiar anal organ. (Pl. **22**, 1–3.)

The members of this family are very active runners and, with certain exceptions, are not seen to fly very much, except when disturbed. They may be picked out quite readily, with a little experience, from the other "small fry" in the sweep-net, or on a window, by their curious and highly distinctive, rapid scrambling mode of progression, which can, perhaps, best be described as "scuttling". Sometimes small species of Dolichopodidae, Sphaeroceridae, Lonchopteridae and Platypezidae give a passable imitation of this gait, but a pocket-lens will soon expose the deception, since the wing-venation and general build of Phoridae (Pl. **22**, 1, 2) together are hardly likely to be confused with any other family. The large third antennal segment may be rounded, lemon-shaped, or even elongate-conical with a bulbous base; it is nearly always thickly but shortly pubescent, as also is the arista. A noteworthy feature of the males is the curious anal tube, which in some genera protrudes very prominently (Pl. **22,** 3). Characters provided by the male genitalia make the males easier to identify than the females.

Megaselia (A.) *meconicera* (Speiser) and some other Phoridae may occur in swarms, Scott[10], Colyer[11], Oldroyd[12]; c.f. Asteiidae and Chloropidae (pp. 228 and 241).

In order to make a brief survey of the family, it will be convenient to divide it into three broad groups, the first of which includes those genera where the tibiae bear distinct, isolated major bristles. In this group, Vein 2 may or may not be present as the upper branch of the apical fork of Vein 3, and both sexes are winged. An example is furnished by *Phora aterrima* F. (Pl. **22,** 2), a very common, deep velvety-black fly, with glassy-clear wings in which the black anterior veins show up strongly. It is to be found, in summer, running actively about on broad leaves such as those of rhododendron or laurel, while the males are not infrequently to be seen hovering in sunlight, usually under a branch of a tree or bush. Not a great deal appears to be known of the life-history; *aterrima* is quite often discovered in the nests of Ants, but the association is not established as other than fortuitous.

Larvae of some of the larger species in our first group are known to develop in dead snails and other carrion, while others are addicted to fungi, including the cultivated mushroom (*vide* Colyer[14] and Hussey[15]). Puparia of a *Conicera* species have been discovered in large numbers on exhumed human bodies, interred two or three years previously, and the evidence has pointed to several generations having developed in the coffins (*vide* Colyer[13]).

The second group consists, on the whole, of much smaller species, with the tibiae bare of isolated bristles, but usually clothed with numerous short, weak bristles or merely cilia.

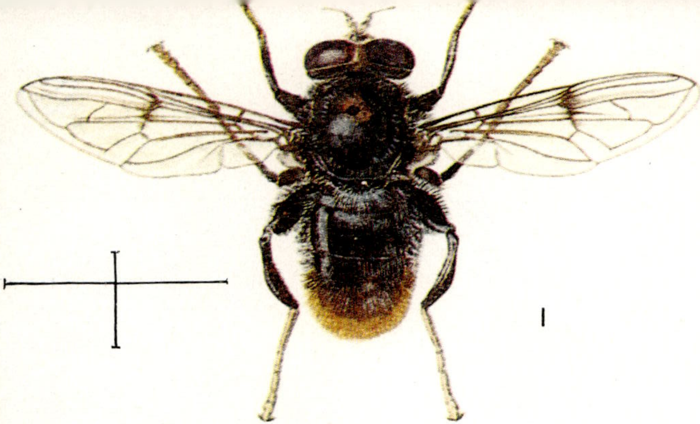

1

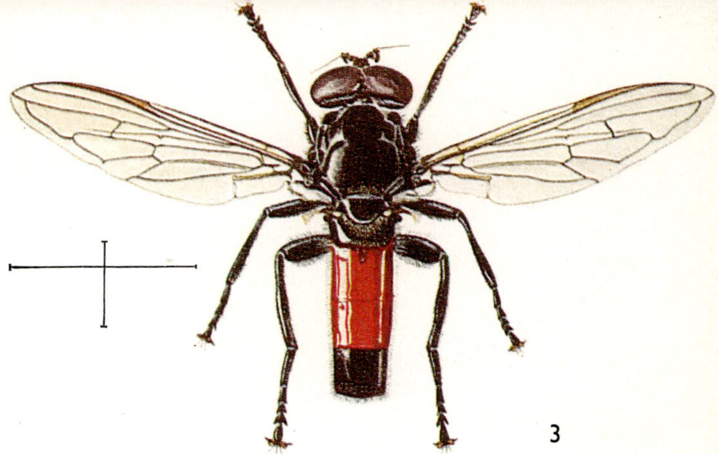

3

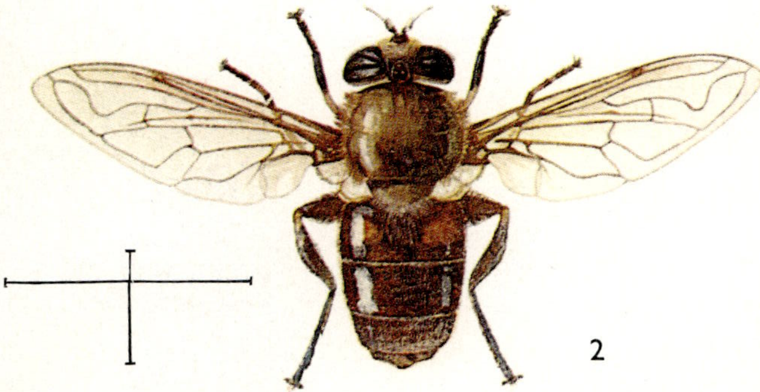

2

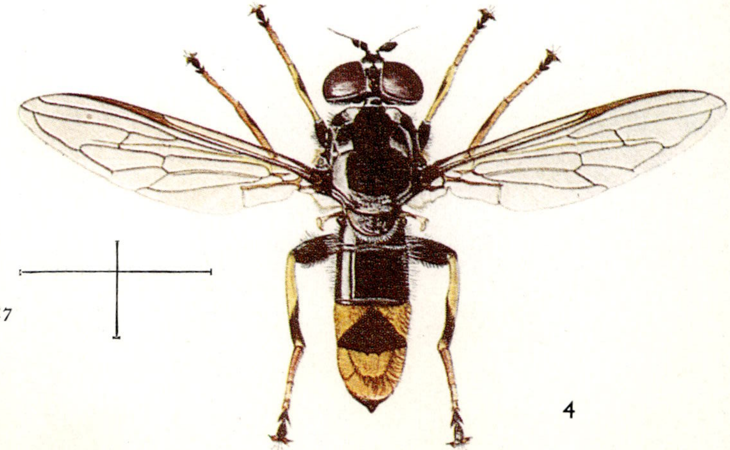

4

PLATE 17

SYRPHIDAE-II

1. *Criorrhina ranunculi* Pz., ♂, p. 170.
2. *Eristalis tenax* L., ♀, p. 163.
3. *Xylota lenta* Mg., ♂, p. 168.
4. *Xylota sylvarum* L., ♀, p. 169.

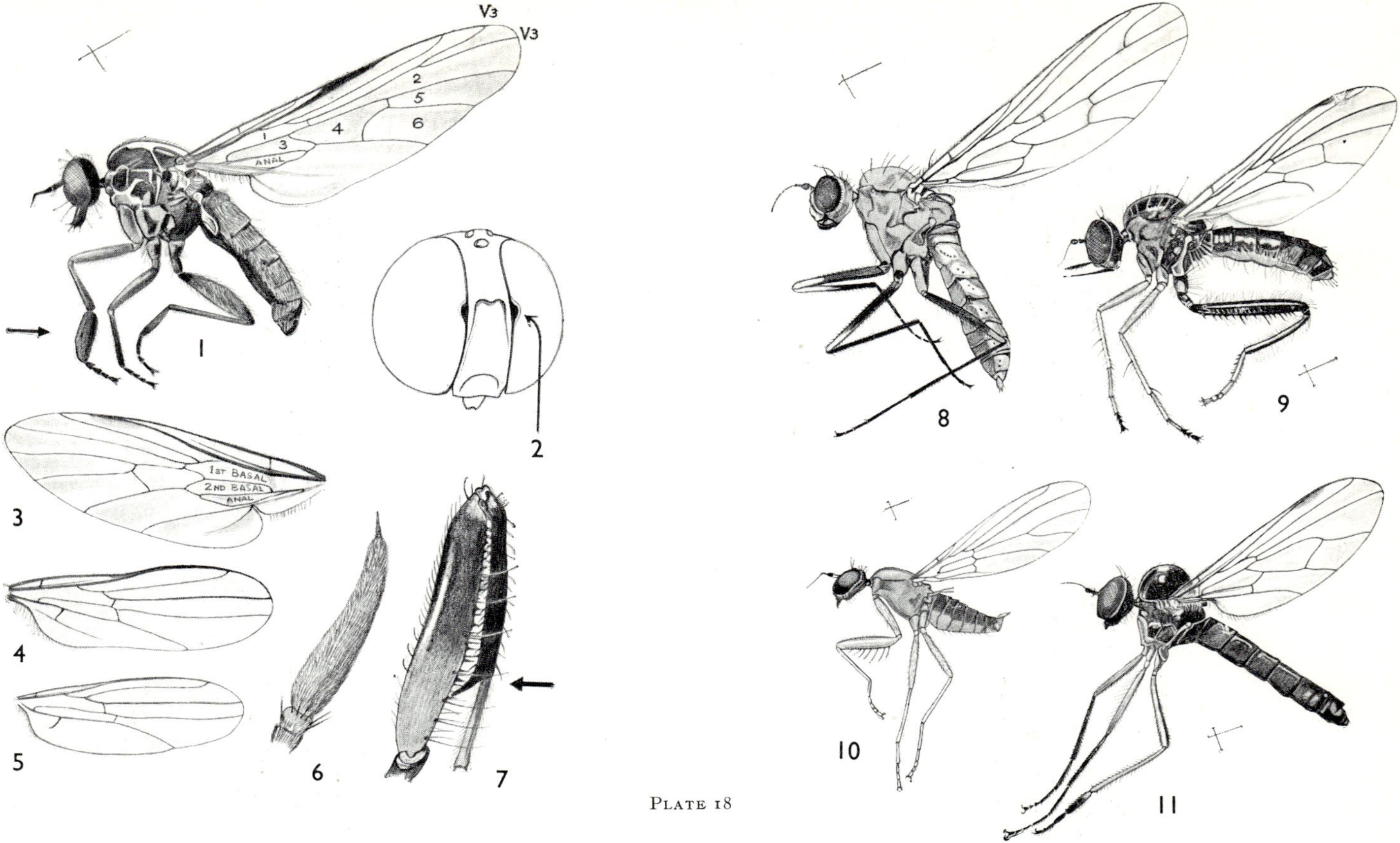

PLATE 18

EMPIDIDAE-II

1. *Hilara maura* F., ♂, p. 125. 2. Empid "eye-nicks", p. 124. 3. *Rhamphomyia sulcata* Mg., wing, p. 126. 4. *Platypalpus*, wing, p. 127. 5. *Tachypeza nubila* Mg., wing, p. 127. 6. *Oedalea holmgreni* Zett., antenna, p. 128. 8. *Clinocera stagnalis* Hal., ♀, p. 129. 9. *Hybos femoratus* Mueller, ♂, p. 128. 10. *Phyllodromia melanocephala* F., ♀, p. 129. 11. *Ocydromia*

Again, both sexes are winged and Vein 3 may have an apical fork or not. In this group is included the complex genus *Megaselia*, containing over 170 species, from which we select, as an example, *M. rufipes* Mg. (Pl. **22**, 1), an exceedingly abundant species, frequently to be seen on windows, even in the depth of winter, if the indoor temperature be favourable. It is very variable in size, ranging from about 1½ mm. to 3 mm. in length ; the male, illustrated in the plate, may be recognised at once by the curious blunt-ended hairs on the sides of the abdomen, in which respect it resembles only one other British species, which is rare ; the female, unless taken in copula, might be confused with several other closely allied species. This fly has a very catholic taste in habitats, locations such as milk-curd in neglected milk-bottles, human excrement, old cheese, and concentrated soap solution having been recorded. In fact, any sort of decaying animal or vegetable matter seems to be acceptable ; no doubt this adaptability has a considerable bearing on the frequent appearance of the species in dwellings.

There is some evidence in the records that the larvae of some of the species in our first two groups, although normally saprophagous, are able to parasitise other insects, such as the larvae of Coleoptera or Lepidoptera, or even fully-developed grasshoppers. Whilst some authorities have considered that such records have in all probability related to fortuitous infestation of sick or moribund hosts, present opinion is that genuine facultative parasitism may well form part of a general polyphagous habit. We possess specimens of *rufipes* which, we are reliably informed, were bred from apparently healthy larvae and pupae of the Cerambycid beetle, *Criocephalus ferus* Muls., and have seen two females of *Diploneura pilosella* Schmitz, which had been bred from larvae found in a mutilated, but still living earthworm.

Pseudacteon formicarum Verrall, which also falls in our second group, is a very tiny species, about ¾–1 mm. long, with

Vein 3 not forked. It is associated, in its life-history, with ants, the larva being an internal parasite. If the nests of certain species of ants, including the common yellow *Lasius flavus* F., whose hummocks are familiar objects in rough pastures, be disturbed in summer, the females of this tiny Phorid species may, occasionally, be seen to hover over the scurrying workers and, now and then, dart down to deposit an egg between two segments of an ant's body with their sharp ovipositors. According to Donisthorpe,[4] the fly will pursue ants on the hand or even dart at the hand itself if scented by acid secretions from them.

There are only two British species of the genus *Aenigmatias* (= *Platyphora*), and these constitute our third group. *A. lubbocki* Verrall is a parasite of the ants, *Formica fusca* L. and *F. picea* Nylander. For a long time, the winged male was known as *Platyphora lubbocki* Verrall, but the wingless female, which looks remarkably like a tiny cockroach, was described as *Aenigmatias blattoides* Meinert; the association between the two was not realised until positively demonstrated by Donisthorpe,[4] to whose works reference should be made for information concerning other Dipterous " guests " of ants. Several other species of Phoridae have been found in the nests of ants, but since they also occur frequently in other locations, the association is probably coincidental. *Aenigmatias* males, which are always winged, are quite unlike most other Phoridae, being very short, broad and flattened in body, with the head nearly as wide, from the dorsal aspect, as the thorax and fitting close to it. Vein 1 and Vein 3, which is not forked, are evenly curved and more or less parallel, joining the costa about half-way along the anterior edge of the wing.

Phorid larvae are generally similar in form to the Cyclorrhaphous type (Fig. 10F), i.e. narrowed anteriorly and truncate posteriorly, with a vestigial head and mouthparts consisting mainly of a pair of hooks. Some species possess a number of processes on the body which give them an odd appearance;

generally, however, these processes are present only in the form of minute tubercles or spines. The pupae often reveal traces of these larval adornments in greater or less degree ; in addition, they bear two more or less elongated protuberances connected with the anterior spiracles. In form, the puparia are variable, many being somewhat kite-shaped, with the dorsal and ventral surfaces curved or arched transversely ; this arching is more pronounced in the dorsal thoracic region and falls away toward the hind extremity.

In connection with the systematic position of the family, to which reference has already been made, the Cyclorrhaphous character of the larvae should be considered together with the manner of emergence of the adults from the pupa, which takes place in various ways. A complete piece of the puparium is thrown off, in some cases, either dorsally or ventrally ; in other cases, the piece thrown off splits longitudinally (cf. p. 35).

There is no up-to-date work dealing with the identification of the British species of Phoridae, but progress may be made with the aid of Wood,[5] Lundbeck,[6] Malloch,[7] Collin,[8] Schmitz [9] and papers by Colyer[16].

LITERATURE CITED

1. de Meijere, J. C. H. Ueber die Larve von *Lonchoptera*. *Zool. Jahrb. Syst.* **14**. 1900.
2. Lubbock, Sir J. On the development of *Lonchoptera*. *Trans. ent. Soc. Lond.* (3) i : 338–44. 1862.
3. Collin, J. E. The British Species of *Lonchoptera*. *Ent. mon. Mag.* **74** : 60–6. 1938.
4. Donisthorpe, H. St. J. K. The Guests of British Ants. Routledge, London. 1927.
5. Wood, J. H. On the British Species of *Phora*. *Ent. mon. Mag.* **42** : 186–96, 262–6. 1906. **44** : 164–74, 215–18, 253–4. 1908. **45** : 24–9, 59–63, 113–20, 143–9, 191–5, 240–4. 1909. **46** : 149–54, 195–202, 243–9. 1910.

   Wood, J. H. Notes on British *Phora* (Corrections and Additions). *Ent. mon. Mag.* **48** : 94–9, 166–81. 1912.

   Wood, J. H. Notes on British *Phora* (Further additions). *Ent. mon. Mag.* **50** : 152–4. 1914.

6. Lundbeck, W. Diptera Danica. Vol. VI. Wm. Wesley & Son, London. 1922.
7. Malloch, J. R. Scottish Phoridae, with tables of all the British Species and notes of localities. *Ann. Scot. Nat. Hist.* **1910**: 15–21, 87–92. 1910.
8. Collin, J. E. On various new or little-known British Diptera, including species bred from the Nests of Birds and Mammals. *Ent. mon. Mag.* **75**: 134–54. 1939.
9. Schmitz, H. On the Irish Species of the Dipterous Family Phoridae. *Proc. Roy. Irish. Acad.* **44**(B)9: 173–204. 1938.
Schmitz, H., in Lindner, E. Die Fliegen der Palaearktischen Region. **33.** Phoridae (not complete). Schweizerbart, Stuttgart. 1938–66.
10. Scott, H. Swarming of *Megaselia meconicera* Speiser (Dipt., Phoridae) in houses. *Ent. mon. Mag.*, **87**: 269–70. 1951.
11. Colyer, C. N. "Swarming" of Phoridae (Diptera). *J. Soc. Brit. Ent.* **5**: 22–7. 1954.
12. Oldroyd, H. The natural history of flies. Weidenfeld & Nicolson, London. 1964.
13. Colyer, C. N. The "coffin" fly, *Conicera tibialis* Schmitz (Dipt., Phoridae). *J. Soc. Brit. Ent.*, **4**: 203–6. 1954.
Colyer, C. N. More about the "coffin" fly, *Conicera tibialis* Schmitz (Diptera, Phoridae). *Entomologist.* **87**: 129–32. 1954.
14. Colyer, C. N. A new species of *Magaselia* (Dipt., Phoridae) from Britain; notes on fungicolous Phoridae. *Ent. mon. Mag.*, **90**: 108–12. 1954.
Colyer, C. N. Phorid flies damaging cultivated mushrooms; *Megaselia* (*M.*) *plurispinosa* (Lundbeck, 1920), a synonym of *M.* (*M.*) *halterata* (Wood, 1910). *J. Soc. Brit. Ent.* **5**: 166–71. 1954.
15. Hussey, N. W. Identification of Phoridae (Diptera) attacking cultivated mushrooms in Britain. *Ann. Mag. Nat. Hist.* (13) **3**: 599–603. 1960.
16. Colyer, C. N. The problem of *Megaselia* (*Aphiochaeta*) *dahli* (Becker, 1901) and *M.* (*A.*) *humilis* (Wood, 1909) (Dipt., Phoridae). *Ent. mon. Mag.*, **90**: 121–8. 1954.
Colyer, C. N. The identity of *Megaselia* (*Aphiochaeta*) *vestita* (Wood, 1914) (Dipt., Phoridae); a description of the male and redescription of the female. *J. Soc. Brit. Ent.*, **5**: 91–4. 1955.
Colyer, C. N. A new species of *Plastophora* (Dipt., Phoridae) from England; a short discussion of the evolution of the present concept of the genus and a key to the identification of the world species. *Broteria Cienc. Nat.*, **26**: 75–89. 1957.

(Continued on p. 352)

CHAPTER XI

SECTION III. CYCLORRHAPHA ASCHIZA

PLATYPEZIDAE (= CLYTHIIDAE) Flat-footed Flies.
PIPUNCULIDAE (= DORILAIDAE) Big-headed Flies.

PLATYPEZIDAE (= CLYTHIIDAE)

5 genera. 23 species. Small, thinly-haired flies, in most cases characterised by flattened hind tarsi and wing-venation ; 3rd antennal segment short-oval (except *Opetia*, *Agathomyia*, elongate-conical) ; arista apical ; ♂ eyes holoptic ; legs short and strong, basal 3 or 4 segments of hind tarsi dilated (but basal 2 slightly, *Microsania* ; hind tarsi normal, *Opetia*) ; wings large ; Vein 4 forked in *Clythia* ; posterior cross-vein absent in *Opetia*, *Microsania* ; ♂ genitalia curved under abdomen, sometimes prominent and resembling *Dolichopus* (cf. Pl. **13**). (Pl. **22**, 6, 7.)

In life, some species of Platypezidae, especially of the genus *Platypeza* (= *Platypeza*) (Pl. **22**, 6), are quite likely to be mistaken for Phoridae, since they have a very similar habit of running about in active, jerky fashion over foliage, often, in their course, rapidly describing a series of circles. Moreover, both sexes have a habit of hovering near low plants, or under the branches of trees, in a fashion resembling that of *Phora* males. Their wings and bodies, however, often exhibit in sunlight a noticeable reddish iridescence which is not the case with Phoridae. When a specimen is caught, examination with a lens will reveal the quite different wing-venation.

Contrary to the usual state of affairs in Diptera, Platypezid females are often more strikingly coloured or patterned than the males, although the species illustrated, *Platypeza atra* Mg. is perhaps the least divergent of all, the female being deep, dull black in colour and not very unlike the male. In other species, the female may be very distinct ; for example, the male of *P. consobrina* Zett. has a dull black abdomen, while in the female the ground-colour is grey, with dull black bands on the hind margins of the segments. Close examination of our example, *atra*, will reveal, just above the base of the antennae, a vague suggestion of a lunule (Pl. **22**, 7); this character, which is rather more pronounced in *Callomyia* (= *Callimyia*), is one which has not occurred in previous families and is significant in that it heralds, thus early, the gradual transition into the Schizophora.

Vein 4 is not forked in *Callomyia* and *Agathomyia*; in the former, the third antennal segment is somewhat longer than in *Platypeza*, while in *Agathomyia*, it is very long and pointed, i.e. gradually tapering off to the apical arista. Females of the very handsome species of *Callomyia* are very distinct from the males and bear various silvery and reddish markings on the thorax and abdomen; generally, *Agathomyia* species are not so striking.

Microsania species have a peculiar liking for the smoke from bonfires; we are aware of one instance where even the acrid smoke from a burning dump of old rubber proved attractive. Appreciation of this somewhat unnatural predilection may lead the collector toward making the acquaintance of these tiny flies, which are a little over 1 mm. in length. Some notes on their habit are given by Edwards,[1] also by Collin,[2] with a key.

Larvae of Platypezidae are more or less broad-oval and flat, with the sides of the body serrated and bearing thin spines or thread-like processes; these processes also occur on the dorsal surface. The pupae are similar, emergence of the adult taking place by the throwing-off of a piece of the anterior dorsal surface of the puparium. Fungi seem to be the normal development-habitat of this family, consequently the adults are likely to be found thereon. Fonseca [3] records females of *Platypeza rufa* Mg., *P. modesta* Zett., and *P. consobrina* Zett. swarming simultaneously around Common Honey Fungus in mid-October, while males of all three species were found sunning themselves on adjacent foliage. We possess specimens of *Platypeza furcata* Fall., an uncommon species, taken on a Bracket Fungus (*Polyporus*), growing on a tree-stump in a London suburban railway station.

Verrall [4] and Wood [5] may be consulted in regard to the British species of the family; Lundbeck[6] and Parmenter[12] will also be found useful.

PIPUNCULIDAE (= DORILAIDAE).

> 8 genera. 76 species. Small, dark, thinly-haired flies, with large, more or less spherical heads composed almost entirely of the compound eyes; 3rd antennal segment short and rounded, kidney-shaped, or with ventral, outward-curved "beak"; arista dorsal; ♂ eyes mostly holoptic, narrowly separated in ♀; wings much longer than abdomen, usually broad at apex and greatly narrowed basally; venation characteristic, Veins 3 and 4 curved and convergent apically, anal cell long, closed near margin (but Vein 4 curtailed and posterior cross-vein absent in *Chalarus*); ♂ genitalia curved beneath tip of abdomen, not very conspicuous externally; ovipositor with bulbous base and long "beak". (Pl. **22**, 8–11.)

The small members of this family, owing to their unobtrusive ways, may be unnoticed by many collectors and thus thought to be more uncommon than, in fact, they are. Once appreciated, their habit of moving in a sort of slow, delicate, hovering flight between grasses or low vegetation, will attract the attention of the keen observer. In the words of Verrall,[4] "the Syrphidae may be the grandest, but the Pipunculidae are the most exquisite, hoverers in the Diptera, as they can hover easily in a folded net, between the folds, without touching the network".

When at rest, adult Pipunculidae may be recognised by their enormous, highly mobile, more or less globular heads (Pl. **22**, 11), consisting almost entirely of the compound eyes. The frons and face, even in the females, are little more than a narrow line, in which are set the antennae.

Pipunculid larvae are internal parasites of Bugs (Hemiptera) of the sub-order Homoptera. Verrall[4] thought that there is some connection between the large, rare species *Nephrocerus flavicornis* Zett. and our only British Cicada, both of which are found exclusively in the New Forest area of Hampshire. *Chalarus spurius* Fall. is parasitic on a species of Leaf-hopper, while Jenkinson[7] records *Verrallia aucta* Fall. as chasing and sitting upon Frog-hoppers. Further reading on the life-history of Dorilaids may be found in Perkins.[8] In shape, the larvae are broad-oval and flat, capable of considerable extension and attenuation, narrowed at both ends but more so anteriorly. When fully developed, the parasitic larva occupies most of the host's abdomen and faces either way. It escapes through the abdominal integument and usually pupates in the ground; the pupa is provided with a pair of respiratory protuberances. Emergence of the adult is effected by the throwing-off of a piece of the anterior dorsal surface of the puparium.

For the British species of Pipunculidae see Coe[11], Collin[9] and Lundbeck[10] will also be found useful.

LITERATURE CITED

1. Edwards, F. W. *Microsania pectinipennis* Mg. attracted to bonfire smoke and some further records of smoke-flies. *J. Soc. Brit. Ent.* **1** (ii): 31–2. 1934.
2. Collin, J. E. Notes on British species of genus *Microsania* Ztt. *Proc. R. ent. Soc. Lond.* **8**: 146–7. 1933.
3. d'Assis-Fonseca, E. C. M. Four Notes on Diptera. *Ent. Rec.* **60**: 50. 1948.
4. Verrall, G. H. British Flies. Vol. VIII. Gurney & Jackson, London. 1901.
5. Wood, J. H. *Agathomyia viduella* Ztt., a new British fly. *Ent. mon. Mag.* **39**: 271. 1903.

   Wood, J. H. Occurrence in Herefordshire of *Callimyia elegantula* Fall. and *Agathomyia boreella* Ztt. *Ent. mon. Mag.* **41**: 5–7. 1905.

   Wood, J. H. *Agathomyia elegantula* Fall., a correction; *Agathomyia zetterstedti* Ztt., a species new to Britain. *Ent. mon. Mag.* **46**: 45–6. 1910.

6. Lundbeck, W. Diptera Danica. Vol. VII. Wm. Wesley & Son, London. 1927.
7. Jenkinson, F. H. *Verrallia aucta* and its Host. *Ent. mon. Mag.* **39**: 222–3. 1903.
8. Perkins, F. H. Leaf-hoppers and their Natural Enemies. *Hawaiian Sugar Planters Assocn. Div. Ent. Bull.* **1.** Pt. 4. 123–57. 1905.
9. Collin, J. E. The British Species of the *Sylvaticus* group of *Pipunculus* (Diptera). *Ent. mon. Mag.* **56**: 270–6. 1920.
Collin, J. E. The British Species of the Genus *Verrallia* (Diptera). *Ent. mon. Mag.* **67**: 234–6. 1931.
Collin, J. E. The British Species of the *Rufipes* group of *Pipunculus* (Diptera). *Ent. mon. Mag.* **73**: 209–18. 1937.
10. Lundbeck, W. Diptera Danica. Vol. VI. Wm. Wesley & Son, London. 1922.
11. Coe, R. L. Handbooks for the Identification of British Insects. Vol. X. Pt. 2(c). Pipunculidae. Royal Entomological Society of London. 1966.
12. Parmenter, L. *Agathomyia falleni* Zett. (Dipt., Platypezidae); an addition to the British List. *J. Soc. Brit. Ent.* **4**: 122–3. 1953.

CHAPTER XII

SYRPHIDAE . . . Hover-Flies, Drone Flies, Flower Flies.

54 genera. Nearly 250 species. Small to large, mostly bristleless, often brightly-marked flies, with the habit of hovering greatly developed ; antennae mostly pendulous, occasionally elongated and porrect, with dorsal arista (but short terminal style in *Callicera*) which is usually bare, but sometimes plumose (e.g. *Volucella*) ; ♂ eyes mostly holoptic, but in some cases well separated (e.g. *Helophilus*) ; hind femora and tibiae sometimes greatly thickened and curved (e.g. *Helophilus, Mallota*) ; ♂ tarsi sometimes with secondary sexual characters (e.g. *Platychirus*) ; wings with " vena spuria " (false vein) lying between Veins 3 and 4 (Fig. 7C) and not joined to a vein ; secondary margin, formed by veins, more or less parallel to actual hind-margin of wing ; ♂ genitalia not usually prominent. (Pl. **16**: **17**: **20**: **21**: **23**: **24**: **26**, 1, 3–6.)

A WALK in the country at any time from April to October, if the sun is shining, is certain to produce an encounter with some of the members of this highly-interesting family, which are remarkable for their powers of hovering. Forest glades and rides will reveal them suspended, apparently motionless, at varying heights from the ground, where they often shine like little jewels in shafts of sunlight piercing the leafy roof. Males of some of the larger species make a splendid sight, in the glorious sunshine of an early spring morning, high above the blackthorn blossom ; at one moment, they will be poised, dark against the background of sky, with wings vibrating so swiftly that they are virtually invisible ; the next, they will have vanished. A second later, they will, almost miraculously it seems, have reappeared in the very same spot where last they were seen. Occasionally, they will make a threatening dart at another male which ventures too near, and return from the sortie to take up the same position.

Both sexes of smaller species, some of which are extraordinarily elongated and attenuated in form, will be seen hovering above, or settled upon flowers ; some early Syrphids congregate in considerable numbers upon Blackthorn and Hawthorn blossom ; others, throughout the summer months, assemble on Compositae and Umbelliferae ; while some may be found, basking on sunlit foliage, with the abdomen making little rhythmic movements reminiscent of wasps and bees. This latter habit is but one of the ways in which Syrphids resemble Hymenoptera ; their shape and colouring is often such that the observer, until he becomes familiar with their styles of flight and behaviour, may need to look very carefully at free specimens in order to note the presence of only one pair of wings and the dipterous type of antennae. Curiously enough, a number of Syrphidae are associated, in their immature stages, with the particular species of wasp or bee which they resemble.

With a captured specimen, there is often a ready means by which the collector may decide whether it belongs to this family ; a " vena spuria " (false vein) lies between the third and fourth veins and is not joined to a vein at either end ; it is, in reality, a thickening of the wing-membrane, which varies in intensity in different species. Another noticeable feature of the wings, especially when the membrane is glassy-clear, is the apparent hind-margin formed by the veins, which runs parallel to the actual hind-margin ; the latter is often hardly visible.

External forms are extremely diverse, which makes it difficult to generalise ; it will be necessary, in order to assess the range of the family, to consider a fairly large number of representative species, and for this purpose we may utilise characters provided by the antennae and wing-venation (*vide* Verrall [1]),* to reduce them to manageable groups.

* The nomenclature adopted by this author is also followed in this chapter.

Group I.

Antennae moderately short, pendulous; arista bare or pubescent; anterior cross-vein before middle of discal cell; cell bounded by Veins 1 and 2 not closed. (Pl. **16,** 1, 2, 4: **20,** 1: **21,** 1, 3–6 : **23,** 1–4, 7, 8, 11–13: **26,** 1, 4.)

A large number of species in this group are black or brownish-black, with orange or yellow bands or crescents on the abdomen and often yellowish markings on the legs. *Syrphus ribesii* L. (Pl. **16,** 1) is an abundant species, almost identical with the equally abundant *S. vitripennis* Mg.; the female of the latter may however be differentiated by the hind femora, which are almost all blackish. *Syrphus balteatus* Deg. (Pl. **23,** 13), the occasional appearance of which, in large swarms with allied species, in coastal districts, especially the Thames Estuary area, has led to the suggestion of possible migration from the Continent, exhibits a second band on the abdominal segments. In other closely-related species, the entire pale abdominal bands become more or less divided until, at the other end of the range, we find lunules or crescents as in *Syrphus luniger* Mg. (Pl. **23,** 12) and *Catabomba pyrastri* L. (Pl. **16,** 2) and the abdomen may be much narrower than the rather oval shape of *ribesii.* The larvae of many of these species are known to be predatory upon aphids (such as the familiar greenfly of roses) and are more or less of the form illustrated on Pl. **26,** 4. *C. pyrastri* furnishes a typical life-history; the female deposits her tiny eggs, less than 1 mm. long, on the underside of a leaf, or on a plant-stem where aphids are feeding or likely to feed: three or four days later, a minute white grub, about 2 mm. long, emerges and commences to hunt its prey. When it comes into contact with an aphis, it plunges its mouthparts into its body and proceeds to suck the juices, holding the prey away from the surface upon which it was resting, whilst retaining its own hold with its hind extremity. This technique, which calls to mind a performing seal balancing a ball upon its nose, renders the victim's struggles of no avail, and in a very short time the husk of the luckless aphis is cast away Starting with a consumption, on the first day, of perhaps three or four aphids, this voracious creature proceeds in the course of the next ten days or so to dispose of an almost incredibly large number, its appetite increasing all the time; by this time its colour has changed to green, with a pale stripe down the middle of the body. The larva is now somewhat slug-like and progresses by using its mouthparts to secure a hold, pulling up its hind part, and then stretching out its head for another grip, the whole movement being a series of rippling undulations. Being sightless, it occasionally halts to wave its head from side to side in quest of its prey. When fully-grown, the larva attaches itself to a leaf or stem by its mouth-parts and shrinks to a brownish, pear-shaped pupa. After a further ten days, the adult fly forces off a piece of the puparium and emerges to feed on pollen and nectar. Larvae of *pyrastri* have also been bred from Lepidopterous pupae, which indicates that prey other than aphids may be taken; in fact, the larva of *luniger* has been recorded as preying on that of the Lackey Moth.

Other genera which are said to have similar life-histories are *Pipiza*, *Pipizella*, *Paragus*, *Cnemodon*, *Melanostoma*, *Sphaerophoria* and *Baccha*; collectively they are thought, in their depredations upon aphids, to outdo even those doughty friends of the gardener, the larvae of the well-known Ladybird beetles, and of the Lace-wings (Neuroptera). *Melanostoma* species are rather small, dark, narrow-bodied flies, about 7–8 mm. long, with reddish-yellow or indistinct bluish spots on the abdomen; the males have a much narrower abdomen than the females. Both *M. scalare* F. (Pl. **23**, 3) and *M. mellinum* L., which is very similar, are very common on the flowers of low plants or Hawthorn. Larvae of *mellinum* have been recorded as attacking adult Muscid flies. Sexual dimorphism also occurs in *Sphaerophoria scripta* L. (Pl. **23**,

7, 8), which is about 10 mm. long, and black with brilliant yellow markings on the abdomen and longitudinal yellow stripes on the sides of the thorax. *Baccha elongata* F. (Pl. **23,** 1) is about 9 mm. and of even more attenuated shape; this frail insect is not easily seen, and is usually to be found hovering in damp situations shaded by vegetation.

Other types of life-history occur in this group, as for instance in the large genus *Chilosia*, the larvae of which have been found living in stems and roots of plants and in fungi. The adults vary greatly in size, from about 11–12 mm. down to 5–6 mm., and are generally of a dark brownish-black or black colour, either dull or shining, sometimes with dense tawny or dark pubescence, or sometimes bare; the wings are mostly clear, but dark markings or clouding occur in one or two cases. Males of some species will often be found, particularly in early spring, hovering at eight to ten feet from the ground in sunshine, while the females are more likely to be found basking on broad leaves, with their wings either folded over the abdomen, or partly extended after the fashion of the Skipper butterflies. Species of this genus have a peculiar feature—slightly raised strips between the face proper and the compound eyes form characteristic "eye-margins" and the jowls droop in a fashion which calls to mind the lips of a mastiff or bloodhound. *C. illustrata* Harris (Pl. **21,** 1) is one of the larger and more striking species; two other large species which may be found in early spring at Sallow are *C. grossa* Fall. and *C. albipila* Mg. Both have clear wings; *grossa* is covered with long, dense pale hairs and has dark antennae, while *albipila* has a lovely reddish-tawny pile, especially dense on the thorax, interspersed with several long black or brown hairs, and orange antennae. Several of the medium-sized species are exceedingly common. *C. albitarsis* Mg. is about 9 mm. long, shining black, with a short oval abdomen and front tarsi with noticeably pale middle segments, and a very common sight on Buttercups and Marshmarigolds in May and early June. *C. variabilis* Pz. is also shining black, but with an elongated abdomen, and is fond of basking on large leaves near the ground, with its wings half-outspread, in May and June ; this species has a peculiar habit of taking short, rapid flights and returning, after a second or two, to the precise place from which it started.

The larvae of *Platychirus* species are thought to belong to the type which develops in decaying organic matter in damp or wet situations ; the adults are remarkable for the fact that the anterior tarsi of the males are curiously modified (Pl. **21,** 5), they are small dark narrow-bodied flies with various reddish, yellowish, or, in certain females, bluish markings on the abdomen. *P. scutatus* Mg. is said to have been bred from rotten fungi.

Closely allied to *Platychirus* are the two British species of *Pyrophaena*, in which the abdomen is somewhat differently shaped (Pl. **21,** 3). In the species illustrated, *P. granditarsa* Forster, both the anterior and middle metatarsi are much dilated and the abdomen bears an extensive fiery-red area ; it is not uncommon in damp situations, but *P. rosarum* F., which is black, with two squarish, basal orange spots on the third abdominal segment, and has simple tarsi, is not so often met. Little seems to be known of their life-history.

Larvae of the very common, but odd-looking species, *Rhingia campestris* Mg. develop in oxdung ; an interesting account of its life-history and early stages, with figures, is provided by Coe.[2] The eggs are laid, not directly on the dung itself, but on grasses, clover and other low plants overhanging it ; when the larvae hatch, they either crawl down the leaf, if it is touching the dung-pat, or drop on to it and seek a crevice in the drying crust, through which to enter the moist part. Both larvae and pupae are difficult to detect, since particles of dung adhere to the integument and make an admirable camouflage. In the adult, a prolongation of the upper mouth-edge forms a pronounced " snout " (Pl. **23,** 11), which, together with the greyish or bluish-grey thorax and reddish abdomen, should make the identification of this fly a simple matter. *R. campestris* is very commonly to be seen on flowers and leaves of low plants, particularly Ground-ivy and Red Campion, on the borders of meadows and woods; the other British species, *R. rostrata* L., which may be distinguished by the absence of the dark, median line on the abdomen, is extremely rare. Of somewhat similar colouring to *Rhingia*, i.e. with black or bluish-black dorsum and rusty-red abdomen, are the closely-allied species of *Brachyopa*, which are rather smaller and bear a not inconsiderable resemblance to certain species of Dung Fly (*Scatophaga*, p. 252); the larvae develop in sap flowing from tree-wounds.

Not a great deal appears to be known of the immature stages of the remaining examples of the group. Larvae of the striking *Xanthogramma ornatum* Mg. (Pl. **16,** 4) have been reared from heaps of turf; a less common species of the same genus, *X. citrofasciatum* Deg., has much narrower yellow bands on the abdomen, and linear instead of triangular basal spots; it will be found in May and early June in some of the older woodlands. The rare *Doros conopseus* F. (Pl. **26,** 1) is interesting on account of its extraordinary likeness to certain Digger Wasps and Ichneumons (Hymenoptera), also to species of Conopidae (p. 177). We have taken two males at rest on Reeds (*Arundo*) on 19 June 1949 at Leigh-on-Sea. *Leucozona lucorum* L. (Pl. **20,** 1) is a handsome and quite common fly closely related to *Syrphus*. Another Syrphid of unusual shape is *Neoascia* (= *Ascia*) *podagrica* F. (Pl. **23,** 2), which is about 5–6 mm. long and much in evidence round nettles and low plants from early spring till quite late in the year; it occurs quite commonly in gardens, except in the very heart of big towns, and the males have habits similar to those of *Syritta pipiens* L. (p. 167).

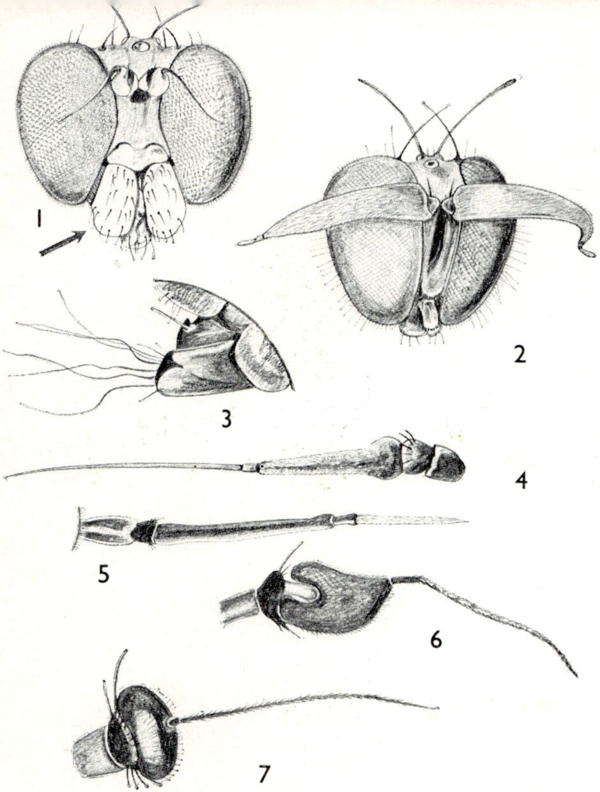

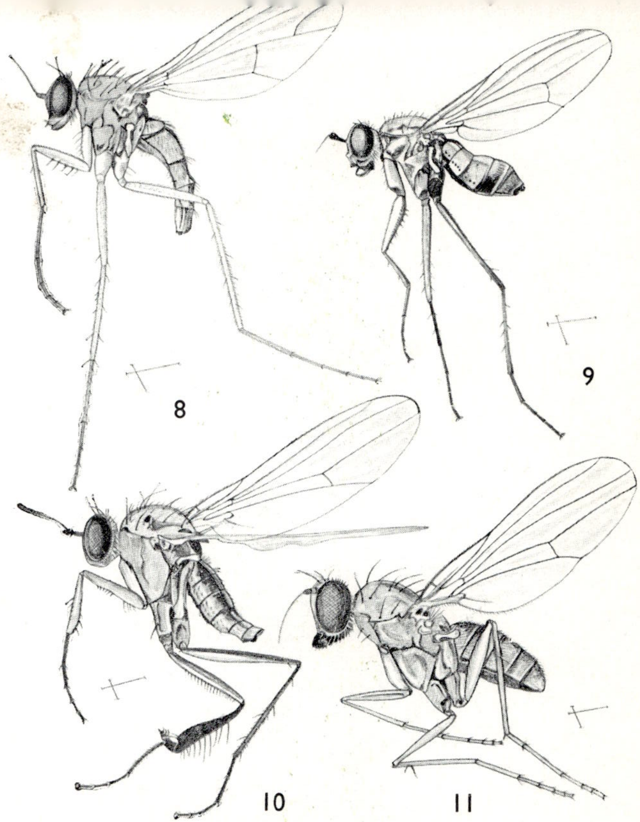

PLATE 19

DOLICHOPODIDAE-II

1. *Thinophilus flavipalpis* Zett., head, p. 135. 2. *Xiphandrium appendiculatum* Zett., head, p. 136. 3. *Aphrosylus raptor* Walker, ♂ genitalia, p. 135. 4. *Aphrosylus raptor* Walker, antenna, p. 135. 5. *Rhaphium longicorne* Fall., antenna, p. 133. 6. *Syntormon pallipes* F., ♀ antenna, p. 136. 7. *Chrysotus neglectus* Wied., antenna, p. 137.

8. *Sciopus platypterus* F., ♀, p. 138. 9. *Hydrophorus oceanus* Macq., ♀, p. 134. 10. *Campsicnemus scambus* Fall., ♂, p. 138. 11. *Medeterus flavipes* Mg., ♀, p. 136.

Plate 20

SYRPHIDAE-III

1. *Leucozona lucorum* L., ♂, p. 169.

3. *Calliprobola speciosa* Rossi, ♀, p. 169.

Group II.

As Group I, but arista plumose, and cell bounded by Veins 1 and 2 closed. (Pl. **24**, 2, 3, 4.)

Species of the single genus contained here, *Volucella*, are large, handsome flies, splendid hoverers and swift in flight; the larvae perform a scavenging function in the nests of wasps and bees. Fabre [3] relates the life-history of a species associated with wasps. The female enters the nest, unharmed by them, and lays her eggs on the papery, outer surface, where they adhere. These eggs are about 2½ mm. long and glistening-white; the larvae which hatch from them are often unable to maintain their hold upon the paper slopes and fall to the vault below the nest, where the wasps throw their dead. This is very convenient for the *Volucella* larvae, since they alight on a store of food. Some of the fly-larvae, however, manage to enter the interior of the nest, where they roam about, quite unchecked. Normally, the wasp-larvae do not excrete in their cells until on the point of pupating; the fly-larva inserts itself alongside the body of the wasp-larva and provokes the emission of excremental liquid, upon which it sups. When fully-grown, the *Volucella* larva is a dirty, reddish-brown in colour, the body being somewhat narrowed anteriorly and covered with fleshy corrugations and excrescences. Pupation takes place in the subterranean midden. It seems that other flies which probably perform a scavenging function are allowed to come and go, unmolested, but insects which have a nefarious design upon the helpless wasp-larvae, are speedily despatched by the formidable jaws or stings of the guardians of the nest.

As regards our British species, *Volucella bombylans* L. (Pl. **24**) resembles in appearance certain species of Humble-Bee perhaps more than any other Syrphid; what is more, it is subject to considerable variation, so that more than one kind of bee is imitated; two of these forms are shown in the plate. Although the larvae are usually associated with the nests of Humble-Bees of the genus *Bombus*, they have been bred from that of the wasp, *Vespula germanica* F. *V. pellucens* L. is a large, shining black fly, about 15 mm. long, with a conspicuous, broad whitish band across the base of the abdomen and a dark blotch across the middle of the wings, the bases of which are suffused with orange. Its larvae and pupae are found mostly in the nests of the Common Wasp, *Vespula vulgaris* L. The males of *pellucens* make a fine sight, when hovering, as is their wont, at about seven or eight feet from the ground in clearing and glades ; the females are not infrequently to be found at the flowers of Bramble and Dog Rose. Deep orange replaces the whitish band in *V. inflata* F. which is much less common and rather local. *V. inanis* L. (Pl. **24,** 2) is associated with the nests of the Hornet, *Vespa crabro* L. and is not so common as *bombylans* and *pellucens* ; a very similar, but much larger fly, *V. zonaria* Poda. should be watched for, since it is a prize of some importance to the collector in this country. Formerly considered extremely rare, it seems to have settled itself in England in recent years, mainly along the south coast, although it has been recorded from such an inland locality as Wimbledon Common. Fraser [4] has bred out *zonaria* and *pellucens* from larvae found in the same nest of the Common Wasp, *Vespula vulgaris* L. in Bournemouth, and has figured one of them ; they are broad-oval in shape and the segments are armed with lateral spinous processes. The *zonaria* which was photographed for the frontispiece was one of many feeding on ivy blossom.

Group III.

Antennae moderately short, pendulous ; anterior cross-vein near or beyond middle of discal cell ; Vein 3 with a deep curve. (Pl. **16,** 3 : **17,** 2 : **20,** 4 : **21,** 2 : **23,** 14, 16 : **26,** 5.)

The larvae of most of the species in this group are of the type popularly referred to as " rat-tailed maggots " ; they are peculiarly adapted to life in liquid or semi-liquid media, such as water or mud, from which they obtain their sustenance by the ingestion of minute particles of decaying organic matter. Typical habitats include gutters and old, neglected water-butts where leaves and other rubbish have been allowed to accumulate. Ponds in or near old-fashioned farmyards, in which the water is fouled with stable effluents, and which recedes in drought periods to expose evil-smelling mud, are almost certain to harbour some of these larvae, probably several species at once. The long, telescopically-extensible breathing-tube (Pl. **26,** 5) enables these creatures to inhale atmospheric air whilst submerged to a considerable depth beneath the surface. When fully grown, the larvae leave the water and seek a drier location in earth or mould in which to pupate. An account of the " rat-tailed maggot " is given by Miall.[5]

One of the commonest species with this type of larva is *Eristalis tenax* L. (Pl. **17,** 2) ; it is popularly known as the " Drone Fly " on account of its remarkable resemblance to the male or " drone " of the Hive Bee. The yellowish or yellowish-brown markings on the basal part of the abdomen are variable and may, in some specimens, be hardly perceptible, as in the example illustrated. *E. tenax* may be readily distinguished by the bands of dark hairs across the eyes and by its bare arista, from its common relative, *E. pertinax* Scop., which is roughly the same size and of very similar appearance, but has the arista plumose, the eyes without defined bands, and the anterior tarsi orange. In addition, *pertinax* usually gives the impression that its abdomen is more tapered than that of *tenax*. *E. intricaria* L. is a handsome fly, covered with a dense, tawny and black pile ; the arista is plumose and the tibiae have a characteristic broad white band on the basal half. Resemblance to various small Humble-Bees is heightened by the hairs at the apex of the abdomen, which may be reddish or whitish.

The males of all the three foregoing species are magnificent hoverers, usually at six to nine feet from the ground, in woodland clearings and glades, or above Blackthorn or Hawthorn blossom; the last-named species is particularly agile and provides a good test of the collector's dexterity with the net. All of them emit a piteous-sounding, high-pitched humming wail when entrapped in the folds, a peculiarity shared to some extent by all the members of the family. A similar sound, though not at quite so high a pitch and of more even tenor, is often made by Syrphids, when free and at rest.

E. arbustorum L. (Pl. **23,** 16) and the closely-similar *E. nemorum* L. are two very common black-and-orange species, about 11–12 mm. long, in which the arista is plumose, at least on the basal half. In fresh specimens, they may be differentiated by the shining black central line on the face of *nemorum* and its absence in *arbustorum,* but this character can be confusing if specimens of the latter have had the short, dust-like pubescence rubbed off that part of the face. *E. horticola* Deg., a similar species, can usually be distinguished by its slightly larger size, and the presence of a transverse clouding across the middle of the wings. All of these species are in evidence from about April until September and may often be seen in considerable numbers on flowers, particularly umbels and Ragwort.

A small, crumpled-looking, black species very unlike the foregoing is *E. sepulchralis* L.; it is about 8–9 mm. long, with its eyes, which are separated in the male, covered with dark, irregular blotches, and is remarkable in being one of the few Syrphidae in which the "vena spuria" is so reduced as to be practically indistinguishable. This fly is almost invariably found around farmyard ponds or on herbage near muddy water containing decaying matter. Very similar to *sepulchralis* but usually a little larger is *E. aeneus* Scop., in which the eyes of the male are holoptic; it is normally found in coastal localities, particularly in the south, although it has been taken on bombed sites in the City of London.

Helophilus pendulus L. (Pl. **16,** 3) is very common and often

Plate 21

SYRPHIDAE-IV

1. *Chilosia illustrata* Harris, ♂, p. 158.
2. *Merodon equestris* F., ♂, p. 166.

3. *Pyrophaena granditarsa* Forster, ♂, p. 159. 4. *Pyrophaena granditarsa*, ♂ fore tarsus. 5. *Platychirus peltatus* Mg., ♂, p. 159. 6. *Platychirus peltatus*, ♂ fore tarsus.

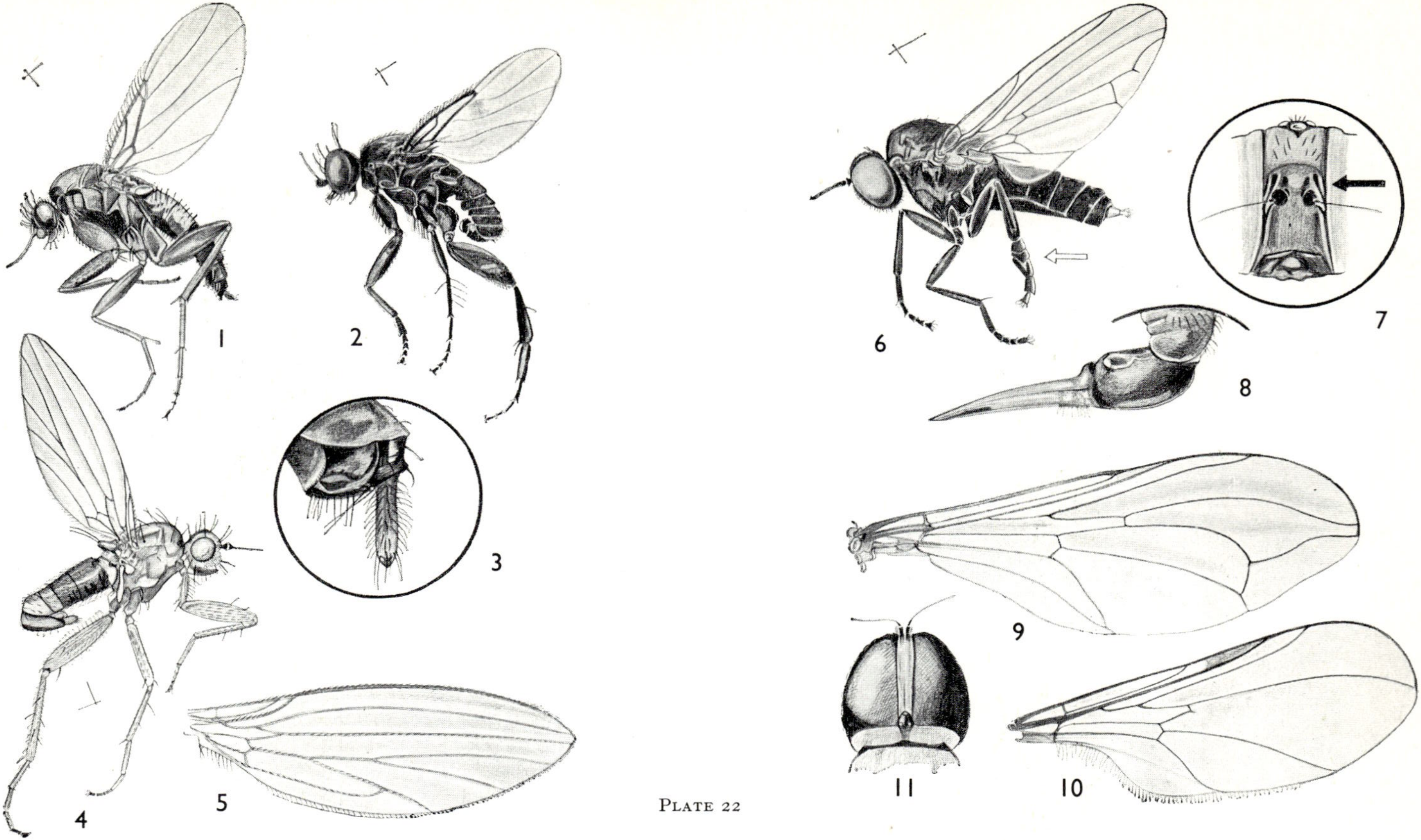

PLATE 22

PHORIDAE

1. *Megaselia rufipes* Mg., ♂, p. 145. 2. *Phora aterrima* F., ♂, p. 144. 3. *Diploneura*, ♂ genitalia and anal tube, p. 144.

LONCHOPTERIDAE

4. *Lonchoptera lutea* Pz., ♂, p. 142. 5. *Lonchoptera lutea* Pz., ♀ wing, p. 142.

PLATYPEZIDAE

6. *Platypeza* (=*Clythia*) *atra* Mg., ♀, p. 149. 7. *Platypeza* (=*Clythia*) *atra* Mg., front view of head, p. 150.

PIPUNCULIDAE

8. *Pipunculus* (=*Dorilas*), ovipositor, p. 151. 9. *Pipunculus* (=*Dorilas*), wing, p. 151. 10. *Chalarus spurius* Fall., wing, p. 151. 11. *Dorilas*, ♀ head, p. 151.

to be found on umbels and Ragwort, or resting on waterside foliage, while the males also have a habit of hovering above ditches and pools, about a foot or so from the surface of the water. Two closely-related species, *H. trivittatus* F. and *H. hybridus* Loew, also occur in similar surroundings ; *trivittatus* may be distinguished from the others in having a reddish or yellowish, instead of black vertical line on the face, while *hybridus* differs from *pendulus* in having the basal third, instead of two thirds of the hind tibiae orange. *H. trivittatus* is more likely to be found in flat coastal districts or estuaries, but sometimes may occur inland with the other two species. *H. versicolor* F., which is not unlike the three previous species, but only about two-thirds of their size and ruddier and more hairy, is not uncommon. Certain other species of the genus *Helophilus* differ from those already mentioned in having linear or parallel-sided abdomens, rather than oval ; an example is provided by the fairly common *H. transfugus* L. (Pl. **23,** 14). *H. lineatus* F. is about the same size, but distinguished at once by its long, pointed " snout ", formed by the projection of the upper mouth-edge. *H. vittatus* Mg., in which the pale abdominal lunules of *transfugus* and *lineatus* are replaced by pairs of greyish-yellow longitudinal spots, is uncommon and more or less confined to flat coastal or estuarine localities, such as the Thames Estuary area ; we have found it, for instance, on the Isle of Sheppey and at Benfleet.

Myiatropa florea L. is very much like a brighter and more yellow version of *Eristalis tenax* L. and is roughly the same size, as a rule, although very much smaller specimens are occasionally encountered. Here, as in *Helophilus* the cell bounded by Veins 1 and 2 is open, which serves to differentiate it from *Eristalis* ; while the thorax bears on the dorsum unusual pale transverse markings. The species is common and widely distributed ; immature stages have been recorded from pools of water standing in hollow beech-stumps and from the saturated decaying parts of fallen beeches. The rather handsome and uncommon *Mallota cimbiciformis* Fall. (Pl. **20,** 4), also resembles *tenax*, but as will be seen from the plates, may be distinguished by wing-venation and the build of the hind femora and tibiae; Coe[14] describes its breeding in an old horse chestnut tree in Hyde Park.

Our last example from the group is *Merodon equestris* F. (Pl. **21,** 2), the "Large Bulb Fly" or "Narcissus Fly", which is a serious pest to the horticulturalist. This is another of the Syrphidae which resembles Humble-Bees; in fact, it may be said to resemble several kinds of Humble-Bee, owing to its extensive colour-variation. There are a number of fairly well-defined forms which turn up regularly, all of which mate freely with one another, with the consequent production of intermediate forms (cf. *Volucella bombylans* L., p. 161). The larva of this species is very different from the "rat-tailed" type of those we have previously considered in this group; it is short, stout and fleshy, with no prominent breathing-tube, and develops, usually singly, in the bulbs of Narcissus and allied plants, which become soft, rotten and quite useless; it has also been recorded from onions. Considerable losses have been caused by *equestris* in parts of this country where there is extensive bulb cultivation, such as Cornwall.

Courtship procedure, in *equestris*, is most interesting; we are reminded of an occasion when we were able to witness the process from start to finish. A female, feeding on a flower-head of Water Hemlock, was under observation when a splendid male suddenly appeared, as if by magic, about two inches above her and hovered for a few seconds, almost motionless save for the slight intermittent flicker of his rapidly vibrating wings. Then he began a series of spectacular, slow vertical helicopter-like ascents and descents, with the tip of his abdomen curving downward on the descents, whilst the female arched her abdomen upward. These evolutions were varied by the male occasionally making a lightning, circular flight round the spot and returning to carry on the up-and-down process. After about ten minutes, the male suddenly came right down and made contact; there ensued a quick flurry of wings and legs and then the coupled pair dropped off the flower-head into the short vegetation below; the mating was completed in approximately a further fifteen minutes.

Group IV.

As Group III, but Vein 3 without a noticeably deep curve. (Pl. **17**, 1, 3, 4: **20**, 2, 3: **23**, 5, 6, 9, 10: **26**, 3.)

Two very striking flies, *Sericomyia borealis* Fall. (Pl. **23**, 9) and *S. lappona* L. have long-tailed larvae which show some relationship to the "rat-tailed maggots" previously mentioned; they develop in decomposing peat sods and peaty, moorland pools. The elegant, wasp-like adults may not infrequently be encountered on wild, boggy moors such as occur on Dartmoor and in the New Forest area. Both have bluish-black or brownish-black bodies, with yellow or yellowish legs and abdominal markings; the three interrupted, pale transverse abdominal bands in *lappona*, are linear and rather whitish. *S. lappona* is more scarce than *borealis*, and both may occur in wooded districts as well as the type of locality already mentioned; in fact, we have seen both in a Surrey wood at no great distance from London.

Of the Syrphid larval types which inhabit decaying organic matter, we have touched upon the "rat-tailed maggots" and the "short-tailed" form (cf. *Rhingia*, p. 159) which is normally to be found in drier and more solid media, such as dung, compost-heaps and rotting vegetation generally. Some more species of the latter type occur in this group. *Syritta pipiens* L. (Pl. **23**, 5) is a very common example, about 8 mm. long, with very thick hind femora and curved tibiae; the legs vary considerably in colour, in some specimens being almost all black, while in others there are extensive orange markings, especially on the hind femora. Parmenter [6] comments on the habit, in the males of this species, of " shadowing " or making threatening darts at males of the same species, while ignoring other species, which suggests that they appropriate and defend a definite " territory ". One male was seen to patrol a limited area, containing a heap of vegetable refuse, where the females were likely to come for the purpose of oviposition, and to perform a series of peculiar flight-evolutions above one which arrived (cf. *Merodon*, p. 166). Reference has already been made to the males of larger species, such as *Eristalis*, making darts at other males, when hovering over Blackthorn and Hawthorn blossom; here it is possible that the behaviour is governed by the probability of females visiting the bloom to feed.

Tropidia scita Harris (Pl. **23,** 6) also belongs to the " short-tailed " section; it is about 8 mm. long, and black with orange-red markings on abdomen and legs. This species often occurs in considerable numbers in water-meadows or marshy places in such localities as the Norfolk Broads, the Cambridgeshire Fens and the Thames Marshes; sometimes it may be found close to fresh-water pools near the seashore. It has a characteristic habit of flying in and out between the stems of low plants and reeds.

Rotten wood or sap from tree-wounds seem to constitute the development-habitats of the larvae of a number of species in this group. When basking, with its wings folded over the abdomen, *Xylota segnis* L. (Pl. **23,** 10), which is about 12 mm. long and black with a very wide sandy or orange band on the abdomen, could easily be taken for one of the Hymenoptera. Sometimes very much smaller specimens than the average size quoted may be taken, and in doubtful cases, it may be distinguished from the other species in the genus by the row of prominent spines under the femora in both sexes. *X. lenta* Mg. (Pl. **17,** 3), which will be known at once by the conspicuous blood-red band on the abdomen (in *segnis* it is sandy or orange), is very local, but apparently widely distributed in the midland and southern counties. *X. sylvarum* L. (Pl. **17,** 4), with its gold tail, is another very distinct species and not at all uncommon in woodlands. Smaller species of the genus, all with much the same characteristic shape, may be met with occasionally on old, rotten tree-trunks and stumps. They are of a blackish colour, with small squarish or indefinite pale markings on the abdomen, which vary a good deal within each species.

Calliprobola speciosa Rossi (Pl. **20,** 3) is, to our mind, one of the most handsome and exotic-looking of our British flies; it is decidedly scarce and apparently confined to the old forests of the south, such as Windsor Forest and the New Forest. We have taken it in both localities, on umbels and hovering round beech stumps; it is said to be attracted to the sap flowing from tree-ulcers and the larva has been recorded from a rotten beech. The species is on the wing in May and June.

Chrysochlamys cuprea Scop. (Pl. **20,** 2), which has characteristic long, black bristly hairs along the margins of the dorsum and scutellum, is also attracted to sap from tree-wounds, in which its larvae develop. It is fairly common and widely distributed and may be distinguished from its very rare and closely-similar relative, *C. ruficornis* F., which we have taken in Epping Forest, by its blackish arista and the fact that the dull bands on the abdomen are actually on the hind-margins of the segments, whereas in *ruficornis* the arista is somewhat reddish and the dull bands are clearly before the hind-margins.

Several bee-like species in this group also breed in rotten wood. *Brachypalpus bimaculatus* Macq. is a dark-brownish, rather hairy fly, with a bare arista and the hind femora thickened and twisted; on the ventral surface there are two rows of short bristles. *B. bimaculatus* is extraordinarily like certain species of Mining Bee of the genus *Andrena*.

Acquaintance with the species of *Criorrhina* is one of the more exciting things which happen to the collector. Mostly in evidence in early spring, they are large, handsome densely hairy flies, which closely resemble Humble-Bees or Carder Bees in appearance and are remarkable for their speed and agility of flight. Although they may occasionally be taken low down, while at rest on Blackthorn, Hawthorn or Sallow bloom, usually a long-handled net, a very quick eye and considerable dexterity are required to capture them. In *C. ranunculi* Pz. (Pl. **17,** 1), which occurs in the south, the hairs at the tip of the abdomen vary from reddish to whitish, thus giving a good imitation of the Large Red-tailed or Large Yellow-tailed Humble-Bees.

C. floccosa Mg. is variable in size, specimens in our collections ranging from 13 mm. to 15 mm.; with its dense tawny or pale yellow hairs it resembles species of Carder Bees. It occurs in the south and south-west. *C. asilica* Fall. has a narrow abdomen bearing rather inconspicuous yellowish transverse bands, and looks like an elongated version of the Hive Bee; it is not so uncommon as generally thought, and may not infrequently be found on early Hawthorn blossom. We have taken all the species within a short distance of London, and have even found *floccosa* in a small park in a built-up North London suburb.

Pocota apiformis Schrank (= *personata* Harris) may be distinguished by the fact that the face is extended very little below the eyes, when regarded in profile, whereas in the species of *Criorrhina*, it is prolonged considerably to form a pronounced " snout ". *P. apiformis* is an early spring species which develops in rotten trees and is very rare; it is recorded from poplar and beech in old forests, and has occurred fairly recently in Epping Forest and Windsor Forest.

Eumerus tuberculatus Rond. and *E. strigatus* Fall. are the " Lesser Bulb Flies " which, like *Merodon* (p. 166), develop in cultivated bulbs; sometimes several larvae are found in a single bulb. Wing-venation is very characteristic (Pl. **26,** 3): the two species mentioned are distinguishable only on minor features. Both have been a serious pest in the Fen areas of

East Anglia; they occasionally, also like *Merodon*, attack non-cultivated plants related to the normal host-plants, such as Wild Hyacinth.

Group V.

Antennae long and porrect; Vein 3 normal, i.e. without a rearward-projecting veinlet. (Pl. **23,** 15: **24,** 1: **26,** 6.)

Species of the genus *Chrysotoxum* form another series of wasp-like flies, with their black bodies, brilliant yellow abdominal arcs, or " bows ", long porrect antennae and yellow legs. *C. cautum* Harris (Pl. **23,** 15), which is about 13 mm. in length, is quite common in certain localities on umbels; the males are interesting by virtue of the fact that the genitalia are not symmetrical, as may be seen by looking at the insect from directly behind (Pl. **26,** 6). *C. octomaculatum* Curtis and *C. verralli* Collin are somewhat similar but smaller, about 10–11 mm. long, while *C. festivum* L. (Pl. **24,** 1) is a much darker-looking species altogether, in which the pale arcs do not extend over the sides of the abdomen. *C. bicinctum* L. is about 9–11 mm. long and differs from all the others in having only two bows, a wide one on the second segment and a narrow one on the fourth. *Chrysotoxum* species are on the wing from about June to August and develop in the nests of ants.

Callicera aenea F. is a beautiful but rare fly, with a blackish thorax and golden-green abdomen, on which there is fairly long tawny pubescence; the antennae are particularly noteworthy on account of the fact that the third segment bears a terminal style about half as long as the segment itself. The Scottish species *C. rufa* Schumm. has been reared by Coe[14] from larvae found in a water-filled rot hole in a pine tree.

Group VI.

Antennae long and porrect; Vein 3 with rearward-projecting

veinlet; dorsal arista; scutellum armed with two blunt processes.

The single genus *Microdon* is represented here; none of the British species are common. They are broad flat flies, much resembling Hive Bees, with the basal segment of the antennae greatly elongated, the second short and the third of medium length. The basal segments are only slightly divergent, but the remainder strongly so; the effect is somewhat reminiscent of *Stratiomys* (cf. Pl. **8**). *Microdon* larvae are like little short slugs, with a strongly convex dorsal and flat ventral surface; so unlike other Dipterous larvae are they that they were first described as Mollusca. They live as scavengers in ants' nests, where they subsist on the pellets of the hosts, which take no notice of them as they creep about in the normal position; their dorsal surface seems to provide adequate protection. If a *Microdon* larva, however, is unfortunate enough to be found on its back by the ants, it is then vulnerable to attack and the ants will despatch it. Donisthorpe [7] gives an account of *Microdon* species in association with British Ants; the commonest appears to be *M. mutabilis* L., which has been recorded from the nests of several kinds of ant, including the Yellow Meadow Ant, *Lasius flavus* F. We have found *M. eggeri* Mik in old woodlands of Surrey and Sussex and have seen adults emerging from the nest of *Lasius niger* L. in a rotten tree-stump, in early May.

Larval habits of the Syrphidae may be summarised as follows:—

(*a*) Aphidivorous (e.g. *Catabomba, Syrphus, Melanostoma,* etc.).

(*b*) Frequenters of decaying organic matter (e.g. *Eristalis, Helophilus, Syritta, Rhingia,* etc.).

(*c*) Frequenters of sap from tree-wounds or of rotting wood (e.g. *Xylota, Brachypalpus, Pocota, Chrysochlamys,* etc.).

(*d*) Frequenters of plant stems and roots, or fungi (e.g. *Chilosia, Merodon, Eumerus,* etc.).

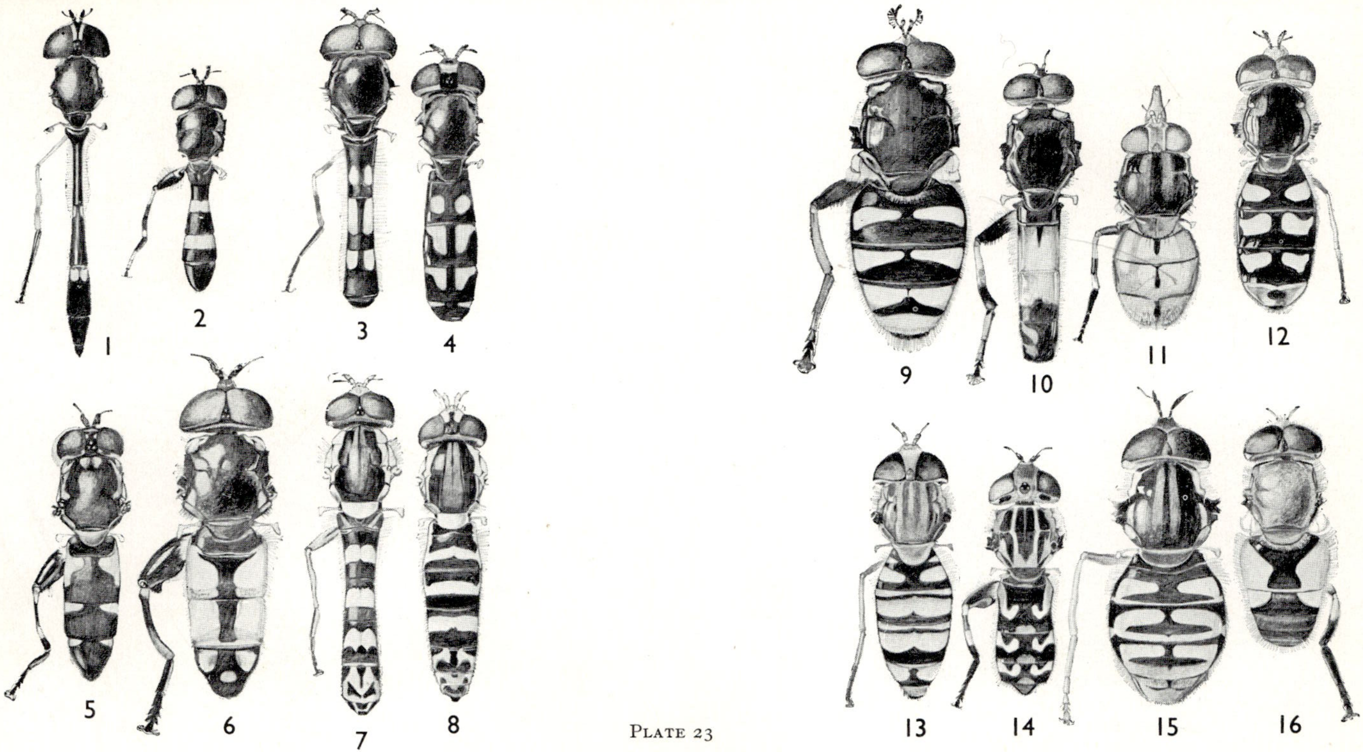

PLATE 23

SYRPHIDAE-V

1. *Baccha elongata* F., ♀, p. 158. 2. *Neoascia podagrica* F., ♂, p. 160. 3. *Melanostoma scalare* F., ♂, p. 157. 4. *Melanostoma scalare* F., ♀, p. 157. 5. *Syritta pipiens* L., ♀, p. 167. 6. *Tropidia scita* Harris, ♂, p. 168. 7. *Sphaerophoria scripta* L., ♂, p. 157. 8. *Sphaerophoria scripta* L., ♀, p. 157.

9. *Sericomyia borealis* Fall., ♂, p. 167. 10. *Xylota segnis* L., ♂, p. 168. 11. *Rhingia campestris* Mg., ♀, p. 159. 12. *Syrphus luniger* Mg., ♂, p. 156. 13. *Syrphus balteatus* Deg., ♀, p. 156. 14. *Helophilus transfugus* L., ♀, p. 165. 15. *Chrysotoxum cautum* Harris, ♂, p. 171. 16. *Eristalis arbustorum* L., ♂, p. 164.

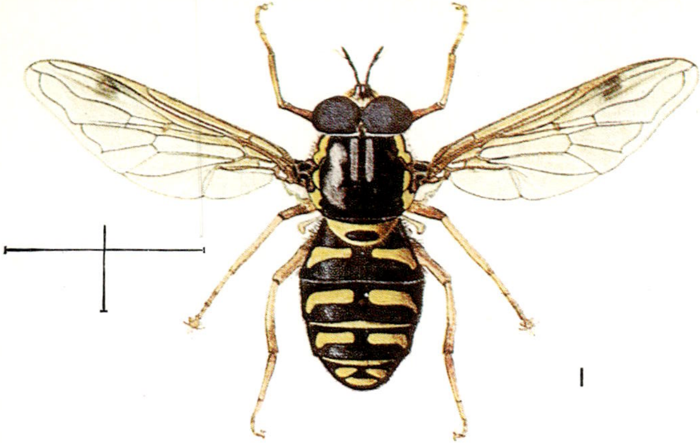

1

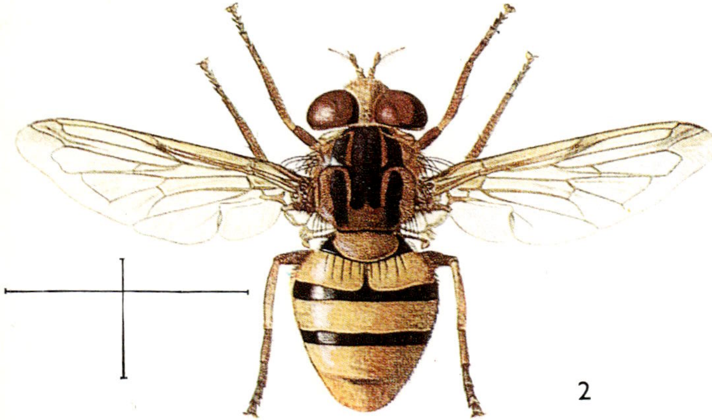

2

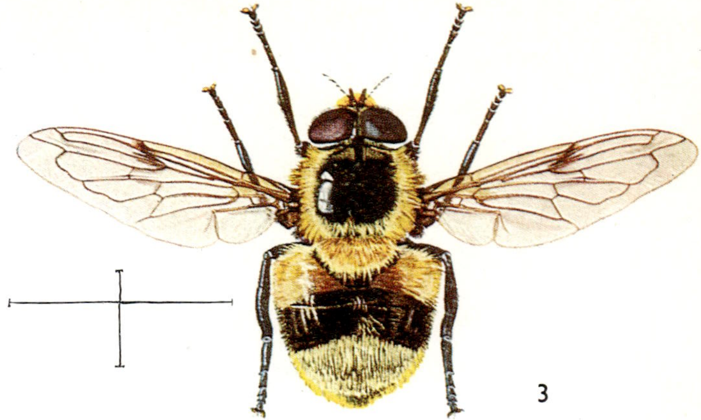

3

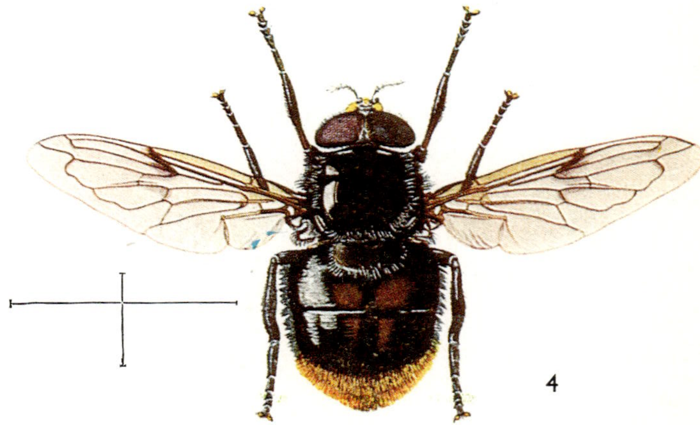

4

PLATE 24

SYRPHIDAE-VI

1. *Chrysotoxum festivum* L., ♂, p. 171.

(*e*) Dwellers in the nests of bees, wasps and ants (e.g. *Volucella, Microdon*).

In structure, they may be classified thus :—

(*a*) the aphidivorous type (Pl. **26,** 4) (e.g. *Catabomba*).

(*b*) the short-tailed inhabitant of fairly solid decaying matter (e.g. *Rhingia, Syritta*).

(*c*) the "rat-tailed maggots" (e.g. *Eristalis, Helophilus*).

(*d*) the boring larva (*Merodon*).

(*e*) the fleshy scavenger (*Volucella*).

(*f*) the mollusc-like scavenger (*Microdon*).

Syrphid pupae are also considerably varied in shape, with either short or long thoracic respiratory horns ; emergence of the adult fly is effected by the throwing-off of a dorsal anterior piece of the puparium.

For the identification of the British species of Syrphidae, see Coe[13] and Verrall[1] and papers by Collin[8] and Coe;[9] additional reading on the family will be found in Lundbeck.[10] For the immature stages see Hartley[11] and Dixon.[12]

LITERATURE CITED

1. Verrall, G. H. British Flies. Vol. VIII. Gurney & Jackson, London. 1901.
2. Coe, R. L. *Rhingia campestris* Meigen (Dipt. Syrphidae) ; An Account of its Life-history and Descriptions of the Early Stages. *Ent. mon. Mag.* **78** : 121–30. 1942.
3. Fabre, J. H. The Life of the Fly. Hodder & Stoughton, London. 1927.
4. Fraser, F. C. *Volucella* (Dipt. Syrphidae) larvae sp., breeding in a nest of *Vespula vulgaris* L. (Hym. Vespidae). *Ent. mon. Mag.* **82** : 55–7 ; also **82** : 158, and **82** : 202. 1946.
5. Miall, L. C. The Natural History of Aquatic Insects. p. 198. The Rat-tailed Maggot (*Eristalis*). McMillan, London. 1895.
6. Parmenter, L. Behaviour of *Syritta pipiens* L. (Dipt. Syrphidae). *Ent. mon. Mag.* **80** : 44. 1944.
7. Donisthorpe, H. St. J. K. The Guests of British Ants. Routledge, London. 1927.

8. Collin, J. E. *Eumerus strigatus* Fall. and *E. tuberculatus* Rondani. *Ent. mon. Mag.* **56**: 102–6. 1920.

Collin, J. E. Notes on some Syrphidae. *Ent. mon. Mag.* **67**: 153–9, 177–82. 1931.

Collin, J. E. Notes on Syrphidae. II. The Genus *Sphegina* Mg. *Ent. mon. Mag.* **73**: 182–5. 1937.

Collin, J. E. Notes on Syrphidae. III. The Genus *Brachyopa* Mg. and Description of the ♂ of a New Species of *Xylota, X. xanthocnema* Coll. *Ent. mon. Mag.* **75**: 104–9. 1939.

Collin, J. E. Notes on Syrphidae. IV. The Genus *Neoascia* Willst; *Chrysotoxum octomaculatum* Curt; *C. elegans* Lw. and allied species; *Chamaesyrphus caledonicus* Coll., sp. n., *C. lusitanicus* Sharp, 1903 (nec Mik, 1890) and its nearest allies. *Ent. mon. Mag.* **76**: 150–8. 1940.

Collin, J. E. *Syrphus ericarum*, sp. n. ♂♀. (Dipt. Syrphidae). *Ent. Rec.* **58**: 117–19. 1946.

9. Coe, R. L. Description of the ♀ of *Xylota xanthocnema* Coll. and a second British Record of *Rhingia rostrata* L., and its distinction from *R. campestris* Mg. *Ent. mon. Mag.* **75**: 224–7. 1939.

Coe, R. L. *Brachypalpus eunotus* Lw., new to Britain; its distinctions from *B. bimaculatus* Mcq. and Notes on Synonymy in the Genus. *Ent. mon. Mag.* **77**: 193–7. 1941.

Coe, R. L. *Chrysochlamys ruficornis* F.; its distinctions from *C. cuprea* Scop. *Ent. mon. Mag.* **77**: 165–7. 1941.

Coe, R. L. *Callicera spinolae* Rond. (Dipt. Syrphidae) taken in Britain; its redescription, with notes on *C. aenea* F. and *C. rufa* Schumm. *Entomologist.* **76**: 155–8. 1943.

10. Lundbeck, W. Diptera Danica. Vol. V. Wm. Wesley & Son, London. 1916.

11. Hartley, J. C. A taxonomic account of the larvae of some British Syrphidae. *Proc. Zool. Soc. Lond.* **136**: 505–73. 1961.

12. Dixon, T. J. A key to and description of the third instar larvae of some species of Syrphidae (Diptera) occurring in Britain. *Trans. R. ent. Soc. Lond.*, **112**: 345–79. 1960.

13. Coe, R. L. Handbooks for the identification of British Insects. Vol. X. Pt. 1. Syrphidae. Royal Entomological Society of London. 1953.

14. Coe, R. L. Rediscovery of *Callicera rufa* Schummel (as *yerburyi* Verrall); its breeding habits, with a description of the larva. *Entomologist*, **71**: 97–102. 1938.

Coe, R. L. *Mallota cimbiciformis* Fallén breeding in Hyde Park, London. *Ent. Gazette*, **4**: 282–286. 1953.

KEY TO THE BRITISH FAMILIES OF BRACHYCERA (SECTION II) AND ASCHIZA (SECTION III)

(Partly based on Oldroyd, Handbooks for the Identification of British Insects, Vol. IX, Pt. 1, 38–9)

1. Empodium pad-like, i.e. foot with three pads (Fig. 8C) ...2
   Empodium bristle-like,[1] sometimes absent, likewise pulvilli, i.e. foot with only two pads (Fig. 8B), or none 5
2. Third antennal segment annulated, i.e. showing indications of primitive segments (Fig. 2: Pl. **11**, 2, 4) 3
   Third antennal segment not annulated,[2] i.e. showing no indication of primitive segments; usually with style or arista (Fig. 4A: Pl. **18**, 6: **19**, 4–7) 4
3. At least the middle tibiae spurred; squamae large;[3] Vein 3 with wide fork embracing apex of wing (Fig, 2: Pl. **9**: **12**: **14**, 3) ....TABANIDAE (p. 101)
   None of the tibiae spurred;[4] squamae mostly small; veins crowded anteriorly; discal cell small, veins issuing apically therefrom faint (Pl. **5**) ........STRATIOMYIDAE (p. 91)
4. Thoracic squamae very large; venation reduced; bulbous flies with small heads (Pl. **9**, 2) ..CYRTIDAE (p. 112)
   Thoracic squamae small; venation complex; slender, long-legged, delicate-looking flies (Pl. **9**, 1)
   RHAGIONIDAE (p. 97)
5. Basal cells long; anal cell long and pointed,[5] sometimes open (Pl. **15**, 7) 6
   Second basal and/or anal cell short[6] (Pl. **18**, 3, 4), or anal cell absent (Pl. **18**, 5); anal cell, if present, closed by recurrent vein (Pl. **18**, 3, 4) or blunt apically (Pl. **18**, 10, 11); never more than 7 cells, including basals and discal, between Vein 3 (posterior branch, if forked) and anal cell (Pl. **18**, 1, 3)....10

[1] Certain species of Dolichopodidae and Empididae have a pad-like empodium, but will be recognised on wing-venation.
[2] Except *Xylophagus* (Rhagionidae), p. 97.
[3] Squamae of Tabanidae not evident in plates, see footnote, p. 101.
[4] Except *Solva* (Stratiomyidae), p. 95 (Pl. **11**, 5).
[5] But note rather short basal cells and anal cell in Clythiidae (Pl. **22**, 6).
[6] Except Hybotinae (Empididae), p. 128 (Pl. **18**, 9).

6. Vein 3 forked (Pl. **15,** 7) 7
   Vein 3 not forked (Pl. **16**: **22,** 6, 9) 13
7. Eight cells, including basals and discal, between Vein 3 (posterior branch) and anal cell (Pl. **15,** 2) 8
   Less than eight cells, including basals and discal, between Vein 3 (posterior branch) and anal cell (Pl. **15,** 1) 9
8. Vertex sunk between eyes; ♂ eyes separated (Pl. **15,** 4) ASILIDAE (p. 117)
   Vertex not sunk between eyes; ♂ eyes holoptic (Pl. **15,** 2) THEREVIDAE (p. 114)
9. Seven cells, including basals and discal, between Vein 3 (posterior branch) and anal cell (Pl. **12,** 3, 4) BOMBYLIIDAE (p. 108)
   Six cells etc., etc. (Pl. **15,** 1) ....SCENOPINIDAE (p. 115)
10. Venation abnormal; Veins 1–3 strong, crowded basally, remainder weak, running in more or less parallel lines from Vein 3 to hind-margin (Pl. **22,** 1, 2) PHORIDAE (p. 143)
    Not thus 11
11. Wing-tips pointed; cross-veins at extreme base (Pl. **22,** 4, 5) LONCHOPTERIDAE (p. 141)
    Wing-tips rounded; at least one cross-vein beyond basal quarter of wing (Pl. **18**: **19**) 12
12. First basal cell long; anterior cross-vein well beyond basal quarter of wing (Pl. **18,** 3–5, 8–11) EMPIDIDAE (p. 123)
    First basal cell very short; anterior cross-vein well within basal quarter of wing; posterior cross-vein prominent (Pl. **19**) DOLICHOPODIDAE (p. 132)
13. Vein 4 turns forward to join Vein 3 before wing-margin; usually, anterior branch of Vein 5 also turns forward to meet Vein 4; the two often running more or less parallel to hind-margin of wing; usually a "vena spuria" (false vein) present between Veins 3 and 4 (Fig. 7C: Pl. **16**) SYRPHIDAE (p. 154)
    Vein 4 may turn forward but does not join Vein 3; no "vena spuria" (Pl. **22,** 6, 9, 10) 14
14. Arista apical; hind tarsi often dilated (Pl. **22,** 6) PLATYPEZIDAE (p. 149)
    Arista dorsal; hind tarsi not dilated; head very large and rounded, almost entirely occupied by compound eyes (Pl. **22,** 11).................. PIPUNCULIDAE (p. 151)

PLATE 25

CONOPIDAE-I

1. *Conops quadrifasciata* Deg., ♂, p. 179. 2. *Sicus ferrugineus* L., ♂, p. 180. 3. *Physocephala rufipes* F., ♂, p. 179. 4. *Myopa testacea* L., ♂, p. 180.

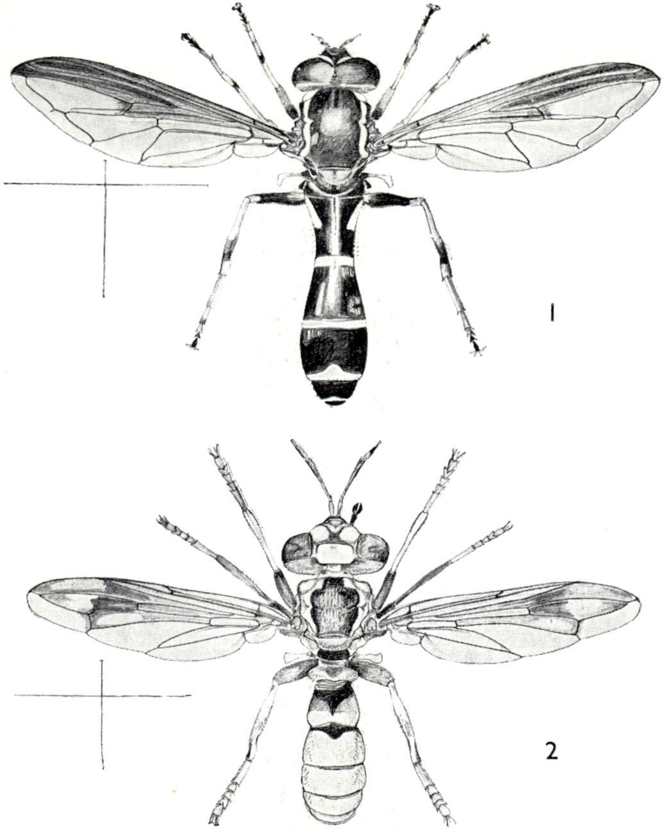

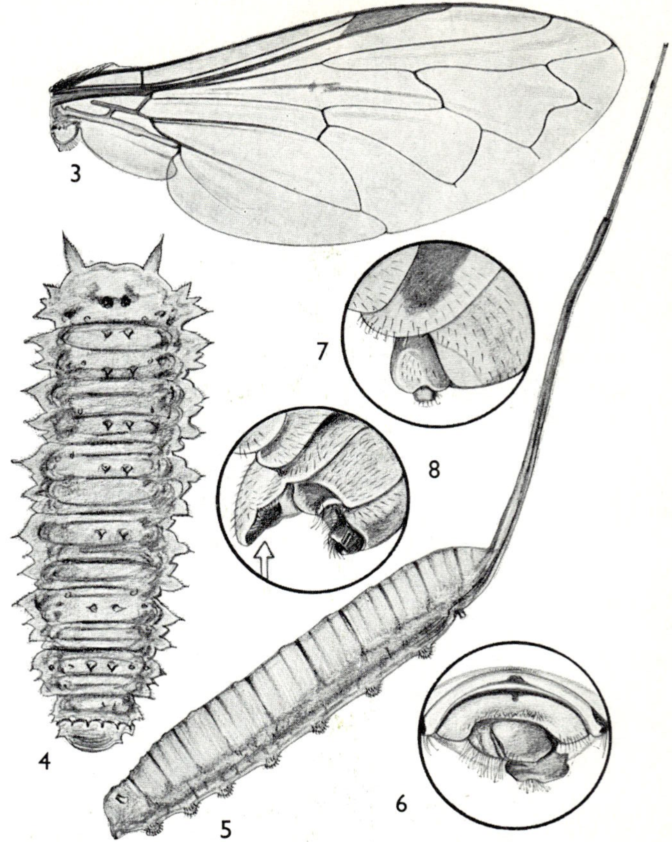

PLATE 26

SYRPHIDAE-VII
1. *Doros conopseus* F., ♂, p. 160.
CONOPIDAE-II

SYRPHIDAE-VIII
3. *Eumerus tuberculatus* Rond., wing, p. 170. 4. Syrphid larva, aphidivorous type, p. 156. 5. Syrphid larva, "rat-tailed maggot", p. 163. 6. *Chysotoxum cautum* Harris, ♂ genitalia from rear, p. 171.
CONOPIDAE-III

CHAPTER XIII

SECTION IV. CYCLORRHAPHA SCHIZOPHORA. THE ACALYPTERATES

The Acalypterates are characterised by a frontal lunule, ptilinal suture, the absence of a definite cleft in the second antennal segment, an incomplete transverse suture on the dorsum, and posterior calli which are not prominent or clearly differentiated from the remainder of the dorsum (Fig. 5). Squamae are mostly, but not invariably, small or vestigial, not concealing the halteres, and the eyes of both sexes are separated.

CONOPIDAE

8 genera. 18 species. Medium-sized to fairly large flies, usually bare or rather thinly haired, often with striking resemblance to Solitary Wasps and other Hymenoptera; head large, as wide, or wider than thorax, from dorsal aspect; antennae long, porrect, clubbed, 3rd segment, with apical style (Conopinae), or short, pendulous, with short, dorsal, jointed arista (Myopinae); mouth-opening wide, with long, slender (except *Leopoldius*) forwardly-directed proboscis, which may be jointed or "elbowed" either at the base only, or also at the middle, in which case the front portion folds back like a pocket-knife; wings sometimes with dark markings; Veins 3 and 4 convergent apically, sometimes forming a stalked cell; anal cell long, pointed, apex mostly close to hind-margin; abdomen often constricted basally, especially in males; genitalia of both sexes (in profile) generally prominent and of unusual form. (Pl. **25**: **26**, 2, 7, 8.)

IN the larval stage, Conopidae are internal parasites of adult bees and wasps. The eggs, which are attached to the host whilst in flight, at least in certain cases, are furnished with tiny, hooked processes at one end, which probably assist respiration by becoming attached to the large tracheal trunks after penetration of the host's cuticle. As a rule, the larvae are to be found in the front part of the host's abdomen, where they are attached by their hind extremity to the tracheae (cf. Cyrtidae,

p. 112, Tachinidae, p. 259). Their mouthparts are much reduced, sustenance apparently being obtained from the blood of the host. Larvae are normally somewhat pear-shaped, with the anterior part narrow, an adaptation for feeding on the thoracic contents *via* the host's " waist ", Smith[4]. The rear spiracles are rather prominent.

Pupation takes place within the larval skin ; the more or less oval puparium exhibits, at the hind extremity, prominent indications of the larval spiracles. The pupa occupies most of the abdominal cavity of the host, where it remains through the winter ; when the adult Conopid emerges, it ruptures the puparium and the enclosing abdominal segments at the same time.

Two sub-families present themselves for consideration. In the Conopinae, the third antennal segment bears a terminal style and ocelli are absent ; traces of a " vena spuria " between Veins 3 and 4, reminiscent of the Syrphidae, are discernible. The third antennal segment, in the Myopinae, bears a dorsal, jointed arista and ocelli are present.

CONOPINAE.

The genus *Conops* contains four large and striking species, in which the antennae are conspicuously clubbed, the second segment being elongated and the third thickened, with a very short terminal style. They are great lovers of sunshine and flowers, especially white or pale-coloured ones ; although fond of basking, motionless, and said by some authorities to be slow fliers, they will depart with great rapidity when alarmed ; in fact, our experience is that their flight is always swift. *C. vesicularis* L. (Pl. **26,** 2) is of a yellow and brown colour similar to that of a Hornet, and distinguishable from the other species of the genus by the fact that the scutellum is brown or reddish-brown instead of black. It is in evidence from about May to August, but mainly in spring and early summer, and in certain localities, is to be found, not infrequently, on

Hawthorn blossom. Two smaller species, *C. quadrifasciata* Deg. (Pl. **25,** 1) and *C. flavipes* L. are more commonly met with, the former distinguishable by its yellow or brownish-yellow hind femora ; *flavipes* has the hind femora black, with perhaps the base and tip yellowish. Both species appear later, generally from about mid-June to August, and are particularly fond of basking on the flowers of Bramble. The fourth species, *C. strigata* Wied., may be known by its black hind femora, yellow face, and jowls bearing black patches ; in the latter respect, it differs from *flavipes*, which has, at most, only a blackish median stripe on the face. *Conopilla* (= *Conops*) *ceriaeformis* Mg. is of very similar appearance to the foregoing species but will be known by its black hind femora coupled with a completely black frons.

Two species of *Leopoldius* (= *Brachyglossum*) have been recorded in Britain and are very rare ; *L. signatus* Wied. has been taken fairly recently in Hertfordshire, on Ivy flowers at the end of September. *Leopoldius* species may be distinguished from the remainder of the Conopinae by the fact that the proboscis is comparatively short and stout, shorter in fact than the width of the head, in profile.

Physocephala rufipes F. (Pl. **25,** 3) is deceptively like a Sand Wasp (*Amnophila*) with its abdomen greatly constricted at the base ; it is not so uncommon as often supposed, occurring in wooded districts from May to September, but chiefly in July and August. In the second species, the larger *P. nigra* Deg., the wide reddish basal abdominal band of *rufipes* is wanting.

Recorded hosts of *Conops* species include Wasps (*Vespula*), Mason Wasps (*Odynerus*), Sand Wasps (*Amnophila*) and Humble Bees (*Bombus*) ; those of *Leopoldius*, Wasps (*Vespula*) ; and those of *Physocephala*, Honey Bees (*Apis*) and Humble Bees (*Bombus*).

MYOPINAE.

Zodion cinereum F. is a small, dark-greyish species, in which the reddish frons and satiny-yellow face and jowls show up in sharp contrast; the proboscis differs from the rest of the Myopinae inasmuch as it is jointed only at the base. We have found this species, which is about 7 mm. long, in late spring and early summer, in the south, at rest upon daisies, where it is somewhat conspicuous by virtue of its dark colouring. *Zodion* is parasitic upon Mason Wasps (*Odynerus*) and Mining Bees (*Halictus*).

The two British species of *Thecophora* are small, greyish flies, about 4–6 mm. long, with the jowls, in profile, much narrower than the height of the eye; the lower part of the face is somewhat silvery and the antennae longer than in the next genus, *Sicus*. They are said to occur near the burrows of Mining Bees (*Halictus*), which makes it not unlikely that these are the hosts.

Sicus ferrugineus L. (Pl. **25**, 2) is somewhat variable in size, from 9 mm. to 14 mm.; the width of the jowls, in profile, is much less than the height of the eye, which serves to distinguish it, together with *Thecophora*, from the next genus, *Myopa*; the antennae are very much shorter than in *Thecophora*, and the size and general colouring very different. *Sicus* is abroad from May to September, but mainly in evidence in July and August, when it is not at all uncommon in, or on the outskirts of woods, very often on umbels or Bramble flowers; it is recorded as a parasite of Humble Bees (*Bombus*).

Myopa species will be known by their deep jowls, which, in profile, are at least equal to the height of the eyes; the antennae are similar to *Sicus* and the abdomen is hardly tapered at the base, while the wings are more or less clouded, especially on the cross-veins, as in our example, *M. testacea* L. (Pl. **25**, 4). Although mainly seen in spring, they are recorded from April to September, occurring on Hawthorn blossom, umbels, dandelions, thistles, etc. Hosts include the Common Wasp,

Vespula vulgaris L., Mining Bees (*Andrena, Colletes*) and Humble Bees (*Bombus*).

Conopid genitalia of both sexes may prove rather confusing at first, since the female abdomen is often equipped with ventral appendages (Pl. **26,** 8) which might easily be mistaken for parts of male genitalia. The underside of the fifth segment has a projection (which may be very pronounced in Conopinae or vestigial as in *Myopa* and *Sicus*), likewise the seventh and eighth segments, the inner surfaces being papillate, i.e. covered with little points, rather like the familiar rubber massage-pads; the whole, at first sight, looks remarkably like male clasping apparatus. Male external genitalia are less prominent, appearing in profile, as projections from the seventh and eighth segments (Pl. **26,** 7). From the dorsal aspect, in Conopinae, the female abdomen is more or less narrow and parallel-sided, whereas in the male it is widened apically, thus presenting an elongated pear-shaped, or Indian Club appearance (Pl. **25,** 1, 3).

The British species may be identified with papers by Smith[1] and Collin.[2] The biology is summarised in papers by Smith[3,4].

LITERATURE CITED

1. Smith, K. G. V. The Irish Conopidae. *Proc. R. Irish Academy* (B) **54**: 203–8. 1952.
   Smith, K. G. V. The Conopidae (Diptera) in the Dale collection with revisionary notes on the British species of *Thecophora* Rond. (=*Oncomyia* Loew). *Ent. mon. Mag.*, **91**: 298–300. 1955.
2. Collin, J. E. The British species of *Myopa* (Dipt., Conopidae). *Ent. mon. Mag.*, **95**: 145–151. 1959.
3. Smith, K. G. V. The distribution and habits of the British Conopidae (Dipt.). *Trans. Soc. Brit. Ent.* **13**: 113–36. 1959. Supplement *Entomologist*, **94**: 237–9. 1961.
4. Smith, K. G. V. The larva of *Thecophora occidensis* with comments upon the biology of Conopidae (Diptera). *J. Zool.*, **149**: 263–76. 1966.

CHAPTER XIV

PLATYSTOMATIDAE
(= PLATYSTOMIDAE)
OTITIDAE (= ORTALIDAE)
PALLOPTERIDAE
} . Pictured-wing Flies.
ULIDIIDAE.
TEPHRITIDAE (= TRYPETIDAE) Large Fruit Flies, Gall Flies.
LONCHAEIDAE.

It will doubtless be noticed from time to time that certain small flies have a curious habit of intermittently opening and closing their wings with a kind of slow, twisting or trembling motion, while walking or resting. This habit of " wing-waving " is characteristic, in some degree, of most of the flies which form the subject of this chapter, and of the Sepsidae (p. 194). The group is also generally characterised by " pictured wings ", which may be banded, spotted or dappled ; a minority have clear wings but show close affinities in other respects. Species with pictured wings will be seen to fall into broad categories as follows :

(i) wings marked with large transverse bands, which may be separate or joined (Pl. **28,** 1, 4, 6);
(ii) wings with large dark spots or patches, probably resulting from reduction of bands (Pl. **28,** 2, 5);
(iii) wings with clouding of cross-veins and/or wing-tips and ends of long veins (Pl. **27,** 1–3) ;
(iv) wings with dark membrane broken up by little clear areas (Pl. **27,** 4).

Another general feature is the more or less elongated, rigid ovipositor-sheaths of the females, the ovipositors themselves being adapted for piercing plant tissues. These sheaths are of two main types, viz. the blunt, wedge-shaped type (Pl. **28,** 2 : **29,** 2), and the narrower type, which may be very short (Pl. **27,** 2) or very long ; as long as, or even longer than the abdomen (Pl. **28,** 1). Male genitalia are not prominent. Larvae of a very large proportion of the species develop in plants. There are no up-to-date keys to the British species, but much assistance may be obtained from Séguy,[1] whose work contains many illustrations and covers all the families listed above. In addition, it contains a very useful index of host-plants and Diptera associated with them. Other foreign works which will provide assistance are quoted under the headings of the families to which they relate.

PLATYSTOMATIDAE.

2 genera. 2 species. Small flies with pictured wings and lower part of occiput swollen ; 3rd antennal segment short and rounded (*Platystoma*) or long and thin (*Rivellia*) ; post-vertical bristles small ; vibrissae absent ; proboscis short and thick ; propleural bristle absent ; three supra-alars ; tibiae without dorsal preapical bristle ; wing-membrane dark with numerous clear spots (*Platystoma*) or clear, with darkened base and tip and two narrow transverse bands between (*Rivellia*) ; costa broken near humeral cross-vein ; *sc* complete ; Vein 1 hairy ; Vein 3 ends before wing-tip ; Vein 6 reaches margin ; Vein 7 absent ; ovipositor with non-retractile sheath. (Pl. **27,** 4: **30,** 4.)

Platystoma seminationis L. (Pl. **27,** 4) and *Rivellia syngenesiae* F. are easily distinguishable from one another on the characters given in the summary ; the former is not uncommon in May and June, on nettles and flowers of low plants, in cool, shady or damp places. The larvae are said to develop in decaying vegetable matter.

OTITIDAE.

8 genera. 17 species. Small to medium flies often with pictured wings ; face with pronounced grooves to accommodate antennae ; 3rd antennal segment more or less pointed or angular ; postverticals of normal size, parallel or divergent ; vibrissae absent ; proboscis short and stout ; propleural bristle present ; 4 supra-alars ; tibiae without dorsal preapical bristle ; costa unbroken ; *sc* complete ; Vein 1 usually hairy ; Vein 3 ending at, or near, wing-tip ; Vein 6 reaching to, or nearly to, margin ; ovipositor with non-retractile sheath. (Pl. **27,** 1, 8: **30,** 7.)

Wings of Otitidae may be faintly or heavily clouded, or spotted, especially on cross-veins or vein-ends at wing-tip ; or may have heavy transverse bands. The species of *Melieria*, from which we select, as an example, *M. omissa* Mg. (Pl. **27,** 1), are fairly common on vegetation, bordering ponds and streams, such as Yellow Flag or Water Hemlock ; or marshy places, sometimes near the seashore. The shape of the third antennal segment should be noted (Pl. **30,** 7).

Dorycera graminum F., which is 5–9 mm. in length, has an unusual appearance, the head, in profile, seeming almost serpentine (Pl. **27,** 8); the structure of the antennae, insertion of the arista and the antennal grooves, or pits, in the face are noteworthy. Three dark stripes adorn the dorsum; the wings, tinged with brown along the anterior margin, are long and broad, about half as long again as the abdomen.

Seioptera vibrans L., a shining blue-black fly, about 5–6 mm. long, with a red frons and clear wings bearing a dark spot near the tip, is included in this family in the Check List. Its wing-venation tends toward that of the Ulidiidae (p. 186), in that Veins 3 and 4 converge apically (cf. Pl. **29,** 4). This fly has a very marked habit of " wing-waving ", with a very agitated trembling motion, which heightens the superficial resemblance to certain Sepsidae already imparted by its shining black colour and the dark apical spot on the wings (cf. Pl. **29,** 5); its greater size will give the first clue to its identity, however. *S. vibrans* is said to be predatory upon small Plant Bugs (Homoptera) and other soft-bodied insects.

Not a great deal appears to be known of the immature stages of Otitidae; the larvae are thought to develop in decaying vegetable matter. The family has been monographed by Hennig[a] and Lyneborg[15] is well illustrated.

PALLOPTERIDAE.

2 genera. 9 species. Small flies with pictured wings conspicuously longer than the abdomen; 3rd antennal segment short, rounded apically; arista bare or short-pubescent (*Palloptera*) or short-plumose (*Toxoneura*); post-verticals parallel or divergent; vibrissae absent, or very weak, hair-like (*Toxoneura*); proboscis short and thick; tibiae without preapical bristle; costa broken at apex of *sc*, which is well separated from apex of Vein 1; *sc* complete; Vein 6 may or may not reach margin; ovipositor with non-retractile sheath. (Pl. **27,** 2, 3, 5: **30,** 6.)

Two very similar, yellowish-brown species, *Palloptera*

trimacula Mg. and *P. arcuata* Fall. may be distinguished on detail of wing-markings (Pl. **27,** 2, 3) ; they are not uncommon from about May to September, on waterside vegetation and in woodland clearings, respectively. Niblett [3] records *P. umbellatarum* F., a dark-grey species with heavily-marked wings, from the heads of the Carline Thistle and Spear Thistle ; in the latter case, the heads already inhabited by larvae of the Tephritid, *Urophora stylata* F. *P. neutra* Pand., which is not included in the Check List, is a rather larger species, about 5½–6 mm. long, with prominent reddish frons, greyish thorax, reddish-brown abdomen, stigma and cross-veins narrowly clouded, and a small triangular cloud just before the wing-tip ; we have found this by sweeping dead reeds beside a New Forest stream in late March. All the foregoing have at least the posterior cross-vein clouded ; two further ones have both cross-veins unmarked. *P. saltuum* L. is a reddish-brown species with a dark continuous band along the costal margin which broadens out to a transverse band at the wing-tip ; it has occurred to us on a number of occasions on umbels in damp places in July, and is particularly addicted to " wing-waving ". *P. ustulata* Fall. is a yellowish-grey species, with somewhat similar wings, except that the costal band is hardly noticeable and the apical cloud extended some distance along Veins 2 and 3 as narrow prongs ; it is common and widely distributed.

Toxoneura (= *Ocneros*) *muliebris* Harris is easy to recognise by its unusual wing-marking (Pl. **27,** 5); it is about 4 mm. long and not uncommon on low plants, even in gardens ; we have found it on windows in London suburbs, where its " wing-waving " was particularly noticeable, the wings being spread wide and in continuous trembling motion.

Larvae of *Palloptera* species and of *Toxoneura* have been recorded on the Continent as living under bark in detritus caused by burrowing-beetles and their larvae. Collin[12] keys the British species.

ULIDIIDAE.

3 genera. 3 species. Small, uncommon flies; 3rd antennal segment short and more or less rounded; postverticals parallel or divergent; vibrissae absent; proboscis short and strong with prominent labella; tibiae without pre-apical bristle; costa unbroken; *sc* entire, apex well separated from apex of Vein 1; Vein 1 bare or hairy; Veins 3 and 4 convergent apically (*Physiphora, Ulidia*) or not (*Homalocephala*); anal cell prolonged to a sharp point (*Physiphora, Ulidia*) or blunt (*Homalocephala*); Vein 6 not reaching margin; ovipositor with non-retractile sheath. (Pl. **29**, 4.)

Physiphora (= *Chrysomyza*) *demandata* F. (Pl. **29**, 4) occurs on horse-dung and manure-heaps, where the larvae develop; decaying vegetable matter is also recorded as a habitat. *Ulidia erythropthalma* Mg. has facial depressions to accommodate the antennae, which are absent in *Physiphora*; it occurs on excrement and flowers in damp places. *Homalocephala albitarsis* Zett. differs from the two foregoing species in having a blunt anal cell. All three are rarely recorded, but it is possible that they are not infrequently overlooked. The family is monographed by Hennig.[4]

TEPHRITIDAE (= TRYPETIDAE)

31 genera. Over 70 species. Small flies with pictured wings; 3rd antennal segment mostly short and rounded apically, occasionally longer and somewhat pointed at lower anterior corner; postverticals parallel; lower fronto-orbitals inwardly curved, nearer eye-margin than upper fronto-orbitals; vibrissae absent; proboscis short with prominent labella (e.g. *Philophylla*) or long and thin, labella turned back (e.g. *Myopites, Ensina*); tibiae without pre-apical bristle; costa broken at apex of *sc*, which is complete, but fades apically after a right-angled bend; another costal break near humeral cross-vein, sometimes indistinct; anal cell blunt apically (e.g. *Urophora*) or pointed (e.g. *Philophylla*); ovipositor with non-retractile sheath of varying type. (Pl. **27**, 6, 7, 9, 10: **28**: **29**, 1–3.)

Larvae of Tephritidae, which are short-oval in shape when fully developed, spend their existence in various parts of plants and are often responsible for the deformities generally referred to as "galls". The illustrated works of Connold [5] and Swanton [6] may be consulted in regard to galls of British plants generally; many of these are caused by Diptera. A series of papers by Niblett [7] deals particularly with British Trypetidae and their host-plants. The shape of the galls is often highly characteristic and, coupled with the identity of the plant, may give a good indication of the species responsible.

Larval habits may be broadly divided into four groups as follows:

Group I. Developing in flowerheads. (*Urophora, Myopites, Trypeta, Xyphosia, Trupanea, Ensina, Noeeta, Tephritis,* etc.)

This group affects mainly the flowerheads of Compositae, such as Thistles, Sow-thistles, Fleabanes, Hawkweeds, etc.; the larvae live in the tissues of the flowerhead and often pupate there, while others leave and pupate in the ground.

Group II. Leaf-Miners. (*Philophylla, Euribia, Prionimera,* etc.)

This group includes the Celery Fly, a well-known pest.

Group III. Developing in stems or roots. (*Oxyna, Spathulina, Cerajocera, Dithryca,* etc.)

Group IV. Developing in fruits. (*Rhagoletis, Anomoia, Ceratitis,* etc.)

The grouping of the genera into two sub-families, Myopitinae and Trypetinae, adopted in the Check List, has recently been revised by Collin,[8] with keys. For our purpose of selecting a few representative species of the family, the order of the genera in the Check List has been followed.

Urophora cardui L. (Pl. 28, 1), a banded-wing species, makes Field and Spear Thistles its hosts, and unlike others in the genus, galls the stem, not the flowerhead. The gall is about the size of a hazel-nut, or larger, and a number of larvae may develop in it; such galls are readily noticeable on the dead thistle stems in winter, and one may often see on them evidence of the attempts of birds to extract the larvae within. *U. stylata* F., another common banded-wing species, in which the wing-bands are rather faint, is associated with the flowerheads of Musk and Field Thistles. It is remarkable in that the female has an ovipositor which is half as long again as the abdomen.

The two British species of *Myopites* are about 3–4 mm. long and have a long, thin, backwardly-bent proboscis and, in profile, a characteristically trapezoid head (Pl. **27,** 9). They affect the flowerheads of *Inula* species, such as Fleabanes, Golden Samphire, etc.

Ceratitis capitata Wied. (Pl. **29,** 3), whose larvae are well-known in the Mediterranean region, where they occur in cherries, plums, peaches, apricots, pears and oranges, is occasionally taken in this country; we have, in fact, taken a specimen in a North London suburban park. A curious feature in the male is the manner in which a pair of the fronto-orbitals are broadly dilated near the apex. Sexual adornment also occurs in the male of *Vidalia* (= *Stemonocera*) *cornuta* Scop.; here the frons is produced on either side into a pair of horns, which bear on their upper surface a series of long, black, upwardly-directed blunt-ended spines. *V. cornuta* mines the leaves of Hemp Agrimony.

Larvae of *Rhagoletis* (= *Zonosema*) *alternata* Fall., which is about 4–6 mm. long, with lightly banded wings, occur in the "hips" or fruits of wild roses.

Euribia (= *Spilographa*) *zoë* Mg. (Pl. **28,** 5, 6) exhibits sexual dimorphism in respect of wing-markings; the larvae mine leaves of Groundsel, Ragwort, Mugwort, and Tansy.

Orellia (= *Goniglossum*) *wiedemanni* Mg., a handsome species, about 4–5 mm. long, with heavily-barred wings, develops in the berries of White Bryony, while the larvae of *Anomoia* (= *Phagocarpus*) *permunda* Harris (Pl. **29,** 1) affect the "haws" or fruits of Hawthorn. *Prionimera* (= *Acidia*) *cognata* Wied., a banded-wing species, mines the leaves of Coltsfoot and Butterbur.

Philophylla heraclei L. (Pl. **28,** 4), a banded-wing species, has, in life, iridescent, deep metallic green eyes, which give it a very striking appearance. This is the Celery Fly, well-known to the gardener, on account of the damage it causes to celery and parsnips; its wild hosts include various Umbelliferae, such as Angelica, Hogweed, etc. Two well-defined forms of the adult are commonly met with, one having the body and wing-markings light reddish-brown, and the other, very dark brownish-black. The eggs are laid, as a rule, underneath the leaf, the female inserting her ovipositor into the tissue, so that the hatching larva is already in place to commence its mining. Sometimes, however, the eggs are to be found on the surface. Young larvae emerge after about a week and make an irregular-shaped, frequently communal mine, which may occupy a considerable space and gives a blotchy appearance to the leaf; the mines look rather like water-blisters with the larval excrement showing through the tissue. Growth of the plant is retarded or brought to an end through the shrivelling of the leaves, due to the effect of the mines. After feeding up for two or three weeks, the larvae, now about 7 mm. long, and glistening greenish-white with the black line of the gut showing through, pupate in the mine, or in the ground, after falling from the leaf. The pupa is oval and wrinkled, about 5 mm. long and of a light brown colour. There are two generations each year, one about May, emerging from over-wintered pupae, and another about August or September; in the second generation, the adults emerge about three weeks to a month after pupation. Larvae are to be found in the leaves as late as December.

Cerajocera (= *Ceriocera*) *cornuta* F. has, in the male, a remarkable horn-like projection from the second antennal segment, furnished with short, spine-like bristles; the larvae develop in flowerheads of Great Knapweed, while the stems of the same plant are affected by *C. microceras* Hering, the male of which has a similar antennal adornment.

Species of *Trypeta* (= *Orellia*) affect the flowerheads of various Thistles, Knapweeds and Goat's-beard; the very common *T. ruficauda* F. (Pl. **29,** 2) may be found commonly on the Marsh Thistle in June and July. *Xyphosia miliaria* Schrank (Pl. **28,** 2) seems to favour the Field Thistle, but both species have been recorded from several different hosts; it will be noticed also that the ovipositor-sheaths are both of the blunt-wedge type.

Spathulina (= *Campiglossa*) *grandinata* Rond., a dappled-wing species, develops in the stems of Golden Rod, causing gall-like swellings.

Oxyna parietina L. (Pl. **28,** 3), in the larval stage, inhabits the stems of Mugwort and other species of *Artemisia*, where the larvae feed up in the central pith, and pupate. If the dried, woody stems of infested plants be split down the middle in winter, the little dark-brown pupae will often be discovered in some numbers; complete stems should be retained and the flies will emerge about May. It is interesting to compare the size of the fully-developed fly with that of the tiny aperture from which it escapes. *O. flavipennis* Lw., which is very similar to *parietina*, is said to form galls about the size and shape of a pea near the top of the stem in Yarrow, while Niblett has found similar galls caused by this fly on the roots; a number of galls may be present, close together, on the same plant.

Ensina sonchi L. is a common little Tephritid in which the wing-markings are at times so reduced as to be almost imperceptible; it is a dull yellow-and-black or greenish-black fly about 3½ mm. long and affects the heads of Sow-thistles, Goat's-beard, Groundsel and Hawkweed Oxtongue (*Picris*), etc. As in *Myopites*, to which genus it shows not inconsiderable affinity, the head is very distinctive in profile; the frons extends downwards for a good distance, and the proboscis is long, thin and turned back (Pl. **27,** 10).

The species of *Tephritis* are mostly dappled-winged, but in some cases the little clear areas in the dark membrane have coalesced to leave dark patches so arranged that, at first sight, they give the impression of normal transverse bands or heavy patches. Larval development takes place in flowerheads ; the identity of the host-plant will often give a good clue to the species. For example, *T. bardanae* Schrank frequents Burdock ; *T. vespertina* Lw., Cat's-ear ; and *T. conjuncta* Lw., Ox-eye Daisy ; the two latter may sometimes be encountered very early in the year by sweeping dead bracken ; we have taken them thus in mild days in February and March, with snow still lying deep on the ground, which indicates over-wintering in the adult stage.

In *Trupanea* (= *Trypanea*) *stellata* Fuessly and *T. amoena* Frauenfeld, the dark patch is confined to an irregular star-shaped marking occupying the apical third of the wing. *T. stellata* is recorded from the heads of Hawkweeds, Ragwort, Sea Aster, etc., and *amoena* from Sow-thistle, Star-thistle, Willow Lettuce, etc.

A dark central patch, with " rays " extending to the margin of the apical half of the wing (Pl. **27,** 6), distinguishes *Noeeta pupillata* Fall., the larvae of which will be found in swollen heads of Hawkweeds, Cat's-ear, etc. The wings of *Dithryca* (= *Ditricha*) *guttularis* Mg., which affects the stem-bases and roots of Yarrow, are somewhat similar, except that the clear patches around the apical margin are more or less circular, not narrowed as in *pupillata*.

The family has been monographed by Hendel.[9]

LONCHAEIDAE.

3 genera. 29 species. Small, shining metallic bluish-black or greenish-black flies, with clear wings ; 3rd antennal segment long ; postverticals divergent ; one orbital, or sometimes a very small lower one as well ; vibrissae absent ;

occiput concave ; proboscis short ; tibiae without preapical bristle ; costa broken at apex of *sc*, which is complete and runs more or less parallel to Vein 1 ; Vein 1 reaches margin, but is very faint apically ; ovipositor with non-retractile sheath. (Pl. **30**, 1, 5.)

The common *Lonchaea chorea* F. (Pl. **30**, 1) is a shining, blue-black fly, which has, in life, striking red eyes and is not unlike certain species of Muscidae in appearance, particularly *Ophyra leucostoma* Wied. and *Hydrotaea bimaculata* Mg. ; unlike them, however, it is rather sluggish in its movements. Its life-history has been studied by Cameron.[10] The adult flies may be seen from very early until quite late in the year ; larval development is recorded as taking place in dung, sugar-beet, and beneath the bark of trees ; we have bred it from pupae found in the last-named situation. On the Continent, Lonchaeid larvae are said to develop in fruits or to cause plant-galls.

The family is monographed by Morge[11]. Collin[13] and Hackman[14] are also useful for British species.

LITERATURE CITED

1. Séguy, E. Faune de France. **28.** Muscidae Acalypterae et Scatophagidae. Lechevalier, Paris. 1934.
2. Hennig, W., in Lindner, E. Die Fliegen der Palaearktischen Region. **46–47.** Otitidae. Schweizerbart, Stuttgart. 1939.
3. Niblett, M. Diptera bred from flower-heads of Compositae. *Ent. Rec.* **58** : 121–3. 1946.
4. Hennig, W., in Lindner, E. Die Fliegen der Palaearktischen Region. **45.** Ulidiidae. Schweizerbart, Stuttgart. 1940.
5. Connold, E. T. British Vegetable Galls. Hutchinson, London. 1901.
   Connold, E. T. Plant Galls of Great Britain. Adlard & Son, London. 1909.
6. Swanton, E. W. British Plant Galls. Methuen, London. 1912.
7. Niblett, M. Some Notes on British Trypetidae. *Ent. Rec.* **46** : 66–9. 1934.
   Niblett, M. British Trypetidae, additional Notes. *Ent. Rec.* **48** : 13–14. 1936.

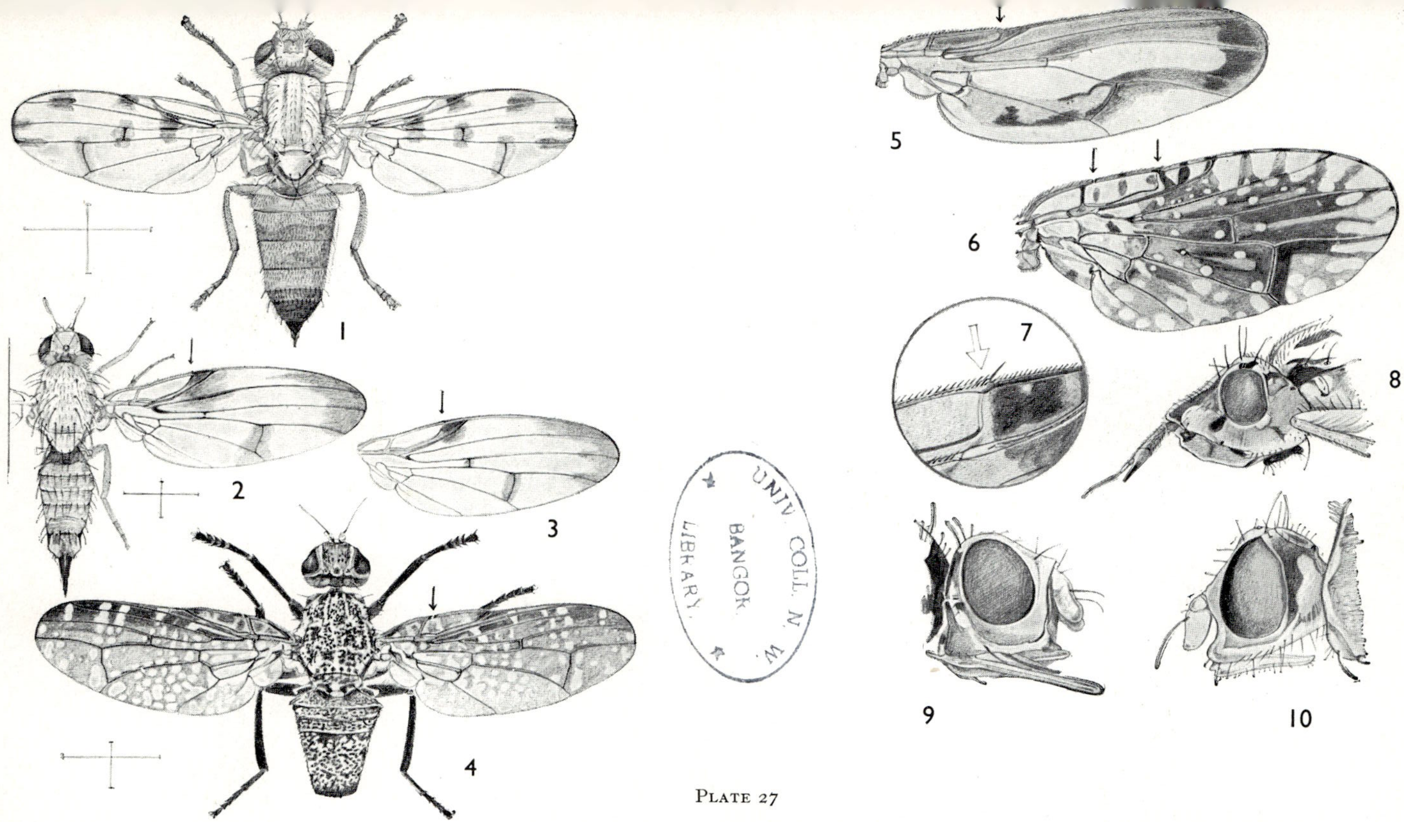

PLATE 27

OTITIDAE-I

1. *Melieria omissa* Mg., ♀, p. 183.

PALLOPTERIDAE-I

2. *Palloptera trimacula* Mg., ♀, p. 185. 3. *Palloptera arcuata* F., wing, p. 185.

PLATYSTOMATIDAE-I

4. *Platystoma seminationis* L., ♂, p. 183.

PALLOPTERIDAE-II

5. *Toxoneura* (=*Ocneros*) *muliebris* Harris, wing, p. 185.

TEPHRITIDAE-I

6. *Noeeta pupillata* Fall., wing, p. 191. 7. Trypetid sub-costa and costal break, p. 186. 9. *Myopites blotii* Brébisson, head, p. 188. 10. *Ensina sonchi* L., head, p. 190.

OTITIDAE-II

8. *Dorycera graminum* F., head, p. 184.

PLATE 28

TEPHRITIDAE-II

1. *Urophora cardui* L., ♀, p. 187. 2. *Xyphosia miliaria* Schrank, ♀, p. 190. 3. *Oxyna parietina* L., ♂, p. 190. 4. *Philophylla heraclei* L., ♀, p. 189. 5. *Euribia* (=*Spilographa*) *zoë* Mg., ♂, p. 188. 6. *Euribia* (=*Spilographa*)

Niblett, M. Notes on Food-plants of the Larvae of British Trypetidae. *Ent. Rec.* **51** : 69–73. 1939.

Niblett, M. Emergences of Trypetidae. *Ent. Rec.* **54** : 66–9. 1942.

Niblett, M. British Trypetidae. *Ent. Rec.* **58** : 6–8. 1946.

Niblett, M. A List of Trypetidae (Diptera) from North-east Surrey. *Ent. Rec.* **59** : 28–32. 1947.

8. Collin, J. E. The British Genera of Trypetidae (Diptera) with notes on a few species. *Ent. Rec.* **59** : Suppl. 1–14, Corr. 36. 1947.

9. Hendel, F., in Lindner, E. Die Fliegen der Palaearktischen Region. **49.** Trypetidae. Schweizerbart, Stuttgart. 1927.

10. Cameron, A. E. On the Life-history of *Lonchaea chorea* F. *Trans. ent. Soc. Lond.* **1913** : 314–22. 1913.

11. Morge, G. Monographie der palaearktischen Lonchaeidae. *Beitr. Ent.*, **9**: 1–371, 909–45; **12**: 381–434. 1959 and 1962.

Morge, G. Die Lonchaeidae und Pallopteridae Österreichs und der angrenzenden Gebeite. *Sonderdruck aus Naturkundliches Jahrbuch der Stadt Linz*, **1963**: 123–312. 1963.

12. Collin, J. E. The British species of the genus *Palloptera* Fln. (Diptera). *Ent. Rec.* **63**: suppl. 1–6. 1951.

13. Collin, J. E. A revision of the British (and notes on other) species of Lonchaeidae (Diptera). *Trans. Soc. Brit. Ent.*, **11**: 181–207. 1953.

14. Hackman, W. The Lonchaeidae (Dipt.) of Eastern Fennoscandia. *Notul. Ent.*, **36**: 89–115. 1956.

15. Lyneborg, L. Danske acalyptrate fluer. 2. Psilidae, Platystomidae og Otitidae (Diptera). *Ent. Medd.*, **32**: 367–88. 1964.

CHAPTER XV

SEPSIDAE.
PIOPHILIDAE . . . . Cheese Skippers, etc.
PSILIDAE . . . . . Carrot Rust Fly, etc.
MEGAMERINIDAE.
MICROPEZIDAE (= TYLIDAE) Stilt-legged Flies.
TANYPEZIDAE.

In addition to the special works cited under family headings, Séguy [1] may be consulted in regard to all the families covered by this chapter.

SEPSIDAE.

7 genera. 23 species. Small, shining, metallic black or brownish-black flies, often with violet or greenish reflections, more or less spherical heads and abdomen basally constricted, giving an ant-like appearance; 3rd antennal segment short-oval; arista bare, at most short-pubescent; postverticals divergent; occiput very convex, making profile of head more or less circular; proboscis short, palpi vestigial; vibrissae absent, or present (*Saltella*); ♂ fore-legs often with secondary sexual characters, the femora and tibiae being deformed and armed with spines or thorny protuberances; tibiae with or without preapical bristle; wings often with dark apical spot; costa unbroken; *sc* entire, apex well separated from apex of Vein 1; Veins 3 and 4 tend to converge; cross-veins somewhat approximated; Vein 6 short and straight; Vein 7 absent; ♂ genitalia fairly prominent. (Pl. **29**,5: **30**, 2, 8–12, 14.)

THE habit of "wing-waving", mentioned in the previous chapter, also occurs most noticeably in the Sepsidae, the species of which are distinguishable from the similar small, shining black species of the next family, Piophilidae, by the very convex occiput (cf. Pl. **30**, 2, 3) and unbroken costa (cf. Pl. **30**, 8, 13, 14), among other characters. Moreover, the curious deformities of the male legs in the Sepsidae are absent in the Piophilidae.

A dark spot at the apex of Vein 2 characterises the genera

Sepsis and *Sepsidimorpha* ; the single British species of the latter, *S. pilipes* v.d. Wulp, will be known, in the male, by the long hairs on the inner side of the mid-femora, and in both sexes, by the absence of marginal bristles on the abdominal segments, which are present in *Sepsis* (Pl. **29,** 5). Several species of this genus are very common, particularly *S. punctum* F., the species illustrated in the plate mentioned, *S. violacea* Mg., *S. fulgens* Mg. and *S. cynipsea* L. While the females are not easy to identify, unless taken in copula, the males are strongly characteristic in regard to the ornamentation of the fore femora and tibiae (Pl. **30,** 9–12). *Sepsis* species may swarm in vast numbers on vegetation; Hobby[12] lists the recorded occurrences.

The remaining genera have the wings unmarked and from them the two uncommon species of *Saltella* (= *Pandora*) may be separated by the absence of the second basal cell, i.e. Vein 4 rises in a sharp curve from Vein 5 just beyond the apex of the anal cell (Pl. **30,**14) and by the presence of well-developed vibrissae. We have taken the rather variable *S. scutellaris* Fall. by sweeping low vegetation in a Norfolk marsh.

In *Meroplius* and *Nemopoda*, each with a single species, there is a well-developed mesopleural bristle ; an orbital bristle is present in *Meroplius*, but absent or vestigial in *Nemopoda* (Pl. **30,** 2). *N. nitidula* Fall. (= *cylindrica* F.) is very common and widely distributed, occurring on excrement, umbels and, in our experience, very often on bracket-fungi on fallen trees ; it is very frequently to be seen running or basking on foliage.

The last two genera have no mesopleural bristle. In *Themira*, the postverticals and orbitals are well-developed and the abdomen little constricted at the base, from the dorsal aspect. *T. putris* L. (Pl. **30,** 8) is an extremely common, shining black fly in which the male genitalia are very conspicuous ; tufts of long, shining black hairs arise from the preceding sternite. Similar adornments occur in some of the other species of this genus, likewise modifications of the male fore-femora and tibiae, similar to those in *Sepsis*. In the remaining genus, *Enicta* (= *Henicita*), the postverticals and orbitals are poorly developed and the abdomen somewhat constricted basally; the mid-tarsi of the males are flattened and dilated.

Sepsid larvae are of the generalised Cyclorrhaphous type (cf. Fig. 10F), rather narrow and elongated, and develop in carrion and decaying organic matter generally, but most frequently in excrement, around which the adults sometimes occur in vast numbers. Particularly-favoured locations are the semi-liquid perimeters of farmyard manure and litter heaps, or the drain-channels from pig-sties or stables. Brindle[7] keys the known larvae.

There is no up-to-date work in English on the British species, but a monograph by Duda[2] may be consulted and papers by Andrewes[9] and Laurence[10] are useful.

PIOPHILIDAE.

3 genera. 9 species. Small, shining black flies, often with bluish reflections; 2nd antennal segment with prominent, erect anterior bristle; 3rd segment short and rounded, arista bare, or practically so; postverticals divergent; one orbital or none; vibrissae strong; occiput flattened; proboscis short and thick; palpi well-developed; ♂ legs simple; costa broken at, or near apex of *sc*, which is not greatly separated from point where upper margin of Vein 1 merges with costa; *sc* entire; Veins 3 and 4 parallel or slightly divergent; Vein 6 curved, fades before margin; Vein 7 absent; genitalia not prominent in either sex. (Pl. **30**, 3, 13.)

The principal genus, *Piophila*, with seven species, is distinguished by a single pair of dorsocentral bristles. *P. casei* L. is a common pest of stored animal-protein foods, such as cheese and bacon; the larvae are the well-known "Cheese Skippers", which have a habit of leaping by grasping their hind part with their mouth-hooks and suddenly releasing their hold. This manoeuvre is executed by other Dipterous larvae (cf. Cecidomyiidae, p. 83, Clusiidae, p. 216). Other species of the genus are not domesticated and are most likely to be encountered on carrion; for example, we have taken *P. varipes* Mg. on chicken-legs discarded by a fox outside its "earth", and another not uncommon species, *P. vulgaris* Fall., on a dead fox, nailed to a keeper's "gibbet" in the New Forest.

Piophilid larvae are keyed by Brindle[8]; they are very similar in shape to those of Sepsidae, but rather more narrowed posteriorly.

A key to the British species of *Piophila* is furnished by Collin,[3] and Duda [4] may also be consulted.

PSILIDAE.

3 genera. 29 species. Small to moderately large flies of varied form; 3rd antennal segment short, rounded (*Psila*) to very long and thin (*Loxocera*); postverticals divergent or absent; ocellar triangle large; vibrissae absent; tibiae without preapical bristle; no strong bristles on pleura; costa extends as far as apex of Vein 4; broken some distance before apex of Vein 1; whitish or transparent streak marks course of vestigial *sc* after right-angled upward bend to costal break; this streak extends as a transverse weakening or cross-fold over the common stem of Veins 2 and 3 and across the second basal cell; basal part of *sc* very close to Vein 1 and often, therefore, not readily perceptible; second basal cell and anal cell large; genitalia of both sexes mostly inconspicuous (but ovipositor long in *Loxocera*). (Pl. **31,** 2, 3: **32** 1, 2).

Psila rosae F. (Pl. **31,** 2) is the common Carrot Fly or Carrot-rust Fly, an agricultural pest of some significance, which, together with allied species, may be mistaken, in the field, for a member of one of the two preceding families. Close examination will reveal that the profile of the head is different and that the wings bear a number of distinctive characters, such as the crossfold peculiar to the Psilidae (Pl. **31,** 3); the sharp upward bend of the sub-costa and its apical fading shows a resemblance to the Tephritidae, (cf. Pl. **27,** 7). *P. rosae* causes damage, not only to carrots which are its principal host-plants, but also to celery, parsley, parsnips and occasionally to turnips. It has been bred from Hemlock and it is thought that the fly may also develop in certain other wild umbelliferous host-plants, such as Hogweed. Eggs are laid in May and June in the soil around the roots of the plant and hatch in about a fortnight. An entry into the lower part of the root is made by the young maggots, which usually remain with their hind parts protruding. After a month they leave the root and pupate close to it; the adults emerge after varying periods to form a second brood. Pupae developing from this generation over-winter to give rise to the early summer brood of the following year. Infested plants show signs of wilting, with the leaves turning yellowish to coppery colour; carrots are usually badly damaged, with a characteristic "rusty" flaking of the outer tissue, beneath which are excavations caused by the feeding larvae.

The genus *Psila* is divided into sub-genera and *rosae* belongs to the sub-genus *Psila* (sensu stricto); this includes a number of similar species, such as *Psila* (*Psila*) *nigricornis* Mg., which is very much like *rosae* in appearance, habits and distribution. Very unlike *rosae*, however, in size and colour, is a species of another sub-genus, *Psila* (*Pelethophila*) *fimetaria* L. (Pl. **32,** 1), which is about twice as long and shining reddish or reddish brown; the profile of the head is long-triangular, i.e. the face is long and much receding. Another fairly common species, *P.* (*Pelethophila*) *merdaria* Collin, is distinguished from it only by minute differences, including the absence of a dark patch on the outer side of the third antennal segment at the base of the arista. Both species occur on low plants in damp places from May to September.

Even more unlike *rosae* are the species of the genus *Loxocera*, which are striking black-and-red, very narrow flies with the third antennal segment lengthened to an exaggerated degree. *L. albiseta* Schrank (Pl. **32,** 2) is a common species which occurs at similar times and in similar locations to *fimetaria*. In the field, *Loxocera* species, especially the females with their long ovipositors, bear a very deceptive resemblance to Ichneumons (Hymenoptera) but the customary, fussy, agitated, antenna-waving behaviour of the latter is not imitated. In the female of *L. aristata* Pz., the ovipositor is very long and blade-like, half as long as the abdomen.

Species of the remaining genus, *Chyliza*, are less common than the others and recognisable by the fact that the face is concave, in profile. The larvae are thought to be stem and root-dwellers and occasionally to cause galls ; it is not unlikely that the life-histories of most of the species of this family are similar to that of *rosae*.

Psilid larvae are very like those of Sepsidae and Piophilidae, i.e. of the generalised Cyclorrhaphous type and rather long and narrow. Brindle[11] keys the known larvae.

Collin [5] provides keys to the British species of Psilidae, with notes. Wakerley[13] adds a further species.

MEGAMERINIDAE.

Single species, *Megamerina loxocerina* Fall. Medium-sized fly superficially like a *Loxocera* (Psilidae) ; 3rd antennal segment short and rounded ; hind femora with double row of short ventral spines; wings narrow; costa unbroken; *sc* visible throughout and gently curved; apex well-separated from apex of Vein 1; abdomen greatly elongated and attenuated; ovipositor short retractile. (Pl. **50,** 6). We have found this fly running over the leaves of bushes in woods in much the same manner as an Ichneumon Fly which it closely resembles. June and July.

MICROPEZIDAE (= TYLIDAE).

3 genera. 8 species. Small to medium-sized slender flies with long legs and wings ; 3rd antennal segment short and rounded ; face receding (very strongly in *Micropeza*) ; ocellar bristles absent ; vibrissae absent ; proboscis stout ; tibiae without preapical bristle ; wings narrow, alula vestigial ; costa unbroken ; *sc* entire, sometimes close to Vein 1 basally and therefore more perceptible at apex ; Veins 3 and 4 convergent apically ; discal cell separate, or confluent with second basal (*Micropeza*); anal cell not noticeably convex apically ; genitalia of both sexes prominent and of peculiar appearance. (Pl. **31,** 1, 4–6 : **50,** 5.)

The name, " Stilt-legged Flies ", is very appropriate in view of the awkward gait of these insects, perched high upon their long, thin legs. Our example, *Calobata* (=*Trepidaria*) *petronella* L. (Pl. **31,** 1) is black with the thorax grey-dusted, and the frons and legs reddish or reddish-brown ; the peculiar genitalia should be noted (Pl. **31,** 1, 5).

In the genus *Micropeza* (=*Tylos*) the second basal cell is confluent with the discal cell, i.e. there is no separating cross-vein, and the profile of the head shallow-triangular, giving a somewhat sinister, serpentine appearance (Pl. **31,** 6). The two British species are less common than those of *Calobata.*

The single rare species of the third genus, *Rainieria calceata* Fall., differs from the remaining members of the family in the possession of a dark sub-apical transverse band on the wings ; it is about 7–8½ mm. long, black, with some of the fore and hind tarsal segments whitish-yellow, and has been recorded from Windsor Forest in June and July. (Pl. **50,** 5).

Micropezidae are usually to be found on bushes and low vegetation on the banks of rivers and streams or in marshy places generally, where they are predatory upon other insects. *Calobata* species are recorded as preying on aphids and small Tipulidae ; we have seen them with small Chironomids and possess a *C. cibaria* L. taken with an aphid. The proboscis (Pl. **31,** 4) should be compared with that of another predatory Acalypterate, *Ochthera mantis* Deg. (cf. Pl. **35,** 10), and with that of *Scatophaga stercorarium* L. (cf. p. 254).

Little appears to be known of the immature stages, but certain foreign species of the family are said to develop in rotten wood or decaying vegetable matter.

Keys to the British species, with notes, are furnished by Collin ; [6] this paper also covers the next family, Tanypezidae.

TANYPEZIDAE.

A single, rare species, *Tanypeza longimana* Fall., about 6–8 mm. long, black, with a silvery area on the vertex and on either side of the frons ; differs from the species of

Micropezidae in having the upper part of the occiput very concave, the anal cell very convex apically, and four scutellar bristles instead of two, among other characters.

This species is included in the Micropezidae by Collin.[6]

LITERATURE CITED

1. Séguy, E. Faune de France. **28.** Muscidae Acalypterae et Scatophagidae. Lechevalier, Paris. 1934.
2. Duda, O. Monographie der Sepsiden. *Ann. Naturh. Mus. Wien.* **39**: 1–153. 1925. **40**: 1–110. 1926.
3. Collin, J. E. Additions and Corrections to the British List of Muscidae Acalyptratae. *Piophila. Ent. mon. Mag.* **46**: 177–8. 1910.
4. Duda, O. Revision der europaischen und grönländischen sowie einiger sudostasiatischen Arten der Gattung *Piophila* Fall. Konowia. **3**: 97–113, 153–203. 1924.
5. Collin, J. E. The British Species of Psilidae. *Ent. mon. Mag.* **80**: 214–24. 1944.
6. Collin, J. E. British Micropezidae (Diptera). *Ent. Rec.* **57**: 115–19. 1945.
7. Brindle, A. Taxonomic notes on the larvae of British Diptera. 20. The Sepsidae. *Entomologist,* **98**: 137–40. 1965.
8. Brindle, A. Taxonomic notes on the larvae of British Diptera. 21. The Piophilidae. *Entomologist,* **98**: 158–60. 1965.
9. Andrewes, C. H. *Nemopoda pecticulata* Lw. (Dipt., Sepsidae) in Britain. *Ent. mon. Mag.,* **99**: 8. 1963.
10. Laurence, B. R. Two coprophilous Diptera (Borboridae and Sepsidae). *Ent. mon. Mag.,* **88**: 81–2. 1952.
11. Brindle, A. Taxonomic notes on the larvae of British Diptera. 22. Psilidae. *Entomologist,* **98**: 169–73. 1965.
12. Hobby, B. M. A swarm of Sepsidae (Dipt.). *Ent. mon. Mag.,* **93**: 34. 1957.
13. Wakerley, S. B. A new species of *Psila* Meigen (Diptera, Psilidae) from Northern England. *Proc. R. ent. Soc. Lond.* (B) · **28**: 107–8. 1959.

CHAPTER XVI

LAUXANIIDAE (= SAPROMYZIDAE).	
SCIOMYZIDAE (= TETANOCERIDAE)	Marsh Flies.
DRYOMYZIDAE.	
NEOTTIOPHILIDAE	Nest Flies.
THYREOPHORIDAE.	
CHAMAEMYIIDAE.	
COELOPIDAE	Kelp Flies.
HELEOMYZIDAE.	
CHYROMYIDAE (= CHIROMYIIDAE).	

Apart from the special works cited under family headings, Séguy [1] may be consulted in regard to all the families covered by this chapter.

LAUXANIIDAE.

13 genera. Nearly 50 species. Small flies of various colours, sometimes with marked wings and, in life, iridescent purplish or greenish eyes; profile of head greatly variable, face often receding, especially in *Trigonometopus*; 3rd antennal segment also greatly variable in shape, but usually with more or less concave anterior edge; arista bare to plumose; postverticals convergent; two pairs of orbitals (except ♂ *Sapromyza albiceps* Fall.); vibrissae absent; tibiae with preapical bristle; costa unbroken; *sc* complete, visible throughout, apex well separated from apex of Vein 1; second basal cell and anal cell short, latter convex apically; Vein 6 short, not reaching margin. (Pl. **31,** 8, 9, 11–13; **34,** 1.)

ADULT Lauxaniidae are retiring in habit and mostly addicted to shady vegetation in damp places; consequently they are more likely to be noticed in the sweep-net than elsewhere. We have swept one species from the lower branches of Surrey pines. They present a varied appearance, colours ranging through shining aeneous, metallic greenish-black, bluish-grey, reddish-brown, deep yellow to a pale, transparent yellow; in some cases they are parti-coloured. The wings may be marked in a fashion reminiscent of *Palloptera* species (cf. Pl. **27,** 2, 3), or the eyes, in life, may be of a brilliant and beautiful purplish or greenish colour, with lighter reflections ; this is most striking in the pale yellow species. The antennae are particularly variable, especially as regards the third segment and the arista (Pl. **31,** 11–13).

Calliopum aeneum Fall. (Pl. **34** 1) is shining greenish-black with somewhat yellowish-tinted wings and the four hinder tibiae conspicuously reddish or yellowish ; like the closely similar *C. simillimum* Collin, it is common and widely distributed and may be swept from grasses and low vegetation from early summer until quite late in the year ; it is recorded as mining clover-leaves.

Cnemacantha (= *Lyciella*) *decempunctata* Fall. is an example of the yellowish species with marked wings (Pl. **31,** 8). *Minettia lupulina* F., a parti-coloured species, is common and will attract attention to itself in the sweep-net by virtue of its blue-grey thorax, with yellowish abdomen and wings. The third antennal segment of *Lauxania cylindricornis* F. (Pl. **31,** 11) is reminiscent of *Loxocera* (p. 198) and will place this species at once.

Trigonometopus frontalis Mg. has a very sharply-receding face and the eyes lengthened horizontally (Pl. **31,** 9), which give it a distinctive appearance.

Lauxaniid larvae are said to develop in decaying vegetable matter, such as fallen leaves ; those of certain foreign species are leaf-miners.

Collin [2] gives keys to the family, with notes.

SCIOMYZIDAE.

29 genera. Over 50 species. Small to medium-sized flies inhabiting damp or marshy places ; head as wide as, or wider than, thorax, from dorsal aspect ; antennae often elongated and porrect ; 2nd segment sometimes greatly thickened or elongated ; 3rd segment mostly triangular, with anterior edge more or less concave ; but sometimes short-oval ; arista bare to plumose ; postverticals parallel or divergent ; vibrissae absent ; femora bristly, sometimes

spinous, especially hind ones ; tibiae with preapical bristle ; wings often conspicuously longer than abdomen, sometimes with dark markings ; costa unbroken ; *sc* entire, visible throughout, apex well separated from apex of Vein 1 ; 2nd basal and anal cells short ; anal cell convex apically and/or Vein 6 reaches margin ; abdomen mostly long and narrow. (Pl. **31,** 7, 10: **32,** 3, 4: **33,** 1, 2.)

The larvae of Sciomyzidae feed on snails and slugs. The aquatic species are predators and kill their hosts quickly; one larva may kill up to a dozen snails, of different families, during development. The terrestrial species are parasitic, developing from an egg deposited on the host's shell. The young larva enters the snail's shell, feeds and pupates and only leaves when adult.

Since some snails are intermediate hosts of human schistosomiasis in various parts of the world sciomyzid biology has been a subject of intensive research in the past few years, particularly by American workers.

The larvae of some British species are described by Knutson *et al.*[12] and a useful summary is given by Parmenter.[17]

TETANOCERINAE.

Here are contained most of the larger and more striking species ; the sub-family is characterised generally by the absence of a propleural bristle and the presence of a facial " keel " or median ridge. *Tetanocera elata* F. (Pl. **32,** 4) is common, from June to September, on vegetation bordering streams or ponds, or in marshy places. The long-legged *Sepedon sphegeus* F. (Pl. **32,** 3) gives the impression of being longer than it really is, owing to its long, thin antennae ; it will be noticed that the second segment is extraordinarily lengthened. This striking blue-black fly is usually to be found on Yellow Flag in marshy places or slow streams from April to September.

Some species have the wings dappled in a manner reminiscent of *Platystoma* (cf. Pl. **27,** 4), such as *Statinia marginata*

F. (Pl. **33,** 2), or *Trypetoptera punctulata* Scop., in which, also, the body is handsomely patterned in pale blue-grey and brown (Pl. **31,** 10). Others have the wings marked with dark clouds or spots, as in *Elgiva cucularia* L. (Pl. **33,** 1).

DITAENIINAE.

In this sub-family the frons is divided by a prolongation of the ocellar triangle; a propleural bristle and facial keel are present but not always very distinct and the third antennal segment is rather blunt by contrast with the Tetanocerinae. Numerous dark spots mark the wings of the common and distinctive little *Pherbellia schoenherri* Fall. (Pl. **31,** 7) which is about 4–5 mm. long; we have taken it in early February.

SCIOMYZINAE.

Here the third antennal segment is even more rounded apically and quite short; the small and mostly undistinguished-looking species are characterised by a well-developed propleural bristle.

TETANURINAE.

The single British species, *Tetanura pallidiventris* Fall., has, in profile, an almost circular head, with the face nearly perpendicular and the third antennal segment very short and rounded, bearing a sub-apical arista.

The British species are included in Knutson and Lyneborg[13] to which Parmenter[17] and Collin[14] are supplementary.

DRYOMYZIDAE.

3 genera. 5 species. Small to medium-sized flies inhabiting damp or marshy places; 3rd antennal segment short and rounded, arista very short in *Helcomyza*; postverticals parallel or divergent; face more or less concave in profile; vibrissae absent; clypeus * prominent; tibiae with preapical bristle; wings very large, much longer than

* See Glossary, p. 355.

abdomen, sometimes marked ; costa unbroken (armed with series of small spines in *Helcomyza*) ; *sc* entire, apex well separated from apex of Vein 1 ; Vein 1 bare, or hairy (*Neuroctena*) ; second basal and anal cells very large ; Vein 6 reaches margin. (Pl. **34**, 2–5.)

There are two sub-families, Dryomyzinae and Helcomyzinae, the latter with a single British species.

DRYOMYZINAE.

These are yellowish or reddish-brown flies addicted to putrefying matter generally and excrement or rotting fungi in particular, where the larvae develop ; they may be found abroad until very late in the autumn. *Neuroctena anilis* Rond., the sole species of the genus, is variable in size from about 6 to 10 mm., and will be known by the hairs on Vein 1, which are absent in *Dryomyza* ; the cross-veins and tips of Veins 3 and 4 are clouded (Pl. **34**, 4). *Dryomyza flaveola* F. is roughly the same size as *anilis*, but rather paler, i.e. yellowish, in colour and has wings unmarked. The profile of the head is similar in both genera ; the small eyes, deep jowls and concave face should be noted (Pl. **34**, 2).

HELCOMYZINAE.

Helcomyza ustulata Curt. (= *Actora aestuum* Mg.) differs a good deal from the Dryomyzinae in having well-developed dorsocentral bristles before the suture, no apical hairs on the palpi and a rounded keel between the antennae on the upper part of the face. The profile of the head shows marked differences, especially as regards the antennae and arista (Pl. **34**, 3) and the wings bear a series of prominent spines on the costa (Pl. **34**, 5), which is also characteristic of Helomyzidae (cf. Pl. **34**, 7), and Neottiophilidae (p. 207).

H. ustulata is one of the familiar flies of the seashore and will be known among the many seaweed-haunting Ephydrids, Coelopids, and Dolichopodids, by its size and marked silvery-grey colour ; it is rather variable in size from about 6 to

10 mm., but its flight habits are characteristic. When disturbed, it makes a swift, darting flight of a yard or so and will repeat the process continuously before the advancing feet of the observer. The larvae live in mud, sand or seaweed thrown up on the shore and are submerged by each tide; they are cylindrical in shape, not greatly narrowed anteriorly and with a rosette of conical, annulated tubercles surrounding the rear spiracles.

The family is monographed by Czerny.[4]

NEOTTIOPHILIDAE.

Single shining-reddish species, *Neottiophilum praeustum* Mg., about 8–9 mm. long; profile of head somewhat similar to *Neuroctena* (cf. Pl. **34,** 2) but eyes larger and vibrissae present; wings clouded on cross-veins, costal margin and apex; costa with spines, similar to *Helcomyza* (cf. Pl. **34,** 5), broken at apex of *sc*, which runs more or less parallel to Vein 1 and is clearly separated at apex; Vein 1 hairy; anal cell closed by sinuate cross-vein; Vein 6 reaches margin.

This fly is recorded from the nests of birds, including thrushes, blackbirds and various finches, where the larvae are said to suck the blood of nestlings, a habit we shall see again in *Protocalliphora* (p. 282). Neottiophilidae are included in Czerny's[4] monograph of the Dryomyzidae.

THYREOPHORIDAE.

Single yellowish-brown species, *Centrophlebomyia furcata* F., about 6–7 mm. long; profile of head somewhat similar to *Neuroctena* (cf. Pl. **34,** 2) but eyes smaller, a pair of equally strong vibrissae present and the face straight and strongly receding; deep antennal depressions; post-verticals long and divergent; scutellum very long with two short projections bearing bristles, triangular in ♀ and truncate in ♂; legs densely hairy; costa broken at apex of *sc*, which is closely approximated to point where *upper* margin of Vein 1 merges with costa; costa with spines, similar to *Helcomyza* (cf. Pl. **34,** 5); Vein 6 reaches margin.

This species is recorded from the dead bodies of various

animals, where doubtless the larvae develop. The family has been monographed by Sack.[5]

CHAMAEMYIIDAE.

3 genera. 13 species. Mostly very small greyish or yellowish-grey flies; 3rd antennal segment short, with angular lower anterior corner, or very broad and rounded (*Leucopis*); postverticals convergent or absent; vibrissae absent; tibiae without preapical bristle; costa unbroken; *sc* entire, more or less parallel with Vein 1, which is thickened apically and extended so that *lower* margin merges with costa some distance from apex of *sc*; Vein 5 not reaching margin; Vein 7 stronger than vestigial Vein 6. (Pl. **34,** 8, 9.)

Larvae of Chamaemyiidae are predaceous upon aphids, coccids and adelgids, as a result of which propensity they, themselves, sometimes become the victims of aphidivorous Syrphid larvae.

As regards the adults, the species of *Leucopis* will be known, in addition to the difference in the third antennal segment, by the absence of postverticals and orbitals; those of *Parochthiphila* by the possession of four pairs of dorsocentral bristles and *Chamaemyia* (= *Ochthiphila*) by three pairs of dorsocentrals. The third antennal segment in both of the last two genera is very distinctive (Pl. **34,** 8) and coupled with the highly characteristic wing-venation of the family (Pl. **34,** 9) should help to "place" these flies quickly. Chamaemyiidae occur in grasses, especially in damp places near streams, from late spring until autumn; one species, *Chamaemyia flavipalpis* Hal., in particular, is common and widely distributed in seashore locations.

Keys to the British species are provided by Smith[15], but Collin[16] and Coe[6] should be consulted for *Chamaemyia.*

COELOPIDAE.

4 genera. 7 species. Small to medium-sized bristly and/or hairy, flattened flies, inhabiting seashore; face, in profile, very concave (*Coelopa, Malacomyia*) or hardly

PLATE 29

TEPHRITIDAE-III

1. *Anomoia* (=*Phagocarpus*) *permunda* Harris, ♀, p. 188. 2. *Trypeta* (=*Orellia*) *ruficauda* F., ♀, p. 190. 3. *Ceratitis capitata* Wied., ♀, p. 188.

ULIDIIDAE

4. *Physiphora* (=*Chrysomyza*) *demandata* F., ♀, p. 186.

SEPSIDAE-I

5. *Sepsis punctum* F., ♂, p. 195.

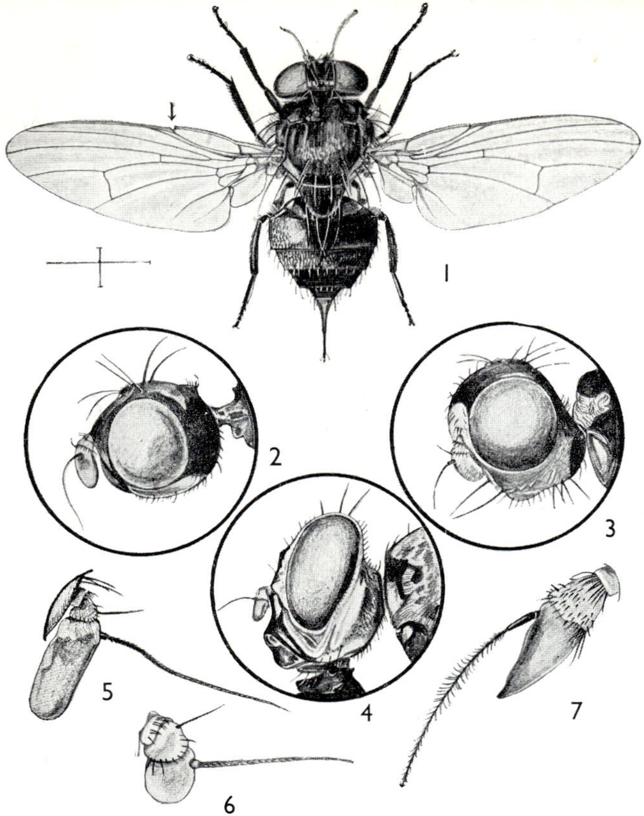

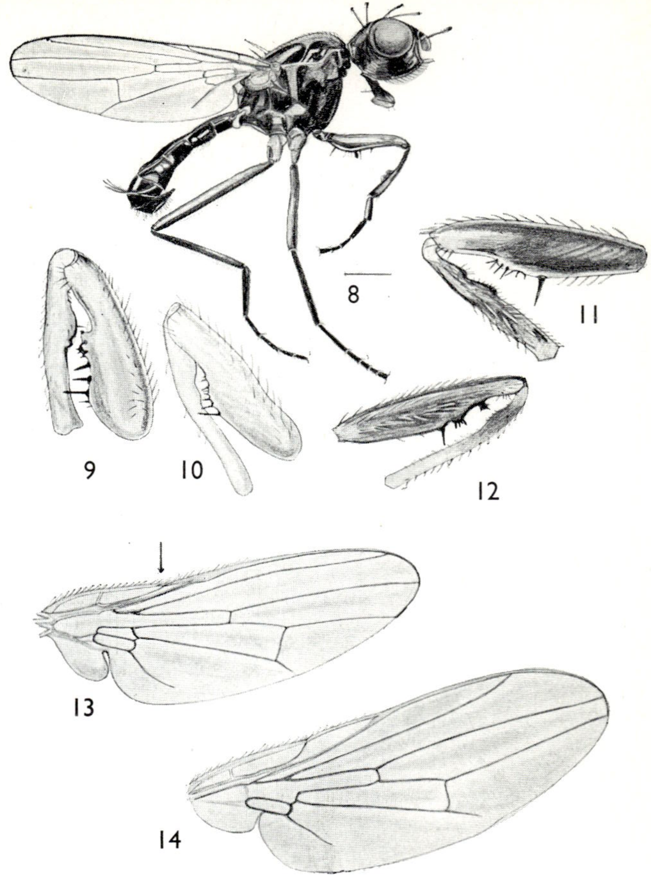

PLATE 30

LONCHAEIDAE

1. *Lonchaea chorea* F., ♀, p. 192. 5. *Lonchaea chorea*, antenna, p. 191.

SEPSIDAE-II

2. *Nemopoda nitidula* Fall. (=*cylindrica* F.), head, p. 195.

PIOPHILIDAE-I 3. *Piophila*, head, p. 196.

PLATYSTOMATIDAE-II 4. *Platystoma seminationis* L., head, p. 183.

PALLOPTERIDAE-III 6. *Palloptera*, antenna, p. 184.

SEPSIDAE-III

8. *Themira putris* L., ♂, p. 195. 9. *Sepsis punctum* F., ♂ fore femur and tibia, p. 195. 10. *Sepsis violacea* Mg., ♂ fore femur and tibia, p. 195. 11. *Sepsis cynipsea* L., ♂ fore femur and tibia, p. 195. 12. *Sepsis fulgens* Mg., ♂ fore femur and tibia, p. 195. 14. *Saltella scutellaris* Fall., wing, p. 195.

at all (*Orygma, Oedoparea*) ; eyes small, jowls deep ; antennal depressions deep ; 3rd antennal segment short and rounded ; postverticals convergent, or divergent (*Orygma*) ; jowls with hairs or bristles but not distinct vibrissae ; legs stout (in *Coelopa* with bristles and dense hairs) ; tibiae with preapical bristle (except *Coelopa pilipes* Hal.) ; apical segment of tarsi dilated ; costa unbroken ; *sc* entire, apex well separated from apex of Vein 1 ; Vein 6 faint apically, but reaches margin ; anal cell not convex apically (except *Orygma*). (Pl. **34,** 6.)

Coelopidae will be found on kelp, or seaweed thrown up on the shore by the tides, in which the larvae develop ; or on nearby flowers.

Coelopa frigida F. (Pl. **34,** 6) and its fellow-species of the genus are very common and widely distributed ; they are in evidence, in suitable weather, throughout the year. *Coelopa* species are in the habit of making mass migratory movements for short distances, hundreds sometimes being seen flying about a foot or so above the shore in one direction ; they are also said to be attracted to aromatic substances, notably trichlorethylene (cf. *Drosophila,* p. 229), and there are recorded instances of large swarms invading seaside chemists' and perfumers' shops. The reason for an invasion of a watch-repairer's shop was not explained until it was noticed that the swarm was particularly attracted to a small bottle of trichlorethylene, used for cleaning delicate mechanism. It is possible that the migratory movements on the shore are in some way connected with the periodical covering of their seaweed habitat by the tide ; further observation on this point would be of considerable interest.

The family is monographed by Hennig , the British species are treated by Burnet[18] and Egglishaw.[19]

HELEOMYZIDAE.

16 genera. Nearly 60 species. Small to medium-sized, yellowish, brownish or grey flies, generally characterised by series of small spines among the costal cilia (except *Borboropsis*) and by the pollinose stripes (vertical plates) on the frons, upon which stand one or two, but not more

than two orbital bristles ; 3rd antennal segment short and rounded ; postverticals convergent ; vibrissae present ; tibiae with preapical bristle ; wings generally large, sometimes marked ; costa broken at apex of *sc*, which is well separated from apex of Vein I ; *sc* entire and visible throughout, or faint apically (Trixoscelinae) ; anal cell convex apically. (Pl. **34**, 7.)

Heleomyzid larvae develop in carrion, excrement and various other kinds of decomposing animal or vegetable matter. Typical habitats include rotting, stored potatoes ; rotten wood ; the primitive cess-pits which still exist in some rural areas ; or the burrows of rabbits and caves frequented by bats, where excrement is plentiful. Larvae of *Heleomyza* species occur in living fungi which decompose as a result of their activities.

Three sub-families present themselves for consideration, viz., Heleomyzinae, Leriinae and Trixoscelinae, the latter forming the family Trixoscelidae of some authors. In the Helomyzinae, the anterior part of the pollinose stripe on each side of the frons is separated from the eye-margin, only one orbital bristle stands upon the stripe and Vein 6 does not reach the margin ; in the remaining two sub-families, the pollinose stripe is in contact with the eye-margin throughout its length and one or two orbitals may be present. A strong propleural bristle is present in the Leriinae, Vein 6 reaches the margin and a pair of ocellar bristles stand on the ocellar triangle almost behind the anterior ocellus, whereas, in Trixoscelinae, at most a hair stands on the propleuron, Vein 6 does not quite reach the margin and the ocellar bristles are in front, or on either side of the anterior ocellus.

HELEOMYZINA.

The species of *Heleomyzia* are small to medium-sized, yellowish-brown or reddish-brown flies, with a humeral bristle; Parmenter [9] refers to the propensity of the males for congregating upon fungi and suggests that they do this to await the arrival of females for mating purposes, in much the same manner as the males of the Yellow Dung Fly (*Scatophaga*) assemble on cow-dung (cf. p. 252); certain Syrphid males, it will be remembered, also have a habit of patrolling or hovering near places where the females are likely to come for oviposition or feeding (p. 168).

LERIINAE.

A number of somewhat undistinguished-looking flies make up the thirteen genera in this sub-family. *Tephrochlamys rufiventris* Mg. (Pl. **34,** 7) is one of the commonest representatives of the family and often to be found on windows indoors, even in winter; like its fellow-species in the genus, it has the thorax dull grey or greyish-brown and in contrast, the abdomen dull reddish. The larvae appear to be able to develop in various kinds of decaying animal or vegetable matter.

TRIXOSCELINAE.

Small flies about 2½–3 mm. long, in some cases with marked wings; two of the British species appear to be associated with coastal localities and one has been bred from a bird's nest.

Collin [10] gives keys to the British species, with notes.

CHYROMYIDAE.

2 genera. 5 species. Small flies with certain resemblances to Helomyzidae, but without series of costal spines, preapical bristle on tibia, or characteristic vertical plates; antennae very short, basal segment often concealed, 3rd segment rounded; postverticals convergent; 2–3 upper fronto-orbitals; face receding; vibrissae poorly developed; palpi short; propleural bristle absent; costa broken before apex of *sc* and Vein 1; *sc* entire, but weakly developed apically, runs more or less parallel to Vein 1 and partly concealed by it; cross-veins approximated; Vein 6 reaches margin (*Aphaniosoma*) or not (*Chyromya*). (Pl. **34,** 10, 11.)

Chyromya (= *Chiromyia*) *flava* L. is a very common little fly about $1\frac{1}{2}$–3 mm. long, semi-transparent yellow in colour, with dark bristles and, in life, beautiful iridescent purple eyes showing greenish reflections; it appears quite often on windows indoors and in the net when sweeping shady vegetation in damp places. It will be noticed that the female abdomen illustrated (Pl. **34,** 11) is distended and pustulate; this appearance occurs in other flies on occasions, and it is perhaps a coincidence that the abdomen of every female *flava* we have taken has been similar.

Larvae of Chyromyidae are said to develop in guano, so that the adults may well be sought near the nests or roosting-places of birds.

The family is monographed by Czerny.[11]

LITERATURE CITED

1. Séguy, E. Faune de France. **28.** Muscidae Acalypterae et Scatophagidae. Lechevalier, Paris. 1934.
2. Collin, J. E. A Short Synopsis of the British Sapromyzidae (Diptera). *Trans. R. ent. Soc. Lond.* **99**: 225–42. 1948.
3. Sack, P., in Lindner, E. Die Fliegen der Palaearktischen Region. **37.** Sciomyzidae. Schweizerbart, Stuttgart. 1939.
4. Czerny, L., in Lindner, E. Die Fliegen der Palaearktischen Region. **38a.** Dryomyzidae. **38b.** Neottiophilidae. Schweizerbart, Stuttgart. 1930.
5. Sack, P., in Lindner, E. Die Fliegen der Palaearktischen Region. **62b.** Thyreophoridae. Schweizerbart, Stuttgart. 1939.
6. Coe, R. L. The British Species of the Genus *Chamaemyia* (Dipt. Chamaemyiidae). *Ent. mon. Mag.* **78**: 173–80. 1942.

   Coe, R. L. *Chamaemyia juncorum* Fall. and *C. herbarum* R–D. (Dipt. Chamaemyidae); a correction to my recent paper on the British Species of this Genus. *Ent. mon. Mag.* **79**, 128–9. 1943.
7. Czerny, L., in Lindner, E. Die Fliegen der Palaearktischen Region. **51.** Chamaemyiidae. Schweizerbart, Stuttgart. 1936.
8. Hennig, W., in Lindner E. Die Fliegen der Palaearktischen Region. **52.** Coelopidae. Schweizerbart, Stuttgart. 1937.

9. Parmenter, L. Notes on the Helomyzidae (Diptera). *Ent. Rec.* **59**: 15–17. 1947.
10. Collin, J. E. The British Species of Helomyzidae (Diptera). *Ent. mon. Mag.* **79**: 234–51. 1943.
11. Czerny, L., in Lindner, E. Die Fliegen der Palaearktischen Region. **53c.** Chiromyiidae. Schweizerbart, Stuttgart. 1927.
12. Knutson, L. V. and Berg, C. O. Biology and immature stages of snail-killing flies. The genus *Elgiva* (Diptera, Sciomyzidae). *Ann. Ent. Soc. America*, **57**: 173–92. 1964.
    Knutson, L. V., Stephenson, J. W. and Berg, C. O. Biology of a slug-killing fly, *Tetanocera elata* (Diptera: Sciomyzidae). *Proc. Malacol. Soc. Lond.* **36**: 213–20. 1965.
13. Knutson, L. V. and Lyneborg, L. Danish Acalypterate flies. 3. Sciomyzidae (Diptera). *Ent. Medd.*, **34**: 61–101. 1965.
14. Collin, J. E. Some overlooked synonymy and additional species in British Sciomyzidae (Diptera), and a note on *Limnophora exsurda* (Anthomyiidae). *Ent. Rec.* **78**: 227–30. 1966.
15. Smith, K. G. V. A short synopsis of British Chamaemyiidae. *Trans. Soc. Brit. Ent.*, **15**: 103–15. 1963.
16. Collin, J. E. The British species of *Chamaeyia* Mg. (*Ochthiphila* Fln.) (Diptera). *Trans. Soc. Brit. Ent.*, **17**: 121–8. 1966.
17. Parmenter, L. The Sciomyzidae (Diptera) in Britain. *Ent. Record*, **78**: 125–8. 1966.
18. Burnet, B. The European species of the genus *Coelopa* (Dipt., Coelopidae). *Ent. mon. Mag.*, **96**: 8–13. 1960.
19. Egglishaw, H. J. Studies on the family Coelopidae (Diptera). *Trans. R. ent. Soc. Lond.* **112**: 109–40. 1960.

CHAPTER XVII

OPOMYZIDAE.
ANTHOMYZIDAE.
CLUSIIDAE.
TETHINIDAE.
EPHYDRIDAE . . . . . . Shore Flies.
CANACEIDAE.
SPHAEROCERIDAE (= BORBORIDAE).

In addition to the special works cited under family headings, Séguy [1] may be consulted in regard to all the families covered by this chapter.

OPOMYZIDAE.

2 genera. 13 species. Small flies with clouded or spotted wings; 3rd antennal segment short, rounded, inclined downward from 2nd segment; arista pubescent (*Opomyza*) or plumose on upper side (*Geomyza*); postverticals absent; upper part of occiput concave; only one orbital, inclined upward; vibrissae absent; one mesopleural and one sternopleural bristle present; tibiae without preapical bristle; wings long and narrow (axillary lobe and alula absent in *Geomyza*); costa broken at apex of *sc*, which is fairly close to apex of Vein 1; Vein 2 very long, convergent towards Vein 3, apically; second basal and anal cells short. (Pl. **35**, 1, 4.)

THE larvae of Opomyzidae develop in the middle shoots and stems of grasses and cereals; two of our British species have been bred from wheat plants.

One of the very common "small fry" which will be found in the sweep-net, especially when working dead leaves and undergrowth in woods, is *Opomyza germinationis* L. (Pl. **35**, 1). Its sluggish gait and habit of plodding its way solemnly up the inside of the opened net, long after most of the other insects have taken to their wings, will, together with its yellowish-brown or reddish-brown colour and marked wings, soon make it a familiar sight.

Geomyza species will be known at once by the absence of the axillary lobe and alula; the wing illustrated (Pl. **35**, 4) is that of the common *G. tripunctata* Fall.

Collin [2] furnishes keys to the British species, with notes.

ANTHOMYZIDAE.

5 genera. 16 species. Very small to small, slender-bodied flies, usually with long, narrow wings; certain resemblances to Opomyzidae; 3rd antennal segment, short, rounded, thickly pubescent, inclined downward from 2nd segment; arista pubescent or short-plumose; postverticals, if present, small, convergent or parallel; two upwardly-inclined orbitals, often a third anterior, not incurved; vibrissae present; no mesopleural but two sternopleurals; fore femora with small ventral spine (except *Striphosoma*); tibiae without preapical bristle; costa with long, fine cilia, broken fairly close to point where *upper* margin of Vein 1 merges with it; *sc* weak, fading apically before reaching break; cross-veins approximated, posterior at middle of wing; second basal and anal cells complete; Vein 6 short. (Pl. **35**, 5.)

The species of Anthomyzidae are mostly under 3 mm. in length, and occur among grasses and other vegetation bordering streams and ponds, or in marshy places; such as Reeds (*Arundo*), Rushes (*Juncus*), etc.; the larvae develop between the closely-fitting leaf-sheaths of the terminal shoots of these plants, and have also been recorded from galls caused thereon by the Chloropid, *Lipara* (p. 241).

Anthomyzidae should not be difficult to recognise on account of the situation of the posterior cross-vein at the middle of the wing, coupled with the pubescent third antennal segment and long costal cilia; the wing illustrated (Pl. **35**, 5) is that of *Anthomyza gracilis* Fall., a common and widely distributed species.

Striphosoma sabulosum Hal. is an aberrant species in which the wings are usually partly atrophied and somewhat strap-like, with incomplete venation; occasionally specimens are taken in which the wings and venation are practically normal.

Keys to the British species, with notes, are furnished by Collin.[3]

CLUSIIDAE.

5 genera. 10 species. Very small to small flies; 2nd antennal segment with an upper and lower bristle, often with a projection over the 3rd segment, which is rounded; arista pubescent, sub-apical (very short and dorsal in *Acartopthalmus*); postverticals divergent; 2–4 fronto-orbitals and sometimes a pair of small, crossed bristles between; vibrissae present (except *Acartopthalmus*); tibiae with or without preapical bristle; costa broken at apex of *sc*, or just after humeral cross-vein (*Acartopthalmus*); *sc* entire, visible throughout, more or less parallel to Vein 1, apex therefore fairly close; anal cell convex apically; Vein 6 short. (Pl. **35**, 6–8.)

Clusiidae mostly occur in the vicinity of rotten wood and fungi, in which the immature stages develop; on one occasion we took four species simultaneously by sweeping over a wet, disintegrating beech stump in the New Forest. The larvae are said to have a similar power of leaping to that of *Piophila* (p. 196) and some species of Cecidomyiidae (p. 83), the hind extremity being grasped with the mouthparts and the hold suddenly released.

The two blackish species of *Acartopthalmus* are about 2 mm. long and have the costa broken near the humeral cross-vein and a dorsal arista. The costal break in the remaining genera is at the apex of *sc*, and the arista is more or less sub-apical. In *Clusiodes*, the cross-veins are very close to one another and before the middle of the wing (Pl. **35**, 6); all the orbitals are upswept, while the second antennal segment projects over the third and has conspicuous upper and lower bristles (Pl. **35**, 7). *Clusiodes* species are dark-coloured; in *Clusia flava* Mg. a yellow species about 4–6 mm. long, the antennae are similar, but the cross-veins more widely separated (Pl. **35**, 8); both genera have a pair of small, crossed inter-frontal bristles between the fronto-orbitals. The anterior pair of fronto-orbitals are convergent (incurved) in *Clusia* and downwardly-directed in *Paraclusia* and *Heteromingia*. Prescutellar acrostichals are present in *Paraclusia* and absent in *Heteromingia*.

The family is monographed by Czerny.[4] Collin[11] is useful.

TETHINIDAE.

2 genera. 6 species. Very small to small, greyish or brownish flies inhabiting the sea-shore; 3rd antennal segment rounded; arista short; postverticals feeble, convergent; 1–4 fronto-orbitals, the upper slightly upswept, often with pairs of small, crossed bristles between; vibrissae present, weakly developed, and a row of bristles around the wide mouth-opening; tibiae without preapical bristle; costa broken close to point where *upper* margin of Vein 1 merges with it; *sc* weakly developed, only visible basally, thence as faint fold confused with Vein 1; cross-veins approximated. (Pl. **35**, 2, 3.)

Tethinidae are small, undistinguished-looking flies about $1\frac{1}{2}$–3 mm. long; little seems to be known of their immature stages. The two species of the genus *Pelomyia* may be distinguished from *Tethina* by the fact that the second basal and discal cells are not distinctly separated, the cross-vein being vestigial; by the absence of acrostichals on the dorsum and of the small, crossed bristles on the frons.

Collin[12] deals with the British species.

EPHYDRIDAE.

30 genera. Over 120 species. Very small to small flies, mainly found near water and marshy places; 2nd antennal segment sometimes with strong, downward-curved spine (Notiphilinae); 3rd segment rounded; arista bare, pubescent or plumose, in the latter case, with the cilia always on the upper surface; postverticals absent, but sometimes a pair of divergent posterior ocellars; mouth-opening very wide and face strongly arched; vibrissae absent or poorly developed, not clearly distinct from remaining bristles round mouth-edge; tibiae without preapical bristle; wings sometimes with dark markings, or dark with clear areas; costa broken twice, once near humeral cross-vein and once near apex of Vein 1; *sc* weak, only visible basally, thence confused with Vein 1; second basal and discal cells confluent; anal cell absent. (Pl. **35**, 9–13.)

In the field, Ephydrids will often be recognised by their characteristic wing-venation (Pl. **35,** 9), the wide mouth-opening and strongly-arched face (Pl. **35,** 12) or the arista plumed on the upper side only (Pl. **35,** 13), coupled with their habitat. They are rarely found elsewhere than in very damp situations, such as the borders of ponds, streams and marshes or on the seashore; they may be found throughout the year, even in mid-winter, if the weather be mild. We have taken a number of species in company with Sciomyzids and *Campsicnemus* (Dolichopodidae) in early February; some Ephydrids are able to run on the surface of the water like the latter.

Adult habits are varied; some are predatory upon small insects or other soft-bodied creatures; others subsist on dead or decaying matter, including excrement. The larvae, which may be either terrestrial or aquatic, are also, in some cases, predatory; some live in decaying matter or excrement and others mine the leaves or live in the stems of terrestrial or aquatic plants.

There are three sub-families for consideration. Notiphilinae are distinguishable by the characteristic curved spine on the second antennal segment (Pl. **35,** 11) or a strong antero-dorsal bristle on the mid-tibiae; Ephydrinae have the mouth-opening very large, the face strongly arched and the eyes bare (Pl. **35,** 12); Hydrellinae have the mouth-opening less wide, the face less strongly arched and the eyes usually hairy (Pl. **35,** 10,13).

NOTIPHILINAE.

Several species of *Notiphila,* which are blackish or dark-greyish flies, are not uncommon on waterside or aquatic vegetation, such as Pondweeds (*Potamogeton*) or Lesser Reed-Mace; the eggs of certain species are said to be laid on the flowers, while the resulting larvae live in the stems. In the case of *N. brunnipes* R-D., the eggs are laid on the flowers of the White Water-lily, which close at sunset and withdraw below the surface of the water, to reappear at sunrise; thus the eggs are alternately immersed and exposed.

The tiny, brilliantly-metallic, bronze-black species of *Psilopa* are common near ponds or streams, especially *P. nitidula* Fall.; a foreign species of this genus spends its immature stages in pools of crude petroleum—an extraordinary instance of adaptability! Larvae of *Discomyza incurva* Fall. have been recorded as developing in a snail.

Dichaeta caudata Fall. is worthy of mention on account of the unusual bristle-arrangement of the abdomen in the male; a series of about 8–10 extraordinarily long bristles extend backwards from the hind-margin of the fourth tergite as a sort of fan, and there are two long bristles extending from the hind extremity of the abdomen.

HYDRELLINAE.

As an example of the carnivorous type, the grotesque *Ochthera mantis* Deg. (Pl. **35,** 9), an uncommon dark, brownish-black fly, is worthy of special notice. The front legs are enormously developed; it will be seen that the coxae and femora are enlarged and dilated, the latter armed with a row of formidable spines; the tibiae, which have a long apical spur, are curved and fold back upon the femora like the blade of a pocket knife, forming a trap from which the unhappy victim has little hope of escaping (cf. *Platypalpus* and *Hemerodromiinae*, Empididae, pp. 127 and 129).

Some years ago, we discovered this species, in mid-April, on grass-stems in a muddy patch surrounding a New Forest spring, which is periodically flooded. Considerable numbers were present, and, in bright sunshine, were flying with rapid, erratic "hops" of two or three feet at a time. When at rest, they were seen to cling, motionless, to the tops of the stems by the middle and hind legs, head uppermost, with the fore-legs extended slightly upwards and the tibiae more or less at a right angle to the femora. One was seen to seize a small black male Chironomid which blundered into the range of the terrible fore-legs and to explore the victim with extended proboscis; an attempt to take both captor and prey was, however, unsuccessful. On further observation, it was noticed that, on the approach of small insects, the stance was rapidly changed so as to present the fore-legs in the required direction, with the body slightly raised, as if expectantly; the effect was extraordinarily sinister and menacing.

Specimens were taken back to our collecting headquarters and kept in captivity for some days in large glass jars, an attempt being made to simulate natural conditions with mud, kept moist, and short grass stems. Small Muscids and Dolichopodids were introduced but, for a day or two, the Ephydrids took very little notice of them, either remaining very aloof on the grass-stems (this time, not in the characteristic attitude noticed in the field, but with head downwards and holding on with all the legs) or resting indifferently upon the mud. Gradually, however, the threatening attitude was again adopted on the approach of their co-dwellers; ultimately, one of the Muscids was trapped by a *mantis*, by means of a lightning spring, both insects being on the mud before and after the capture. In this case, the prey was held down by the outstretched fore-legs of the *mantis*, in a fashion somewhat reminiscent of the manner in which a dog will sometimes hold down a bone with his front paws, preparatory to gnawing it. After a brief exploration with the proboscis, the prey was allowed to escape, evidently worse for wear but apparently not mortally injured; there were several similar instances subsequently. Then we introduced some newly-emerged mosquitoes, *Culex pipiens* L., and *Aëdes punctor* Kirby, which we were rearing from pupae taken locally a few days previously; these were received with considerable enthusiasm. The "stem" attitude was not, however, adopted, possibly due to the limited confines of the jars; all the prey were taken on the mud, by the pinning-down technique. Rapid execution was made, and in a few minutes after capture, only the shrivelled husks of the mosquitoes remained.

In June of the same year, the habitat was again visited in order to conduct further observation on habits under natural conditions, but this time the flies were very much less in evidence. However, prey were seen to be taken on the mud, which was now confined to a very small area adjacent to the source of the spring; sometimes the leaps made by the captors were as much as an inch in extent. By September, the flies had entirely disappeared.

O. mantis seems to be extremely local; prolonged search and sweeping on these visits, within some hundreds of yards of the particular habitat, were unavailing, although the area contained several places of very similar character. Moreover, search in subsequent years has produced no specimens from the original habitat, and only one from a similar location about a mile away.

Some of the little species of *Hydrellia* (=*Hydropota*) are commonly swept from waterside vegetation; we have taken *H. albilabris* Mg., a shining-black fly, just over 1 mm. long, with a silvery face, running on the surface of the water in a lily-pool in early February. *H. griseola* Fall. (Pl. **35,** 13) is a very common little greyish-brown species, the larvae of which mine the leaves and stems of various plants, including Daisy, Ragged Robin, Duckweed, and sundry grasses and cereals.

EPHYDRINAE.

The common coastal species, *Ephydra riparia* Fall. (Pl. **35,** 12), is about 5 mm. long and of a metallic, dark brownish-grey colour, with dull bronze-blue or green reflections; it will be encountered on salt-flats or near brackish or salty pools adjacent to the shore. The larva lives beneath the surface of the water, on aquatic plants; in the pupa, there is a curious excision in the hind segments, by means of which the creature is able to moor itself to a plant, at will; the rear spiracles are situated at the ends of the branches of a long, forked appendage at the hind extremity. Larvae of certain American species of this genus are said to have been used by the Indians as food.

Some of the species of the genera *Scatella, Scatophila* and *Parydra* (= *Napaea*) have the wing-membrane somewhat darkened, with a number of clear areas, which are more clearly visible when viewed obliquely, against the light. In a few species of *Parydra*, a small veinlet branches off from Vein 2 near the apex.

Larvae of *Teichomyza fusca* Macq. are said sometimes to accumulate in such numbers in liquid excremental matter, on the Continent, as to block up sewage- and drain-pipes.

Papers by Collin [6] on the genera *Psilopa, Discocerina, Scatella* and *Limnellia* (= *Stictoscatella*) will be useful to the student of British Ephydridae; the family has been monographed by Becker.[7] Collin[13] treats the genera *Ephydra* and *Hydrellia*.

CANACEIDAE.

2 genera. 2 species. Very small greyish, or greyish-brown flies of the seashore; bases of the antennae widely separated; 3rd segment short and rounded; arista short; postverticals small, divergent; eyes small, lengthened horizontally; ocellar triangle large, extending to anterior margin of the frons; 3–5 fronto-orbitals, divergent; cheeks with bristles, two pairs extending forward, vibrissa-like; jowls deep; mouth-opening wide; costa broken close to point where *upper* margin of Vein 1 merges with it; *sc* entire, but tending to fade, and very close to Vein 1 apically; 2nd basal and anal cells very small. (Pl. **35**, 14, 15.)

Four pairs of strong, black dorsocentrals and other bristles are present on the dorsum of *Canace nasica* Hal.; in *Xanthocanace ranula* Loew, the bristles are whitish and only two pairs of dorsocentrals are present; all are weakly developed.

The family is monographed by Becker.[8]

SPHAEROCERIDAE.

24 genera. Nearly 100 species. Minute to small, usually black or brown flies with characteristically dilated and shortened hind metatarsus, often attracted to excrement; 3rd antennal segment short, rounded; arista dorsal, long, bare or pubescent: postverticals convergent, often indistinct or absent; fronto-orbitals strong, never inwardly-curved, but often small, crossed inter-frontals between; vibrissae present, often indistinct; face, in profile, usually concave, with mouth-edge projecting forward; mouth-opening wide; femora strong, tibiae often with curved apical spur; hind metatarsus usually very short and more or less dilated; wings sometimes reduced; costa broken twice, once near humeral cross-vein, and once close to point where upper margin of Vein 1 merges; apical section of Vein 5 (e.g. *Trichiaspis*), or Veins 4 and 5 (e.g. *Limosina*) sometimes wanting or not pigmented, in the latter case 2nd basal cell confluent with discal cell and anal cell absent. (Pl. **38,** 1–4.)

Every nature-student will doubtless be familiar with the Yellow Dung Flies (*Scatophaga,* p. 252), which foregather on cow-dung, since they are conspicuous enough by virtue of colour, numbers and activity. It is possible, however, that other, less conspicuous companions in their habitat will not have attracted the attention of many observers. Closer observation will often reveal large numbers of very small, sombre brown or blackish flies on or near the pat, which rapidly disperse when alarmed; a sweep-net cast lightly to and fro above the droppings will sometimes produce a large assembly with a variety of species. On examination with a lens, they will usually be found to have the hind metatarsus shortened and dilated (Pl. **38,** 1, 2, 4), a characteristic wing-venation (Pl. **38,** 1–3), in which the apical sections of Vein 5 or both Vein 4 and 5 may be curtailed or faint, together with a wide mouth-opening somewhat reminiscent of the Ephydridae (Pl. **38,** 1, 2). These are Sphaeroceridae (= Borboridae); most of which are dark-coloured, only a very few being reddish or reddish-yellow, and rarely exceeding 5 mm. in length; not a few being exceedingly minute, as small as 1 mm. Many species are addicted to the excrement of various mammals, including cattle and horses, or rabbits and other burrowing animals, the larvae developing therein. Others are to be found under dead leaves, near fungi, in decaying organic matter, or at sap exuding from tree-wounds; Donisthorpe [9] instances four species of *Limosina* as guests of ants. Some species seem to require the proximity of water and certain coastal species are associated with seaweed.

Mites are often found in association with flies, especially those frequenting excrement, but perhaps in no other family is their occurrence more prevalent than in the case of the dung-inhabiting Sphaeroceridae. This will probably not surprise the naturalist who is familiar with other dung-frequenting insects, such as the well-known Dor Beetles (*Geotrupes*), which are almost invariably well loaded with mites. The method and point of attachment varies a great deal, but there seems to be no direct evidence that fatal harm results to the fly, although, in some cases, certain structural modifications and deformities ensue. It would appear that the principal purpose served by the attachment is the transporting of the mites to fresh territory. Sphaerocerid flies, themselves, are said to utilise dung-frequenting beetles in other parts of the world for a similar purpose. Although capable of sustained flight (swarms occasionally take place) they seldom make more than a rapid hop of a few inches, when disturbed, and seem to be adept at self-concealment. Some species can run, as for instance, in the sweep-net, in a rapid and agitated "scuttling" manner reminiscent of the Phoridae (p. 143), from which they may be differentiated immediately by the quite distinct wing-venation.

We may conveniently consider the Sphaeroceridae in three broad groups.

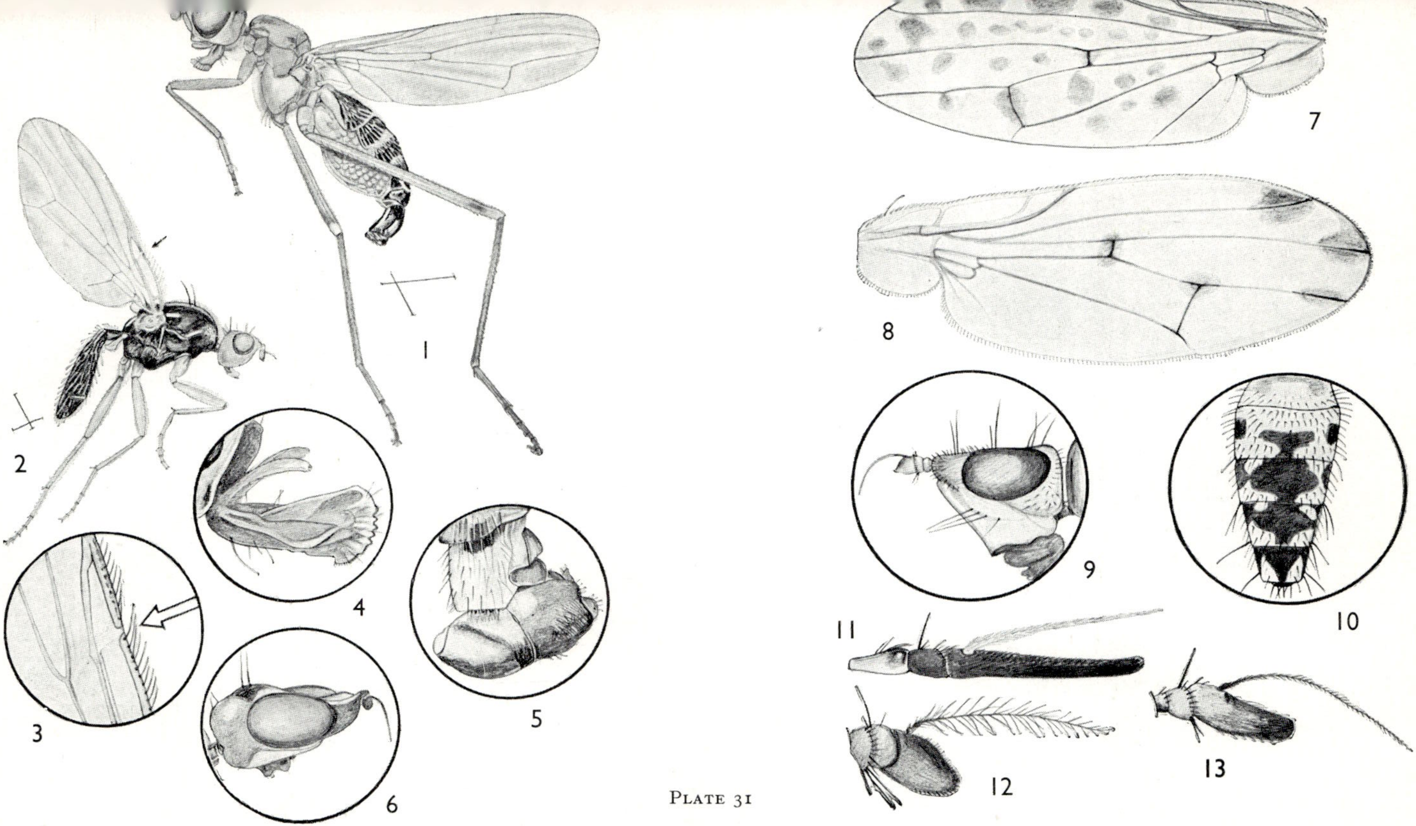

PLATE 31

MICROPEZIDAE (=TYLIDAE)-I

1. *Calobata* (=*Trepidaria*) *petronella* L., ♀, p. 200. 4. *Calobata* (=*Trepidaria*) *petronella* L., proboscis, p. 200. 5. *Calobata* (=*Trepidaria*) *petronella* L., ♂ genitalia, p. 200. 6. *Micropeza* (=*Tylos*) *corrigiolatus* L., head, p. 200.

PSILIDAE-I

2. *Psila* (*Psila*) *rosae* F., ♀, p. 197. 3. *Psila* (*Psila*) *rosae* F., costal break and cross-fold, p. 197.

SCIOMYZIDAE-I

7. *Pherbellia schoenherri* Fall., wing, p. 205. 10. *Trypetoptera punctulata* Scop., abdomen, p. 205.

LAUXANIIDAE-I

8. *Cnemacantha* (=*Lyciella*) *decempunctata* Fall., wing, p. 203. 9. *Trigonometopus*, head, p. 203. 11. *Lauxania cylindricornis* F., antenna, p. 203. 12. *Prorhaphochaeta inusta* Mg., antenna, p. 203. 13. *Calliopum aeneum* Fall., antenna, p. 203.

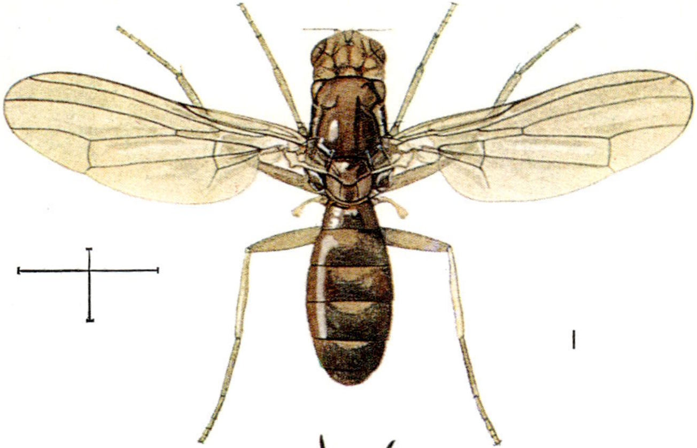

1

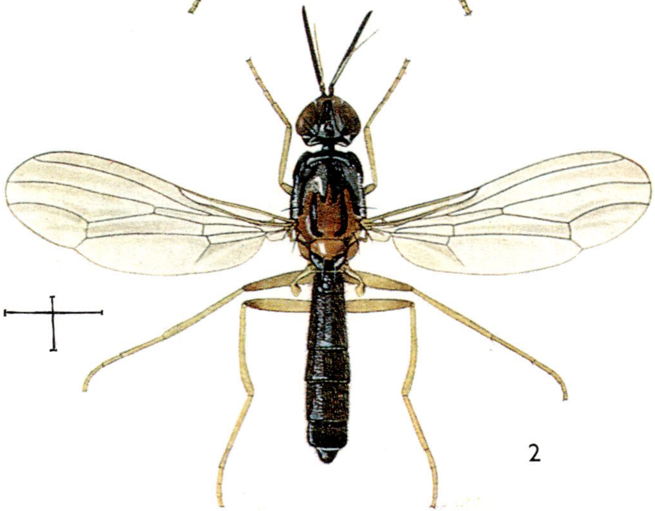

2

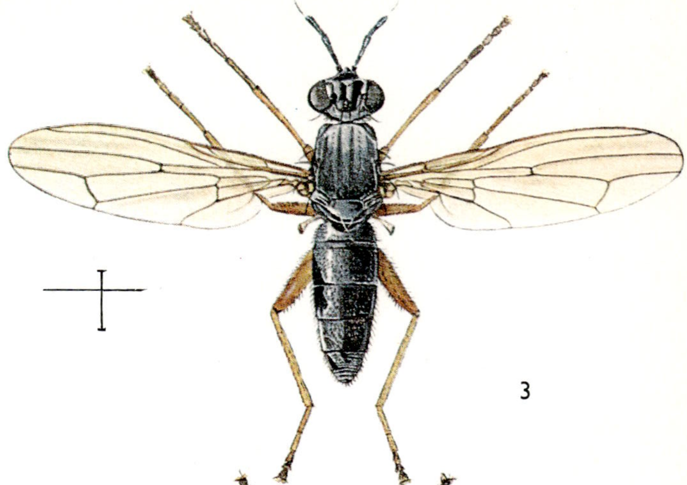

3

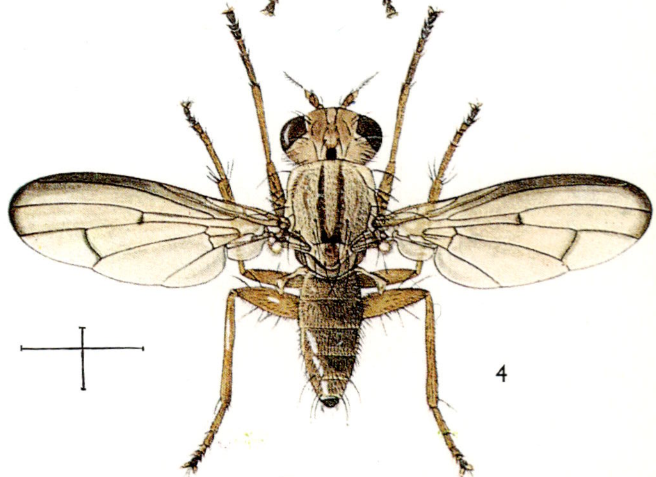

4

Plate 32

PSILIDAE-II

1. *Psila (Pelethophila) fimetaria* L., ♂, p. 198.

SCIOMYZIDAE-II

3. *Sepedon sphegeus* F., ♀, p. 204.

Group I.

Both Veins 4 and 5 either appear to cease at the posterior cross-vein or become very faint thereafter; second basal cell confluent with discal, and anal cell absent; scutellar bristles present. *Paracollinella fontinalis* Fall. (Pl. **38,** 2), a very common and widely distributed woodland fly, will serve as an example of this group, which contains by far the greatest proportion of genera and species.

Group II.

Vein 5 not reaching the margin; second basal and anal cells present; scutellar bristles present. The common *Trichiaspis* (= *Borborus*) *similis* Collin (Pl. **38,** 1) and the very closely similar *T. equina* Fall. and *T. stercoraria* Mg. are all frequenters of horse-dung.

Group III.

Vein 5 reaching the margin (Pl. **38,** 3); second basal and anal cells present; scutellum without bristles, but sometimes toothed, or with warty excrescences on the disc, which may also occur on the dorsum, or even on the frons. *Sphaerocera curvipes* Latr. is a very common frequenter of carrion and other decaying matter, including the excrement of various animals, especially rodents; oddly enough, it does not seem to be addicted to horse-dung. The greatly enlarged and dilated hind femora and tibiae, which appear to be bowed or curved, are sufficiently conspicuous to enable this fly to be recognised; it may positively be known by the characteristic way in which the basal hairs on the ventral surface of the hind metatarsus are in the form of a series of downward-pointing hooks (Pl. **38,** 4).

Sphaeroceridae are to be found at most times of the year, even in the winter months, in mild weather or sheltered situations.

The British species are dealt with by Richards,[10] whose paper contains much interesting matter on biology and habits.

Collin[14] and Laurence[15] should also be consulted. Hackman[16] keys the genus *Copromyza*.

LITERATURE CITED

1. Séguy, E. Faune de France. **28.** Muscidae Acalypterae et Scatophagidae. Lechevalier, Paris. 1934.
2. Collin, J. E. The British Species of Opomyzidae (Diptera). *Ent. Rec.* **57**: 13–16. 1945.
3. Collin, J. E. The British Species of Anthomyzidae (Diptera). *Ent. mon. Mag.* **80**: 265–72. 1944.
4. Czerny, L., in Lindner, E. Die Fliegen der Palaearktischen Region. **54a**. Clusiidae. Schweizerbart, Stuttgart. 1928.
5. Czerny, L., in Lindner, E. Die Fliegen der Palaearktischen Region. **55.** Tethinidae. Schweizerbart, Stuttgart. 1928.
6. Collin, J. E. The British Species of *Psilopa* Fall. and *Discocerina* Mcq. (Dipt. Ephydridae). *Ent. mon. Mag.* **79**: 145–51. 1943.
   Collin, J. E. Some new species of the Dipterous Genus, *Scatella* Dsv., and the differentiation of *Stictoscatella* gen. nov. *Ent. mon. Mag.* **66**: 133–9. 1930.
7. Becker, T., in Lindner, E. Die Fliegen der Palaearktischen Region. **56a.** Ephydridae. Schweizerbart, Stuttgart. 1926.
8. Becker, T., in Lindner, E. Die Fliegen der Palaearktischen Region. **56b.** Canaceidae. Schweizerbart, Stuttgart. 1926.
9. Donisthorpe, H. St. J. K. The Guests of British Ants. Routledge, London. 1927.
10. Richards, O. W. The British Species of Sphaeroceridae (Borboridae, Diptera). *Proc. Zool. Soc. Lond.* **1930**: 261–345. 1930.
11. Collin, J. E. A new species of *Clusiaria* (Dipt., Clusiidae). *J. Soc. Brit. Ent.*, **4**: 138–9. 1953.
    Collin, J. E. *Clusiaria apicalis* Zett., new to Britain. *J. Soc. Brit., Ent.*, **4**: 139–40. 1953.
12. Collin, J. E. British Tethinidae (Diptera). *Entomologist*, **93**: 191–3. 1960.
    Collin, J. E. A revision of the palaearctic species of *Tethina* and *Rhicnoessa*. *Boll. Mus. civ. Stor. nat. Venezia*, **16**: 19–32. 1966.
13. Collin, J. E. The British species of *Ephydra* (Dipt., Ephydridae). *Ent. mon. Mag.*, **99**: 147–52. 1963.
    Collin, J. E. A contribution towards the knowledge of the male genitalia of species of *Hydrellia* (Dipt., Ephydridae). *Boll. Mus. civ. Stor. nat. Venezia*, **16**: 7–18. 1966.

(Continued on p. 353)

CHAPTER XVIII

PERISCELIDAE.
ASTEIIDAE.
DROSOPHILIDAE . Small Fruit Flies, Vinegar Flies.
CAMILLIDAE.
DIASTATIDAE.
AULACIGASTRIDAE.

In addition to the special works cited under family headings, Séguy [1] may be consulted in regard to all the families covered by this chapter.

PERISCELIDAE.

1 genus. 4 species. Very small to small flies, greyish or yellowish-grey with black-and-yellow banded legs; 2nd antennal segment projecting dorsally over 3rd, like a hood; 3rd segment short, rounded; arista plumose; postverticals divergent; occiput concave, somewhat inflated on lower part; vibrissae absent; tibiae without preapical bristle; costa unbroken, reaching as far as apex of Vein 3; *sc* only visible basally; Veins 3 and 4 not convergent apically; Vein 4 bent angularly between anterior and posterior cross-veins, which are approximated; 2nd basal cell present; Vein 6 feeble, therefore anal cell not distinctly complete. (Pl. **38,** 5.)

PERISCELIDAE are uncommon and usually occur at tree-wounds, in the decomposing exudations from which the larvae are said to develop. A curious feature in the wing is a kind of cross-fold, or apparent weakening running from the costa near the base of the wing, across the basal cells to beyond the apex of the anal cell, which is somewhat reminiscent of the Psilidae (p. 197). The wing illustrated is that of *Periscelis* (*Microperiscelis*) *winnertzi* Egger.

The family has been monographed by Duda.[2]

ASTEIIDAE.

2 genera. 5 species. Very small to small flies, with very characteristic wing-venation ; 3rd antennal segment short, rounded ; arista pubescent (*Leiomyza*) or plumose (*Asteia*) ; postverticals divergent ; vibrissae weak ; tibiae without preapical bristle ; costal breaks near humeral cross-vein and apex of Vein 1 very indistinct, represented only by slight attenuation of nervure ; costa extends to apex of Vein 4 ; Vein 2 ends close to Vein 1 in basal half of wing (*Asteia*) or extended near to apex of wing (*Leiomyza*) ; Veins 3 and 4 convergent apically ; posterior cross-vein present, close to anterior cross-vein, before middle of wing (*Leiomyza*), or absent (*Asteia*) ; 2nd basal and anal cells absent. (Pl. **38,** 6, 7.)

Asteia amoena Mg. is a not uncommon little fly, about 2–3 mm. long, and of rather distinguished appearance ; the dorsum and frons are shining black, the pleura and legs yellowish, the abdomen reddish or yellowish, patterned with black, and the face reddish with silvery reflections ; a broad, conspicuous pearly-white band extends above the mouth-edge (Pl. **38,** 7). The immature stages of this fly are said to develop in the debris in hollow trees and similar locations. Normally the adults hibernate in sheltered cavities under bark ; sometimes, however, they congregate in autumn (*vide* Blair [3]) in considerable numbers indoors, on windows (cf. *Thaumatomyia* and other *Chloropids*, p. 241, and *Pollenia*, p. 286). The reason for this occasional abundance and swarming is obscure.

Larvae of *Leiomyza scatophagina* Fall. have been recorded from dried stems of the Common Reed (*Arundo*) together with those of *Lipara* species (Chloropidae, p. 241).

The family is monographed by Duda.[4]

DROSOPHILIDAE.

11 genera. 52 species. Very small to small flies, often attracted to fermenting substances ; 3rd antennal segment short, rounded ; arista usually plumose, with characteristic fork at tip, formed by main stem and last hair (Pl. **38,** 9) ; postverticals convergent ; 2–3 orbitals, not incurved, one

often curved forward, remainder upswept; vibrissae present, sometimes indistinct; tibiae with preapical bristle; costa broken twice, once near humeral cross-vein, once close to apex of Vein 1; *sc* visible basally, fading apically and more or less confused with Vein 1; Vein 1 short; second basal cell confluent with discal cell, or separated (e.g. *Stegana, Amiota, Phortica*); anal cell present. (Pl. **38,** 8, 9.)

Those who have had an opportunity of seeing the manufacture of beer, cider, vinegar or pickles in progress, or who have occasion to visit places where fruit is stored, will almost certainly have noticed numbers of small yellowish or brownish flies of rather fat or bulbous appearance, with bright red eyes, congregating in the places where fermentation is going on, and obviously attracted by the odours which prevail in such situations. They have a slow, ponderous, hovering sort of flight, with the abdomen hanging downward, and give the impression that it is as much as they can do to keep their bodies suspended in the air. Perhaps, also, they will have been noticed in restaurants or public-houses, where they are often attracted to beer or wines and will settle on the rims of glasses. These are Drosophilidae, sometimes referred to as the Small Fruit Flies, or Vinegar Flies. Some species have certain resemblances to the Ephydridae, but are usually not difficult to recognise on account of the characteristic tip of the arista, which is plumose on both sides (Pl. **38,** 9); there are two costal breaks and the second basal cell is usually merged with the discal cell, but an anal cell is present (Pl. **38,** 8).

The common, ferment-loving species, already mentioned, belong to the genus *Drosophila* and include *D. funebris* F., *D. fenestrarum* Fall., *D. fasciata* Mg. and *D. obscura* Fall., all of which are somewhat variable in colour, particularly as regards the abdominal markings. Some of these species will be encountered not infrequently out of doors; we have found *obscura*, for example, on several occasions, at sap exuding from tree-wounds and at rotten fruit. *Drosophila* species have been much used in the study of genetics, on account of the large chromosomes and brief life-cycle.

A cosmopolitan species, *D. repleta* Wollaston, which may be distinguished by its mottled or dappled dorsum (small irregular dark patches round the sockets of the bristles), was first recorded in this country by Coe,[5] who found it in 1942 and the succeeding year, in a number of canteens and a hospital in London; it was established that it was developing in kitchens, in sufficient numbers to constitute a nuisance, the habitat of the immature stages consisting of decomposing vegetables, including onions and cabbages. We found this species to be numerous in the kitchen of a Services canteen in the south-western district of London, during the autumn of 1945 and throughout the winter; it swarmed wherever the temperature was high, such as the vicinity of hot-water sinks and cooking apparatus, and became a nuisance by settling in considerable numbers on exposed food, including bread, cheese and butter, in the adjoining eating-rooms. Since then, we have found it in a number of restaurants and taverns within a radius of a mile of the original location, so that it would appear to be firmly established.

Drosophila species may be distinguished by the plumose arista, confluent second basal and discal cells, strong post-verticals and the fact that the lowest (anterior) of the upswept orbitals is much weaker than those behind it. In the very small species of the sub-genus *Scaptomyza*, only one lower hair is present on the arista, behind the fork at the tip, as against at least two in the species of *Drosophila* proper; the larvae are leaf-miners or develop in vegetable matter generally. The dark grey *D.* (*Scaptomyza*) *graminum* Fall., which is about 2 mm. long, is often swept from grasses; the larvae have been recorded from the stems of Watercress.

Chymomyza and *Leucophenga* have weak postverticals and the anterior upswept orbital as strong as those behind it; in the latter, the foremost orbital is a good distance from the bases of the antennae and the cross-veins are approximated. *Leucophenga maculata* Dufour is a pretty little fly, about 2½–3½ mm. long, the male with a white frons, yellow antennae, thorax with shimmering silvery pruinosity and white lateral stripes, and yellow abdomen with three longitudinal rows of large, blackish spots; the larva has been recorded from a *Polyporus* fungus.

In *Stegana, Protostegana, Amiota,* and *Phortica,* the second basal cell is separated from the discal cell by a cross-vein. We have taken *Phortica variegata* Fall., which is about 3 mm. long, with grey thorax mottled with brown, yellowish abdomen with hind-margins of the segments blackish, and pale tibiae with three dark bands, at sap flowing from a wound in an elm-trunk in the New Forest; the larvae have been recorded in the sap of a Weeping Willow. *Stegana coleoptrata* Scop. has also been recorded at sap exuding from the stumps of felled oaks and poplars, and the larvae from the latter.

Acletoxenus and *Cacoxenus* have a bare or pubescent arista and weak postverticals. Ocellar bristles are absent in *Acletoxenus formosus* Loew, the larvae of which are said to feed on species of "White Fly" (Aleyrodidae, Hemiptera) infesting Hawthorn and Viburnum; the whitish pupae are found on the underside of the leaves among the husks of the victims. In *Cacoxenus,* ocellar bristles are present; the larvae are said to be parasitic upon Carpenter Bees (*Osmia*) and Mason Wasps (*Odynerus*).

Some Drosophilids are said to be attracted to carrion or excrement; there are records of development taking place in the curd which forms in neglected milk-bottles. The larvae of the ferment-loving species seem to be abnormally resistant to the alcohols and other by-products of fermentation which occur in their development-habitats; it is probable that these would militate against the survival of many other kinds of Dipterous larvae. A number of provisions for existence in fluid or semi-fluid media are to be found in the immature stages. The eggs of *Drosophila*, for instance, are furnished with long filamentous processes at one end; in the larvae, the anal segment is equipped with a series of small excrescences enclosing a retractile process, which bears the rear spiracles; the ovoid pupae have a pair of anterior respiratory horns, which are plumose at the apex.

The family is monographed by Duda.[6] The British species are keyed by Fonseca[10] and Basden[11] is very useful.

CAMILLIDAE.

1 genus. 5 species. Very small to small, uncommon, shining black flies allied to Drosophilidae; 3rd antennal segment short, rounded apically; arista plumose, some long pubescence in addition to the long, upper hairs; post-verticals not strongly developed, convergent; vibrissae present; mesopleural bristle present; tibiae without preapical bristle; costa broken twice; *sc* vestigial; second basal cell, anal cell, and Vein 6 absent.

Collin [7] may be consulted in regard to the British species of the single genus, *Camilla*.

DIASTATIDAE.

2 genera. 7 species. Very small to small greyish or brownish-grey flies allied to Drosophilidae; 3rd antennal segment short-oval, pubescent (*Diastata*) or long, reaching almost to mouth-edge (*Campichaeta*); arista pubescent, or short-plumose; 2–3 orbitals, with bases close together, anterior inset and upswept, the second close to eye-margin and curved forward (reverse of arrangement in Aulacigastridae); one pair of strong vibrissae present, distinct from remaining strong bristles round mouth-edge; tibiae with preapical bristle; costa broken twice, once near humeral cross-vein, once near apex of Vein 1; *sc* visible basally, thence confused with Vein 1; anterior cross-vein near base of wing; second basal cell not clearly separated from the discal cell, i.e. the cross-vein not complete; anal cell present. (Pl. **38**, 10.)

The species of *Diastata* may be distinguished from those of *Campichaeta* (= *Thryptochaeta*) by the single inset, upswept orbital (Pl. **38**, 10), mesopleural bristles and short third antennal segment. In the latter there are two inset upswept orbitals, the mesopleura are bare and the third antennal segment is very much longer.

Diastata unipunctata Zett. is a common species about 3 mm. long, with a dark brownish cloud obscuring the posterior cross-vein and the anterior, basal part of the wing, which may be found in marshy places more or less throughout the year, in suitable weather. *D. vagans* Loew is another common species, about the same size, with the wings practically unmarked, which may be swept from dead leaves in woods.

The family is monographed by Duda.[8]

AULACIGASTRIDAE.

Single, shining brown species, *Aulacigaster leucopeza* Mg., about 2½ mm. long, with distinctive head-profile; 3rd antennal segment short, rounded, a little angular anteriorly; arista pubescent; postverticals absent; ocellar bristles indistinct or absent; 2 orbitals, bases close together, anterior curved forward and the inset posterior upswept (reverse of arrangement in Diastatidae); two pairs strong vibrissae present and row of hairs round mouth-edge; mouth-opening very wide, Ephydrid-like; tibiae with indistinct preapical bristle; costa broken at apex of *sc*, which is well separated from apex of Vein 1; *sc* entire, visible throughout; anterior cross-vein very near base of wing; 2nd basal and discal cells confluent; anal cell present; abdominal tergites with transverse furrows. Pl. **38**, 11.)

The larvae of this fly are said to develop in the decomposing sap exuding from tree-wounds; the rear spiracles are borne at the end of a long, retractile process at the hind extremity, which is shortly forked at the apex; the anterior spiracles are borne on short, plumose processes. In the pupa, there are two long, plumose, respiratory "horns".

Further reference may be made to Duda.[9]

LITERATURE CITED

1. Séguy, E. Faune de France. **28.** Muscidae. Acalypterae et Scatophagidae. Lechevalier, Paris. 1934.

2. Duda, O., in Lindner, E. Die Fliegen der Palaearktischen Region. **58a.** Periscelidae. Schweizerbart, Stuttgart. 1934.
3. Blair, K. G. *Asteia amoena* Mg. (Dipt. Asteiidae) swarming on windows. *Ent. mon. Mag.* **82**: 53. 1946.
4. Duda, O., in Lindner, E. Die Fliegen der Palaearktischen Region. **58b.** Astiidae. Schweizerbart, Stuttgart. 1934.
5. Coe, R. L. *Drosophila repleta* Wollaston (Dipt. Drosophilidae), new to Britain, with notes on the species and some account of its breeding-habits. *Ent. mon. Mag.* **79**: 204–7. 1943.
6. Duda, O., in Lindner, E. Die Fliegen der Palaearktischen Region. **58g.** Drosophilidae. Schweizerbart, Stuttgart. 1934–5.
7. Collin, J. E. Five New Species of Diptera. *Ent. mon. Mag.* **69**: 272–5. 1933.
8. Duda, O., in Lindner, E. Die Fliegen der Palaearktischen Region. **58e.** Diastatidae. Schweizerbart, Stuttgart. 1934.
9. Duda, O., in Lindner, E. Die Fliegen der Palaearktischen Region. **58c.** Aulacogastridae. Schweizerbart, Stuttgart. 1934.
10. Fonseca, E. A. A short key to the British Drosophilidae (Diptera) including a new species of *Amiota*. *Trans. Soc. Brit. Ent.*, **16**: 233–44. 1966.
11. Basden, E. B. The distribution and biology of Drosophilidae (Diptera) in Scotland, including a new species of *Drosophila*. *Trans. Roy. Soc. Edinburgh*, **62**: 603–54. 1954.

CHAPTER XIX

AGROMYZIDAE	. . . .	Leaf Miners.
MILICHIIDAE.		
ODINIIDAE.		
CHLOROPIDAE (= OSCINIDAE)	.	Gout Fly, Frit Fly, etc.
BRAULIDAE	. . . . .	Bee "Lice".

In addition to the special works cited under family headings, Séguy [1] may be consulted in regard to all the families covered by this chapter.

AGROMYZIDAE.

About 20 genera. Over 300 species. Very small to small flies, with characteristic larval habit of leaf-mining and stem boring; 3rd antennal segment short, usually rounded, but sometimes with sharp point (e.g. *Cerodontha* **s.s.**); arista pubescent or bare; postverticals divergent; lower fronto-orbitals incurved; vibrissae present; tibiae without preapical bristle; costa broken some distance from point where upper margin of Vein 1 merges with it, continued as far as apex of Vein 3 (e.g. *Phytomyza Napomyza*) or of Vein 4 (e.g. *Agromyza*, ***Liriomyza***); *sc* faint, visible throughout (e.g. *Cerodontha*, *Phytomyza*) or confused with **Vein** 1 apically (e.g. *Agromyza*); posterior cross-vein present (e.g. *Agromyza*) or absent (e.g. *Phytomyza*); anal cell present; in ♀, last (7th) abdominal tergite and sternite fused into non-retractile sheath for ovipositor. (Pl. **38,** 12–15.)

Leaf-mining is a typical habit of the larvae of Agromyzidae; the species generally confine themselves to a particular plant-host, or to a closely-related series of plants, and make characteristic mines which frequently give a very good clue to the identity of the miner. Apart from various *Phytomyza* species which turn up quite frequently when sweeping grasses and low herbage, such as nettles, adult Agromyzids may not be found commonly unless the food plant of a particular species is swept. There is little doubt that a good many more species will be added to the list as time goes on; much has been done by Spencer and Griffiths during the past 14 years. It is certain that there is still scope for the student taking the trouble to rear Agromyzids from mined plants and work out their life-histories.

The family presents some very varied types. In *Phytomyza*, features which will catch the eye are the absence of the posterior cross-vein and the weakening of the posterior veins, which throws up the anterior veins in contrast (Pl. **38,** 12). The wings of *Napomyza* and *Pseudonapomyza* are very similar, but a posterior cross-vein is present, very close to the base of the wing, forming a very short discal cell. In the wings of other genera, the posterior cross-vein is more normally situated, but is almost invariably close to the anterior cross-vein. There are two types of sub-costa: one pursuing an independent course, albeit faintly, more or less parallel to Vein 1 (Pl. **38**, 13), and the other coalescing with Vein 1 just before its apex (Pl. **38,** 14).

The observer will doubtless have noticed certain Holly-leaves in which the normal dark green colour is broken up by a blotch of yellowish or paler green. On examination, such lighter patches will sometimes be found to consist of shallow blisters between the upper and lower surfaces of the leaf, perhaps with a small hole, or appearing to contain a small, seed-like object, which can be seen more easily by holding the leaf against the light. This little object is quite likely to be the larva or pupa of the Holly-leaf Miner, *Phytomyza ilicis* Curt. If a small hole be present, it will mark the exit from which the adult fly has emerged. From leaves without an exit-hole, preserved at home under suitable conditions, the adults may often be bred out. The life-history of this species has been studied in detail by Miall and Taylor.[2]

A number of the larger species of *Agromyza* can be reared from mines on nettle and Boraginaceae such as Hound's Tongue, Viper's Bugloss and Comfrey.

The species of *Cerodontha* s.s. are noteworthy for the curious shape of the third antennal segment, which is pointed apically and bears a short spine-like process (Pl. **38,** 15); the larvae mine the leaves of various grasses.

The family has been monographed by Hendel[3] and the papers of Spencer[11] and Griffiths[12] deal with British species.

MILICHIIDAE.

7 genera. 26 species. Very small to small, greyish-black or shining black flies; 3rd antennal segment short, rounded, sometimes greatly dilated or enlarged (e.g. *Phyllomyza*); postverticals convergent or parallel; ocellar triangle strongly developed, considerably prolonged anteriorly (Carninae) or normal; anterior orbital incurved; sometimes pairs of small, crossed inter-frontals present between orbitals; vibrissae present, often indistinct; proboscis often long, jointed in middle, i.e. labella fold back like blade of pocket-knife, or short and thick, adapted for piercing (*Carnus*); wings large; costa broken twice, once near humeral cross-vein, once near apex of Vein 1; costa extended to apex of Vein 4, or Vein 3 (Carninae); *sc* visible throughout (e.g. *Milichia*), or faint apically, more or less confused with Vein 1 (e.g. *Phyllomyza*); Vein 4 fades apically (Carninae) or normal; second basal and anal cells very small, or absent (Carninae); posterior cross-vein very close to anterior (*Meoneura*) or absent (*Carnus*); Vein 6 present as a shadowy fold. (Pl. **39,** 1, 3, 4.)

The family consists of three sub-families, Milichiinae, Madizinae and Carninae.

MILICHIINAE.

Represented by the single species, *Milichia ludens* Whlbg., a little greyish-black fly, about 2½–3 mm. long, which, according to Donisthorpe,[4] is associated with the Ant, *Lasius fuliginosus* Latr.

MADIZINAE.

Species of *Phyllomyza* are also recorded by Donisthorpe as guests of ants. An interesting representative of this genus is *P. securicornis* Fall., in which the third antennal segment (Pl. **39,** 1) is enormously developed. *Leptometopa latipes* Mg.

has an extraordinary modification of the hind tibiae (Pl. **39**, 4); it develops in human and other excrement. *Madiza glabra* Fall., a shining black fly about 2½ mm. long, has similar propensities, often occurring in country areas where sanitary arrangements are still of a primitive nature.

According to Séguy,[1] species of *Desmometopa* and *Phyllomyza* are veritable jackals of the insect world; they attach themselves to predatory Diptera, such as Asilids, or even to Reduviid bugs and spiders, by which they are transported until prey is captured; whereupon they proceed, in company with the predator, to sup the exudations from the wounds of the victim.

CARNINAE.

The two genera, *Meoneura* and *Carnus*, are characterised by their wing-venation, in which Vein 4 fades apically (Pl. **39**, 3); the species of both are very minute, about 1–1½ mm. in length. In *Meoneura*, a posterior cross-vein is present, close to the anterior cross-vein; the adults are attracted to decomposing substances, in which the larvae develop. *Carnus hemapterus* Nitzsch, the sole species of the genus, has no posterior cross-vein (Pl. **39**, 3). In the adult stage, it is parasitic upon birds, sucking their blood; the proboscis is much chitinised and strengthened for the purpose. After attachment to a host, these flies shed their wings, their abdomens, meantime, becoming greatly distended; the larvae develop in refuse in the nests. *C. hemapterus* is recorded in association with a number of birds, including the Kestrel, Great Spotted Woodpecker, Blackcap and Starling.

The family has been monographed by Hennig;[5] a paper by Collin [6] on the genus *Meoneura* will be found very useful.

ODINIIDAE.

1 genus. 4 species. Very small to small, greyish-black flies, with broad wings, in which the cross-veins and apex of Vein 1 are clouded, and with dark patches on the abdo-

men ; 3rd antennal segment short, squarish in profile ; arista practically bare ; postverticals divergent ; vibrissae present ; tibiae with preapical bristle ; costa broken close to point where *upper* margin of Vein 1 merges with it ; costa extends as far as apex of Vein 3 ; *sc* visible basally, fading apically ; cross-veins approximated ; second basal and anal cells present ; Vein 6 extended almost to margin. (Pl. **39**, 2.)

The two British species of this family are very uncommon. In the adult stage, they are attracted to the suppurations of tree-wounds; we have swept a male *Odinia boletina* Zett. from a rotten beech-stump in the New Forest. The larva is recorded from a *Polyporus* fungus.

The family is monographed by Hennig.[7] Collin[13] keys the British species.

CHLOROPIDAE.

28 genera. over 80 species. Very small to small, smooth, almost bristleless flies, often black, yellow-and-black, or green-and-black, with characteristically large plate-like ocellar triangle, and reduced wing-venation ; antennae of variable shape ; arista pubescent or bare, sometimes appearing greatly thickened by dense pubescence ; head bristles weakly developed or absent ; eyes often brilliantly coloured in life ; proboscis usually short, or long, thin, folded back from middle (*Siphonella*) ; tibiae without preapical bristle ; costa broken, usually close to point where *upper* margin of Vein 1 merges, but sometimes a little distant (e.g. *Dicraeus*, *Oscinella*) ; *sc* vestigial, only perceptible at base, or absent ; 2nd basal cell confluent with discal cell ; anal cell absent ; Vein 6 weak or absent. (Pl. **39**, 6–11.)

Chloropid larvae are mainly plant-feeders, some species constituting agricultural pests of some importance. Those of *Chloropisca glabra* Mg., however, are said to be predatory upon aphids ; certain species abroad have been recorded as developing in the egg-cases of locusts or egg-cocoons of spiders.

As regards the adults, certain foreign species are attracted to the eye-secretions of human beings and are thought to introduce disease organisms mechanically by means of the mouthparts ; there is evidence to suggest that conjunctivitis and yaws are thus transmitted.

The British species are divided into two sub-families : Oscinellinae and Chloropinae. In the former, the species are predominantly black, and the costa reaches as far as the apex of Vein 4 (Pl. **39,** 9) ; in the Chloropinae, various combinations of yellow, green and black occur and the costa only reaches as far as the apex of Vein 3 (Pl. **39,** 6). Many species of Chloropidae will turn up in the sweep-net, when working grasses and low vegetation in meadows, or on road-verges and waste land ; some will be recognisable at once, as belonging to this family, by the very large ocellar triangle (Pl. **39,** 11) and yellow or greenish-yellow colour, with dark longitudinal stripes on the dorsum and various dark markings elsewhere (Pl. **39,** 6).

OSCINELLINAE.

Oscinella frit L., the "Frit Fly" (Pl. **39,** 9), is a tiny black insect about 1–1½ mm. long, which is a cereal-pest of considerable economic significance in other parts of Europe and in the New World. In this country, it affects barley, wheat, rye and maize, but its preference appears to be for oats, in consequence of which it is also known as the "Oat Fly". The species overwinters in the larval stage, either in various wild grasses, or winter cereals, and pupates in the spring. Adults of the first generation appear in April and May ; eggs are laid on the lower parts of spring-sown cereals and the larvae work their way into the stem, with the result that the central shoot withers. Pupation takes place in the plant and the adults of the second generation emerge from June to August ; this time the eggs are laid on the developing ears, the larvae feeding on the grain. Adults of the third generation, which appear from July to September, lay their eggs on wild grasses and late-sown cereals and so the cycle is completed.

Elachiptera species are noteworthy for the stout, densely-pubescent arista (Pl. **39,** 10) and the frequently warted margin

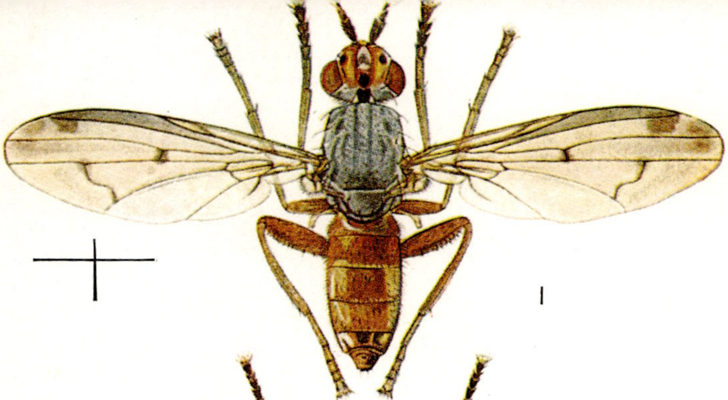

1

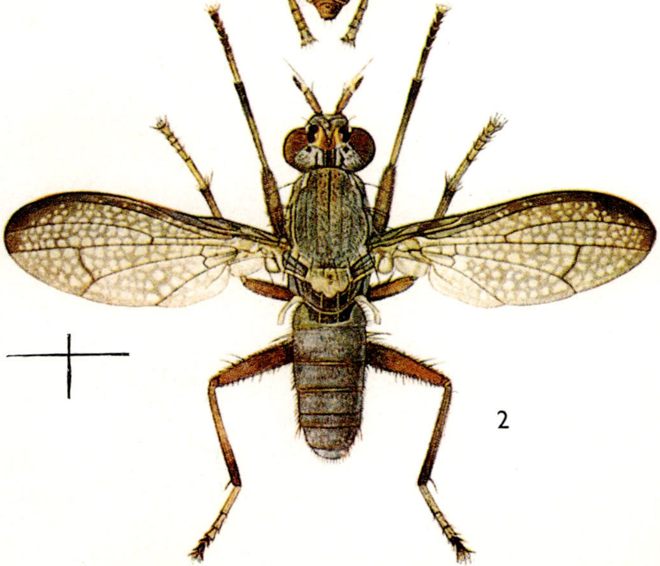

2

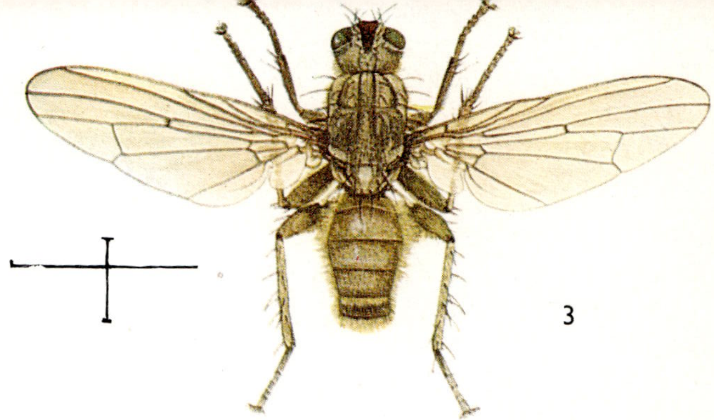

3

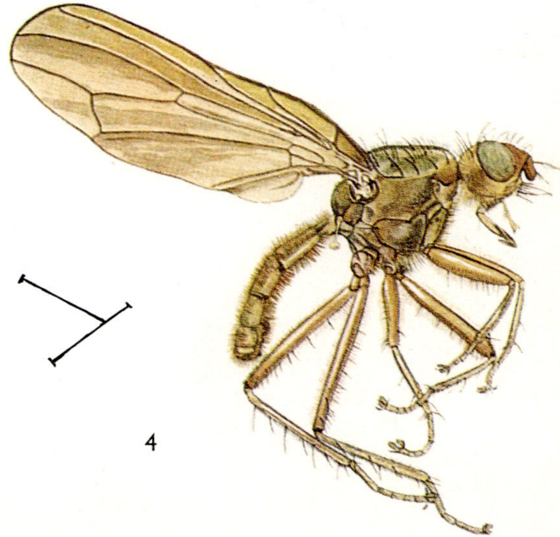

4

PLATE 33

SCIOMYZIDAE-III

1. *Elgiva cucularia* L., ♀, p. 205.
2. *Statinia marginata* F., ♂, p. 204.

SCATOPHAGIDAE-I

3. *Scatophaga* (=*Scopeuma*) *stercorarium* L., ♂, p. 253.
4. *Scatophaga* (=*Scopeuma*) *scybalarium* L., ♂, p. 253.

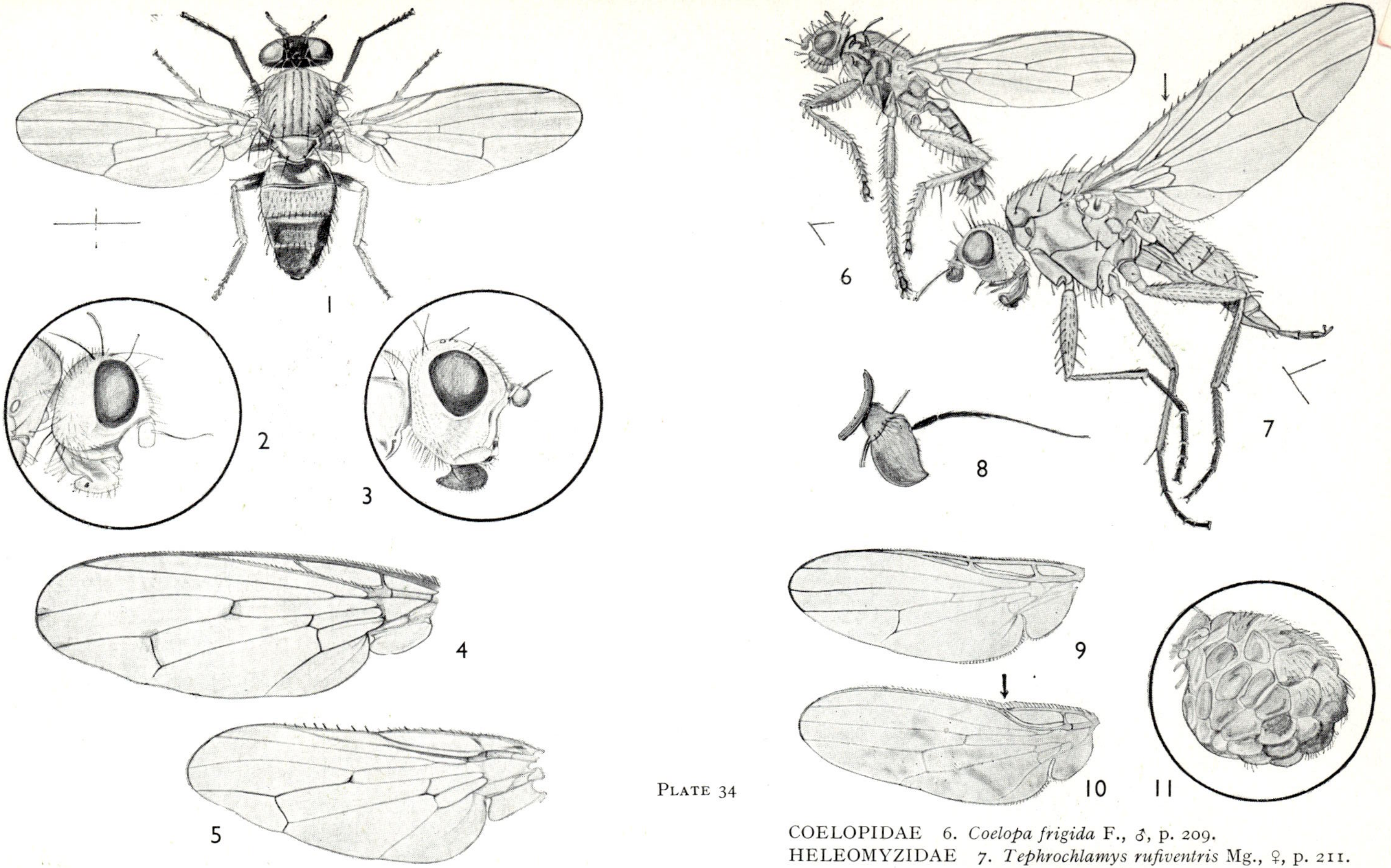

PLATE 34

LAUXANIIDAE-II 1. *Calliopum aeneum* Fall., ♀, p. 203.

DRYOMYZIDAE

2. *Neuroctena anilis* Fall., head, p. 206. 3. *Helcomyza ustulata* Curt., head, p. 206. 4. *Neuroctena anilis* Fall., wing, p. 206. 5. *Helcomyza ustulata*

COELOPIDAE 6. *Coelopa frigida* F., ♂, p. 209.

HELEOMYZIDAE 7. *Tephrochlamys rufiventris* Mg., ♀, p. 211.

CHAMAEMYIIDAE

8. *Chamaemyia flavipalpis* Hal., antenna, p. 208. 9. *Chamaemyia flavipalpis* Hal., wing, p. 208.

CHYROMYIDAE

10. *Chyromya* (=*Chiromyia*) *flava* L., wing, p. 211. 11. *Chyromya* (=*Chiromyia*) *flava* L., ♀ abdomen, p. 212.

of the scutellum, the warts bearing the marginal bristles ; the dorsum is marked with longitudinal lines of little punctures. *E. cornuta* Fall. is a very common and widely distributed species ; while *E. brevipennis* Mg. has greatly reduced wings, which are very little more than half the length of the abdomen. The larvae in this genus develop mainly in the stems of grasses and cereals, but recorded host-plants include turnips and watercress.

Larvae of the British species of *Lipara* cause large cigar-shaped galls in the stems of the Common Reed (*Arundo*), which are often found to contain other Dipterous larvae ; in particular, those of other Chloropids, including *Cryptoneura* (= *Haplegis*) *flavitarsis* Loew, which appears to be a more or less regular inquiline.

CHLOROPINAE.

Platycephala planifrons F., a fairly large species, for this family, at 7–8 mm., is worthy of notice on account of the extraordinary flat-triangular head, elongated antennae (Pl. **39,** 7), dilated hind femora, curved hind tibiae and pitted integument, especially on the frons ; the larvae develop in the stems of the Reed (*Arundo*) and the adults occur, therefore, in waterside situations where this is to be found. Somewhat similar hind femora and tibiae are to be found in *Meromyza* species (Pl. **39,** 8), which are pale-green and black flies, about 3–4½ mm. long, frequently swept from grasses ; they have a characteristic habit of making little jumps or leaps in the net.

Another cereal-pest of some importance is *Chlorops pumilionis* Bjerkander (= *taeniopus* Mg.), otherwise known as the Gout Fly or Ribbon-footed Corn Fly, which attacks mainly wheat and barley, but also develops in couch-grass.

The phenomenon of vast autumn swarms which was noted in the Asteiidae (p. 228) also occurs in the Chloropidae (*vide* Séguy [1]) ; in Britain, the habit, in the open, has been recorded in respect of *Thaumatomyia notata* Mg., *Oscinosoma nitidissima* and *Oscinella frit* L. Mr. A. E. LeGros sent us specimens of *notata* in October, 1948, which he had taken from a huge swarm on a ceiling of a very old building in Winchester; according to his informants, these flies had congregated thus, in the autumn, in precisely the same place for many years. A similar habit is recorded in respect of the Calliphorid, *Pollenia rudis* F. (p. 286).

British species of the sub-family Oscinellinae are covered in a paper by Collin[8]; some Choropinae are covered by Collin[14] and Smith[15]; the family, as a whole, is monographed by Duda.[9]

BRAULIDAE.

A single tiny, wingless species, *Braula coeca* Nitzsch, associated with Honey Bees; length about 1–1½ mm.; head wide; antennae set in lateral grooves; eyes minute; ocelli absent; no frontal or vertical bristles; thorax not clearly differentiated from rounded abdomen; scutellum absent; legs stout; claws inconspicuous, but a pair of comb-like processes present. (Pl. **39**, 5.)

Our last Acalypterate, *Braula coeca* Nitzsch, the "Bee Louse" (Pl. **39**, 5), is a most unusual kind of fly, since neither wings nor halteres are present; superficially, it has a somewhat mite-like appearance. The hairy flattened legs and tarsal combs are well-adapted for clinging to the hairs of Honey Bees, on which these insects are sometimes found in considerable numbers, especially on the thorax. *Braula* has often been included in the Pupipara (Section VI) with certain species of which it has some characters in common (cf. Pl. **50**, 3, 4); in contrast to the reproductory processes of that group, however, it lays eggs. These are deposited in the brood-combs of the hive and the resulting larvae, which are of the normal Cyclorrhaphous type, feed on the stored pollen-paste and pupate in the hive.

The life-history has been studied by Skaife.[10]

LITERATURE CITED

1. Séguy, E. Faune de France. **28.** Muscidae Acalypterae et Scatophagidae. Lechevalier, Paris. 1934.
2. Miall, J. C., and Taylor, T. H. The Structure and Life-history of the Holly Fly (*Phytomyza aquifolii*). *Trans. ent. Soc. Lond.* **1907**: 259. 1907.
3. Hendel, F., in Lindner, E. Die Fliegen der Palaearktischen Region. **59.** Agromyzidae. Schweizerbart, Stuttgart. 1931–6.
4. Donisthorpe, H. St. J. K. The Guests of British Ants. Routledge, London. 1927.
5. Hennig, W., in Lindner, E. Die Fliegen der Palaearktischen Region. **60a.** Milichiidae & Carnidae. Schweizerbart, Stuttgart. 1937.
6. Collin, J. E. Some species of the genus *Meoneura*. *Ent. mon. Mag.* **66**: 82–9. 1930.
   Collin, J. E. Two new species of the genus *Meoneura*. *Ent. mon. Mag.* **73**: 250–2. 1937.
7. Hennig, W., in Lindner, E. Die Fliegen der Palaearktischen Region. **60b.** Odiniidae. Schweizerbart, Stuttgart. 1938.
8. Collin, J. E. The British Genera and Species of Oscinellinae. *Trans. R. ent. Soc. Lond.* **97**: 117–48. 1946.
9. Duda, O., in Lindner, E. Die Fliegen der Palaearktischen Region. **61.** Chloropidae. Schweizerbart, Stuttgart. 1932–3.
10. Skaife, S. H. On *Braula coeca* Nitzsch, a dipterous parasite of the honey bee. *Trans. R. Soc. S. Afr.* (Cape Town). **10**: 41–8. 1921.
11. Spencer, K. A. Notes on the British Agromyzidae. *Ent. mon. Mag.*, **89**: 296–7; **90**: 64–5; **91**: 68–70. 1953–55.
   Spencer, K. A. Observations on the British Agromyzidae. *Ent. Gaz.*, **5**: 183–8; **7**: 51–4; **8**: 21–7; 93–8; **10**: 165–72; **13**: 18–20. 1954–62.
   Spencer, K. A. The British Agromyzidae (Dipt.). *Proc. Trans. S. London Ent. Nat. Hist. Soc.*, 1954–5; 98–108;. 1956.
   Spencer, K. A. Notes on the British species of *Melanagromyza* Hendel (Dipt., Agromyzidae). *Proc. R. ent. Soc. Lond.* (B) **26**: 179–88. 1957.
12. Griffiths, G. C. D. New British species of Agromyzidae and similar titles. *Ent. Gaz.*, **6**: 10–2; 62–4; **7**: 190–3; **10**: 101–6; **12**: 123–6. 1955–61.
   Griffiths, G. C. D. The Agromyzidae of Wood-Walton Fen. *Ent. mon. Mag.*, **98**: 125–158. 1962.
13. Collin, J. E. On the European species of the genus *Odinia* (Diptera, Odiniidae). *Proc. R. ent. Soc. Lond.* (B) **21**: 110–6. 1952.

(Continued on p. 353)

KEY TO THE BRITISH ACALYPTERATE FAMILIES (SECTION IV)

(Partly based on Oldroyd, Handbooks for the Identification of British Insects, Vol. IX, Pt. 1, 39–42)

1. Abnormal flies, very small, wingless, mite-like, living externally on the bodies of bees (Pl. **39**, 5)
   BRAULIDAE (p. 242)
   Normal flies 2
2. Apex of anal cell at or fairly near to wing-margin (Pl. **25**); head very wide; abdomen often basally constricted; proboscis nearly always long, forwardly projecting, sometimes bent back from middle .... CONOPIDAE (p. 177)
   Apex of anal cell well removed from wing-margin 3
3. Costa not interrupted or "broken"[1] before apex of Vein 1 .. 4
   Costa interrupted or "broken" before apex of Vein 1 ... 17
4. Vibrissae strong; palpi rudimentary
   SEPSIDAE (p. 194)
   Vibrissae absent or weak 5
5. Legs long, thin and stilt-like; tibiae without preapical bristle; abdomen long and slender; Veins 3 and 4 convergent apically (Pl. **31**, 1) .... MICROPEZIDAE AND TANYPEZIDAE (pp. 199–200)
   Legs normal, or not all above characters present, together .. 6
6. Tibiae with dorsal preapical bristle[2] 7
   Tibiae without dorsal preapical bristle.................. 11
7. Postvertical bristles well developed, convergent or crossed .. 8
   Postvertical bristles parallel, divergent or (rarely) absent ... 9
8. Vein 6 faint apically, but reaches hind-margin at least as crease or shadow; last tarsal segment broadened; 3–4 fronto-orbitals; anal cell not convex apically; very bristly and/or hairy seashore flies (Pl. **34**, 6) .. COELOPIDAE (p. 208)
   Vein 6 short and blunt, not reaching margin; anal cell convex apically; 1–2 pairs fronto-orbitals; last tarsal segment not broadened (Pl. **34**, 1) .... LAUXANIIDAE (p. 202)
9. Second basal and anal cells long and large; anal cell not convex apically; clypeus[3] prominent (Pl. **34**, 2–5)
   DRYOMYZIDAE (p. 205)

[1] Characters based on costal breaks will require close attention; if doubt is experienced, try alternative.

[2] Except *Coelopa pilipes* Hal. (Coelopidae).

[3] See Glossary, p. 355.

Second basal and anal cells small; anal cell often convex apically 10

10. Hypopleural bristle present; 3–4 orbitals
COELOPIDAE,
g. *Orygma* (p. 208)

No hypopleural bristle; at most 2 orbitals (Pl. **32,** 3, 4: **33,** 1, 2)
SCIOMYZIDAE (p. 203)

11. Hind femora thickened, with 2 rows ventral spines; abdomen long and thin, somewhat constricted basally; anal cell long, parallel-sided .... MEGAMERINIDAE (p. 199)

Hind femora not thickened and not spinous ventrally; anal cell comparatively short, broad, not parallel-sided 12

12. *Sc* only visible basally 13

Sc visible throughout, even if faint apically 14

13. Costa reaches to apex of Vein 3; Veins 3 and 4 not convergent apically; 2nd basal cell present (Pl. **38,** 5)
PERISCELIDAE (p. 227)

Costa reaches to apex of Vein 4; Veins 3 and 4 convergent apically; 2nd basal cell absent (Pl. **38,** 6)
ASTEIIDAE (p. 228)

14. Palpi rudimentary; head more or less spherical; wings often with black spot at tip and/or abdomen somewhat constricted basally (Pl. **29,** 5: **30,** 8) ..... SEPSIDAE (p. 194)

Palpi well developed 15

15. Vein 6 well developed; postverticals divergent, parallel, or absent 16

Vein 6 vestigial; postverticals convergent or absent (Pl. **34,** 9)
CHAMAEMYIIDAE (p. 208)

16. Vein 1 bare on dorsal surface; if hairy, then anal cell sharp-pointed,[1] *sc* and Vein 1 close together, and body-colour shining metallic; Veins 3 and 4 convergent apically (Pl. **29,** 4)
ULIDIIDAE (p. 186)

Vein 1 hairy on dorsal surface; if bare, then *sc* and Vein 1 comparatively widely separated or Veins 3 and 4 convergent apically (Pl. **27,** 1) .... OTITIDAE (p. 183)

17. A costal break near humeral cross-vein only 18

A costal break near apex of *sc* or Vein 1 19

18. Vein 6 reaches margin; wings dappled or banded; small flies, about 3–5 mm. (Pl. **27,** 4)
PLATYSTOMATIDAE (p. 183)

Vein 6 short; wings not dappled or banded; very small flies, about 2 mm. .... CLUSIIDAE,
g. *Acartopthalmus* (p. 216)

[1] But see *Homalocephala*, p. 186.

19. Costal break well separated [1] from point where *upper* margin of Vein I merges with costa............................20
Costal break very close to point where *upper* margin of Vein I merges with costa..................................33
20. *Sc* distinct, separated from Vein I, ending in costa, or abruptly cut off before reaching it 21
Sc weakly developed, vestigial, or absent 29
21. Postvertical bristles divergent, parallel or absent 22
Postvertical bristles convergent or crossed 28
22. *Sc* bent abruptly at right-angle, fading apically (Pl. **27,** 6, 7) TEPHRITIDAE (p. 186)
Sc not abruptly bent at right-angle 23
23. Vibrissae present 24
Vibrissae absent, or very weak, hair-like (Pl. **27,** 2, 3: **27,**5) PALLOPTERIDAE (p. 184)
24. Second basal cell separated from discal cell...............25
Second basal cell confluent with discal cell (Pl. **38,** 11) AULACIGASTRIDAE (p. 233)
25. Vertical plates [2] extended forward, bearing 2–4 fronto-orbitals; anal cell convex apically (Pl. **35,** 6, 8) CLUSIIDAE (p. 216)
Vertical plates short, bearing one fronto-orbital 26
26. Head concave behind; eyes in profile semicircular; male frons narrowed; ovipositor with non-retractile sheath (Pl. **30,** 1) LONCHAEIDAE (p. 191)
Head convex behind; eyes rounded; male frons not narrowed; ovipositor not with non-retractile sheath 27
27. Vein I hairy dorsally; Vein 6 reaches margin NEOTTIOPHILIDAE (p. 207)
Vein I bare above; Vein 6 not reaching margin (Pl. **30,** 3, 13) PIOPHILIDAE (p. 196)
28. Tibiae with preapical dorsal bristle; *sc* usually strong, well separated from Vein I at apex; costa with series of spines [3] distinct from normal cilia (Pl. **34,** 7) HELEOMYZIDAE (p. 209)
Tibiae without preapical dorsal bristle; *sc* rather weakly developed but distinct, almost parallel with Vein I, sometimes concealed by it, apically; costa without series of distinct spines (Pl. **34,** 10).....CHYROMYIDAE (p. 211)
29. Second basal and discal cell confluent; anal cell absent; *sc* and Vein 6 weak or absent; ocellar triangle very large, plate-like (Pl. **39,** 6–11).......CHLOROPIDAE (p. 239)

[1] In doubtful cases, the family is keyed out under both headings.
[2] See Glossary, p. 366.
[3] Except the rare *Borboropsis fulviceps* Strobl.

Second basal and discal cell usually separated; anal cell and Vein 6 always present 30

30. Vibrissae present 31

Vibrissae absent; if bristles at mouth-edge appear vibrissa-like, then only one fronto-orbital present; all upper fronto-orbitals upswept 32

31. Lower fronto-orbitals incurved; postverticals divergent (Pl. **38,** 12–14) .... AGROMYZIDAE (p. 235)

Lower fronto-orbitals absent; postverticals convergent or (rarely) absent; fore femora usually with ventral spine; 3rd antennal segment with long pubescence (Pl. **35,** 5) ANTHOMYZIDAE (p. 215)

32. Mesopleural and sternopleural bristle present; ocellar triangle small; second basal and anal cells small; Vein 2 long, convergent towards Vein 3 apically (Pl. **35,** 1) OPOMYZIDAE (p. 214)

Mesopleural and sternopleural bristle absent; ocellar triangle large; second basal and anal cell long and wide; wings with cross-fold across basal part (Pl. **31,** 2, 3: **32,** 1, 2) PSILIDAE (p. 197)

33. Lower fronto-orbitals incurved 34

No incurved lower fronto-orbitals 36

34. Costa interrupted once, near apex of Vein 1; Vein 6 present, even if short 35

Costa interrupted twice, once near humeral cross-vein and once near apex of Vein 1; Vein 6 only perceptible as shadow (Pl. **39,** 1, 3, 4) .... MILICHIIDAE (p. 237)

35. Tibiae with dorsal preapical bristle; abdomen with five visible segments before genitalia; ovipositor retractile (Pl. **39,** 2) .... ODINIIDAE (p. 238)

Tibiae without dorsal preapical bristle; abdomen with six visible segments before genitalia; seventh segment in ♀ consitutes non-retractile sheath of ovipositor (Pl. **38,** 12–14) AGROMYZIDAE (p. 235)

36. Hind metatarsus shortened and dilated; apical section of Vein 5, or Veins 4 and 5, often curtailed or not pigmented (Pl. **38,** 1–4) SPHAEROCERIDAE (p. 223)

Hind metatarsus not shortened or thickened 37

37. Second basal and anal cells, and Vein 6 absent 38

At least anal cell, and usually Vein 6 present 41

38. *Sc* rudimentary 39

Sc distinct, at least basally 40

39. Ocellar triangle very large, plate-like; no strong frontal bristles; costa interrupted only at apex of *sc* (Pl. **39,** 6–11) CHLOROPIDAE (p. 239)

Ocellar triangle not unusually large; normal frontal bristles present; costa broken twice, once at humeral cross-vein, once near apex of Vein I .....CAMILLIDAE (p. 232)

40. Costa distinctly broken twice; if arista plumose, hairs are above stem (Pl. **35**, 9–13) ....EPHYDRIDAE (p. 217)

Costa indistinctly broken near humeral cross-vein and apex of Vein I; Veins 3 and 4 convergent apically; arista plumose above or on both sides (Pl. **38**, 6)
ASTEIIDAE (p. 228)

41. Costa broken twice, once near humeral cross-vein and once near apex of Vein I 42

Costa broken once, near apex of Vein I.................43

42. Anterior orbital bristle inset and upswept, the second close to eye-margin and forwardly-curved (Pl. **38**, 10)
DIASTATIDAE (p. 232)

Orbitals more or less in line, or if not, then inner anterior bristle downswept (Pl. **38**, 8, 9) ......DROSOPHILIDAE (p. 228)

43. Ocellar triangle very large; antennae widely separated at base (Pl. **35**, 14, 15) ....CANACEIDAE (p. 222)

Ocellar triangle not extending more than halfway down frons; antennae not widely separated at base................44

44. *Sc* entire, visible throughout [1] 45

Sc weakly developed, visible basally, thence as faint fold confused with Vein I; postverticals small and convergent (Pl. **35**, 2, 3) ....TETHINIDAE (p. 217)

45. Postverticals long and divergent; scutellum exceptionally long or flattened, with two short projections bearing stout bristles
THYREOPHORIDAE (p. 207)

Postverticals convergent; scutellum normal (Pl. **34**, 10)
CHYROMYIDAE (p. 211)

[1] In Chyromyidae, the apical part of *sc* may be concealed by Vein I when regarded from directly dorsal aspect, but will be perceptible if the wing is regarded obliquely from the basal end.

CHAPTER XX

SECTION V. CYCLORRHAPHA SCHIZOPHORA. THE CALYPTERATES

The Calypterates are characterised by a frontal lunule, ptilinal suture, a cleft in the second antennal segment (except Gasterophilinae), a complete transverse suture on the dorsum, and posterior calli which are clearly differentiated from the remainder of the dorsum (Fig. 5). Squamae are often, but not invariably, large and concealing the halteres. Male eyes are frequently holoptic.

Limits of the families are not by any means clear-cut. Tachinidae (= Larvaevoridae) and Calliphoridae are separable on the character of a "fan" of hypopleural bristles (Pl. **46,** 6): the remaider, which consists of the Scatophagidae (= Cordiluridae) and Muscidae are not so readily dealt with, since the former have certain characters in common with the sub-families Fucellinae and Coenosiinae of the latter, and moreover with the Acalypterates. However, very broadly, they may be separated thus:

1. Thoracic squama as long as, or longer than, alar squama and never reduced to a very narrow strip; lower, outer corner of anal cell, formed by Vein 6 and cross-vein, never sharply-pointed; abdomen with not more than five visible segments MUSCIDAE (excl. Fucellinae)
   Thoracic squama usually vestigial or reduced to a narrow strip . . 2
2. Abdomen with four or five segments visible; female with crossed bristles on frontal stripe; wings with costal spine close to apex of *sc* . . . . MUSCIDAE (Fucellinae)
   Abdomen with at least five or six segments visible; female without crossed bristles on frontal stripe; wings without costal spine . . . . SCATOPHAGIDAE

The Scatophagidae have, in particular, certain characters which have presented a case for their inclusion in the Acalypterates, especially the reduced thoracic squamae, the number of abdominal segments visible, the dichoptic eyes of the males, and, in many cases, the presence of only one sternopleural bristle; in fact, they have often in the past been so included. Today, they are generally regarded as belonging to the Calypterates and forming a link between the "lower" and the "higher" Schizophora.

SCATOPHAGIDAE . . . Dung Flies, etc.
23 genera. About 50 species. Head more or less rounded in profile; at least four orbitals; ocellar triangle never pointed anteriorly; one or two large vibrissae, sometimes accompanied by smaller bristles posteriorly; proboscis in predatory species strong, labella armed with prestomal teeth for perforating integument of prey; palpi well developed; arista pubescent or plumose; wings long, often tinged, or clouded at cross-veins, in one case spotted with dark band along costal margin (*Ernoneura*); Veins 3 and 4 more or less divergent apically, or very slightly convergent (e.g. *Scatophaga scybalarium* L.); supplementary cross-veins between Veins 3 and 4 (*Pogonoia*) or recurrent veinlets from Vein 2 (*Ernoneura*); 2nd basal and anal cells complete; anal cell never sharply-pointed at lower, outer corner; Vein 6 may or may not reach margin; Vein 7 always in evidence, even if only as a fold; abdomen, often in profile, appearing long and flattened, and with five or six visible segments; male genitalia often very prominent. (Pl. **33**, 3, 4: **42**, 1–5.)

HABITS of both adults and larvae of Scatophagidae are greatly varied. Some species are predatory on other insects, some are dung-frequenters and feeders, or develop in decaying matter generally, while others are plant-feeders or leaf-miners; many types of habitat are affected, some species being apparently confined to the seashore. The larvae are of the generalised Cyclorrhaphous type (cf. Fig. 10F), usually somewhat rigid and devoid of external processes, excepting fleshy protuberances on the truncate anal segment.

The family is divided into five subfamilies here, though Collin[10] in his recent revision recognises only two.

CORDILURINAE.

The species of *Cordilura* are medium-sized, elongated, mostly black flies, with the femora and tibiae armed with very robust bristles; five pairs of dorsocentrals, one sternopleural, and well-developed orbitals are present; Vein 1 is hairy in the apical half, the palpi have a long apical bristle, the third antennal segment is short and the arista long-plumose, with the stem much thickened at the base; male genitalia are very prominent in profile. *C. pubera* L., which is about 8–11 mm. long, occurs commonly from April to September, mainly in damp or marshy places; it may be known from other species of the genus by its four scutellar bristles, together with a black spot on the underside of each of the reddish tarsal segments (Pl. **42,** 3). Another common species, *C. umbrosa* Loew, which is about the same size, has two scutellar bristles, the wings yellowish, often distinctly darkened at the apex, and the tibiae and tarsi yellowish-orange.

In *Parallelomma,* most of the characters applicable to *Cordilura* are present, but the stem of the arista is not greatly thickened basally and the bristles on the femora and tibiae are somewhat frail. *P. albipes* Fall. (Pl. **42,** 1) is variable in colour, very pale specimens not infrequently turning up; in its typical form, which is that illustrated, it is very striking, the dorsum and abdomen being brilliantly black, the frons pale yellow with black orbits and the pleura and legs pale, almost whitish; the dividing line of colour between the dorsum and pleura is carried on to the occiput. *P. albipes* is quite common on foliage of trees and bushes near water or in damp places from April to October.

Chylizosoma differs from *Parallelomma* in having only one instead of two humeral bristles; the larvae are said to mine the leaves of Herb Paris, Lily of the Valley and various species of the Orchid family. A single British species, *C. vittatum* Mg., represents the genus.

The little grey *Cnemopogon apicalis* Wied., about 4½–6 mm. long, has three sternopleurals, Vein 1 hairy, palps without a long apical bristle, two scutellars, and the third antennal segment with pointed lower anterior corner, while the femora are densely-haired (Pl. **42,** 4). The larvae have been found in the galls of *Lipara* species (Chloropidae, p. 241), in stems of the Common Reed (*Arundo*); and in the stem of a Dock (*Rumex*); one has been recorded from a caterpillar of the Fen Wainscot Moth, the immature stages of which develop in the Common Reed.

Amaurosoma species are little greyish flies about 4–6 mm. long, which are allied to *Cnemopogon*, but have Vein 1 bare and very small, narrow palps; we have usually taken the common *A. fasciatum* Mg. by sweeping over dead beech leaves.

SCATOMYZINAE.

The golden-yellow, furry-looking flies which assemble in large numbers on cow-pats are one of the most familiar sights in the country. Usually, they disperse rather reluctantly on the approach of the observer and reassemble rapidly after his departure. For long periods they may remain almost motionless, occasionally making a rapid hop of an inch or two. Sometimes the observer will, if he be watching carefully, notice a sudden flurry of movement, giving the impression that some discovery has been made, which has attracted the attention of several of the flies simultaneously. On closer investigation, in all probability it will be found that the cause of the flurry is a much less conspicuous, dusty grey-green fly of somewhat similar, but generally smaller, build, and without the furry appearance. This is the female of the Common Yellow Dung Fly, *Scatophaga* (=*Scopeuma*) *stercorarium* L., and by the time the watcher has located her, the flurry of golden-yellow males will be dispersing and she will be found to have acquired one of them as her mate. The paired couple will, as a rule, retire a little way from the cow-pat; later, the female will return to lay her eggs within it, and here the larvae will develop, pupation taking place in the earth below. The observer will probably come to the conclusion that the assembly of males on the cow-pats is primarily for the purpose of awaiting the females seeking dung for oviposition; a process which has a parallel in the swarming of Nematocera (p. 55), the "male territories" of Syrphidae (p. 168) and the assembly of *Heleomyza* males on fungi (p. 211).

Species of the genus *Scatophaga* are generally characterised by a single sternopleural, hairy pteropleura or dense, furry pubescence on the body, and absence of prothoracic bristles; the arista may be pubescent or plumose. The familiar golden-yellow male of *stercorarium* (Pl. **33,** 3) has a variety known as v. *merdarium* F., which is less hirsute and of a dull greyish or greenish-grey colour, more like the typical female; all the forms may be known from the other species of the genus by the black antennae, plumose arista and ten or more bristles on the hind tibia. Several of the other species are somewhat similar in appearance and habits. The association with excrement often results in the attachment of various mites to the bodies of the flies. *S. scybalarium* L. (Pl. **33,** 4) is a larger and rather rare species, which occurs in the south-west and is, in the male, of a handsome red-gold colour, quite distinct from the typical male of *stercorarium*, and moreover, has the wings deeply tinged with yellowish-brown. Most *Scatophaga* species are in evidence from very early until very late in the year and are predominantly carnivorous in habit, preying on small, soft-bodied insects, mainly Diptera; sometimes on the dung, and often in other situations, for example, on Blackthorn in the spring; later, on Hawthorn and later still, on Bramble, when prey are foregathering at the blossom; or at fruits or sweet exudations to which prey are attracted; or on foliage, where prey are likely to bask. Lists of the recorded prey of *Scatophaga* have been compiled by Hobby.[1] Interesting notes on the techniques employed by these flies are given by Parmenter [2] and Grensted; [3] it appears that very little, if any, motion on the part of the prey takes place once the proboscis of the *Scatophaga* has made contact and this is performed by a single stroke, without any attempt at trapping the victim by means of the legs. There appears to be no evidence to suggest that poison is injected, so that the almost immediate immobility of the victim must be otherwise accounted for; as a rule, the prey is first seized by the thorax near the neck, the juices being subsequently extracted from the head and abdomen in that order. In captivity, males will prey on females; the process of feeding takes about two to three minutes to complete. The life-history and habits of *stercorarium* have been studied by Cotterell.[4]

Structure of the mouthparts is interesting. Whereas in the Nematocera and Brachycera piercing mouthparts were in the form of rigid needle-like or blade-like organs, here we find a non-rigid retractile proboscis, with somewhat fleshy labella; surrounding the opening of the pharynx are arranged a number of processes known as prestomal teeth (Pl. **42**, 5). The labella, which are furnished with minute collecting channels for liquids, radiating from the aperture, and ingeniously designed to prevent ingress of anything more than minute particles of solid matter, can be withdrawn so as to expose the prestomal teeth for action and allow the entry of larger items of food directly into the aperture. A very full study of the mouthparts of a non-predatory Cyclorrhaphous fly, *Calliphora* (p. 282), has been made by Graham-Smith [5] and reference to this author or to Imms [6] will demonstrate the general principles on which the Cyclorrhaphous mouthparts operate; it will be seen that in this case the prestomal teeth are much less developed and the collecting channels, or pseudotracheae, much more. In certain predatory Muscids also, like *Scatophaga*, the prestomal teeth are highly-developed. In the bloodsucking Muscids (p. 301), the whole proboscis becomes more rigid and non-retractile, the pseudotracheae disappear and greatly developed prestomal teeth are borne on tiny labella, or the labella have disappeared and the teeth work like saws. Bletchly[9] compares the mouthparts of *Scatophaga* with *Calliphora*.

Some seashore species occur in the Scatomyzinae; for example, *Scatomyza litorea* Fall., which is about 5–6 mm. long, of similar character to *Scatophaga*, but with weak, though distinct prothoracic bristles; and *Ceratinostoma ostiorum* Hal., which is about 6–8 mm. long, with bare pteropleura, a strong mid-anterior bristle on the mid-femora, and palpi as long as the proboscis. These flies are of the dull grey or bluish-grey so often associated with seaweed or salt-flat haunting species.

NORELLINAE.

Represented by one genus, *Norellisoma*. Larvae of *N. spinimanum* Fall. are said to occur in the stems of a species of Dock (*Rumex*). Adults are recognisable by the double ventral row of bristles on the fore femora (Pl. **42,** 2) and prominent bristles on the tibiae; male genitalia are very outstanding.

HYDROMYZINAE.

Hydromyza livens F., a greenish or bluish-grey species about 6–8 mm. long, has one sternopleural, bare pteropleura and feeble orbitals; the larvae are said to mine the stems and leaves of the Yellow Water Lily.

CLEIGASTRINAE.

Larvae of *Cleigastra* (= *Clidogastra*) species are thought to be leaf-miners; the genus is represented on the British list by *C. nigrita* Fall., a shining-black fly about 5 mm. long, with three sternopleurals, Vein 1 bare, short third antennal segment, almost bare arista, and small palps.

The family is monographed by Sack;[7] much assistance may also be obtained from Séguy.[8]

LITERATURE CITED

1. Hobby, B. M. A List of the Prey of Dung Flies (Dipt. Cordyluridae). *Trans. ent. Soc. S. of Engl.* **7** (i): 35–9. 1931.
   Hobby, B. M. Prey of *Scatophaga stercoraria* L. (Dipt. Cordyluridae). *J. Ent. Soc. S. of Engl.* **1**: 106–10. 1933.
   Hobby, B. M. Notes on Predacious Anthomyiidae and Cordyluridae (Dipt.). *Ent. mon. Mag.* **70**: 185–90. 1934.
   Hobby, B. M. Predacious Diptera and their Prey. *J. Soc. Brit. Ent.* **1**: 35–9. 1934.
2. Parmenter, L. Predacious Flies and their Prey. *London Naturalist.* **1936**: 51–4. 1936.
   Parmenter, L. Further records of Predacious Flies and their Prey. *Ent. mon. Mag.* **77**: 154–5. 1941.

3. Grensted, L. W. The Technique of *Scatophaga lutaria* F. (Dipt. Cordyluridae) as a predator. *Ent. mon. Mag.* **80**: 296. 1944.

4. Cotterell, G. S. The life-history and habits of the yellow dung-fly (*Scatophaga stercoraria*); a possible blow-fly check. *Proc. Zool. Soc. Lond.* **1920**: 629–47. 1920.

5. Graham-Smith, G. S. Further observations on the Anatomy and Function of the Proboscis of the Blow Fly. *Parasitology.* **22**: 47–115. 1930.

6. Imms, A. D. A General Textbook of Entomology. Methuen, London. 1938.

7. Sack, P., in Lindner, E. Die Fliegen der Palaearktischen Region. **62a.** Cordyluridae. Schweizerbart, Stuttgart. 1937.

8. Séguy, E. Faune de France. **28.** Muscidae Acalypterae et Scatophagidae. Lechevalier, Paris. 1934.

9. Bletchly, J. D. The mouth-parts of the yellow dung-fly *Scopeuma* (= *Scatophaga*) *stercoraria* (L.) (Diptera, Cordyluridae). *Proc. Zool. Soc. Lond.*, **123**: 143–65. 1953.

10. Collin, J. E. A short synopsis of the British Scatophagidae (Diptera). *Trans. Soc. Brit. Ent.*, **13**: 37–56. 1958.

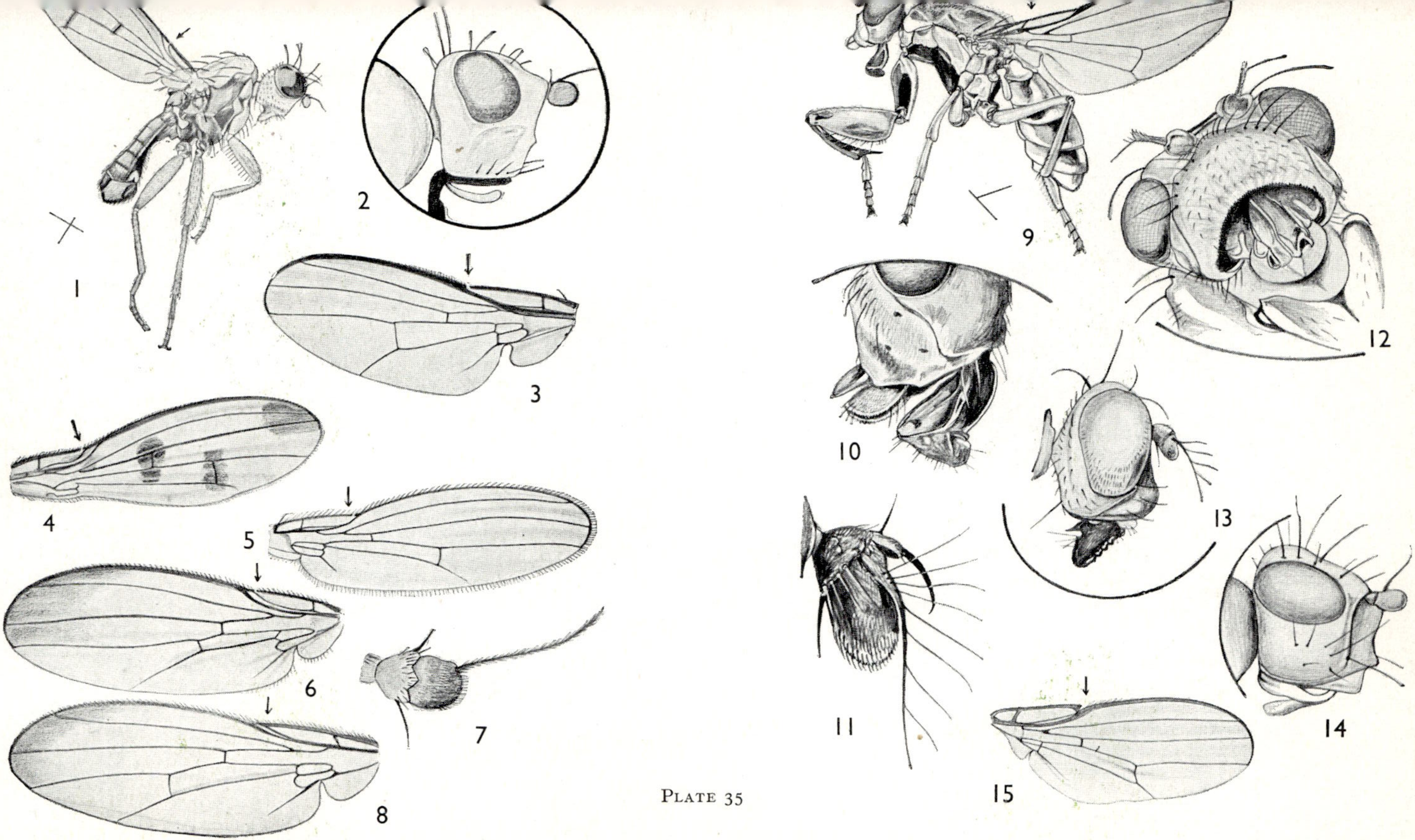

Plate 35

OPOMYZIDAE

1. *Opomyza germinationis* L., ♂, p. 214. 4. *Geomyza tripunctata* Fall., wing, p. 215.

TETHINIDAE 2. *Tethina*, head, p. 217. 3. *Tethina*, wing, p. 217.

ANTHOMYZIDAE 5. *Anthomyza gracilis* Fall., wing, p. 215.

CLUSIIDAE 6. *Clusiodes albimana* Mg., wing, p. 216. 7. *Clusiodes albimana* Mg., antenna, p. 216. 8. *Clusia flava* Mg., wing, p. 216.

EPHYDRIDAE

9. *Ochthera mantis* Deg., ♀, p. 219. 10. *Ochthera mantis* Deg., proboscis, p. 219. 11. *Notiphila uliginosa* Hal., antenna, p. 218. 12. *Ephydra riparia* Fall., head, p. 221. 13. *Hydrellia griseola* Fall., head, p. 221.

CANACEIDAE

14. *Canace nasica* Hal., head, p. 222. 15. *Canace nasica* Hal., wing, p. 222.

Plate 36

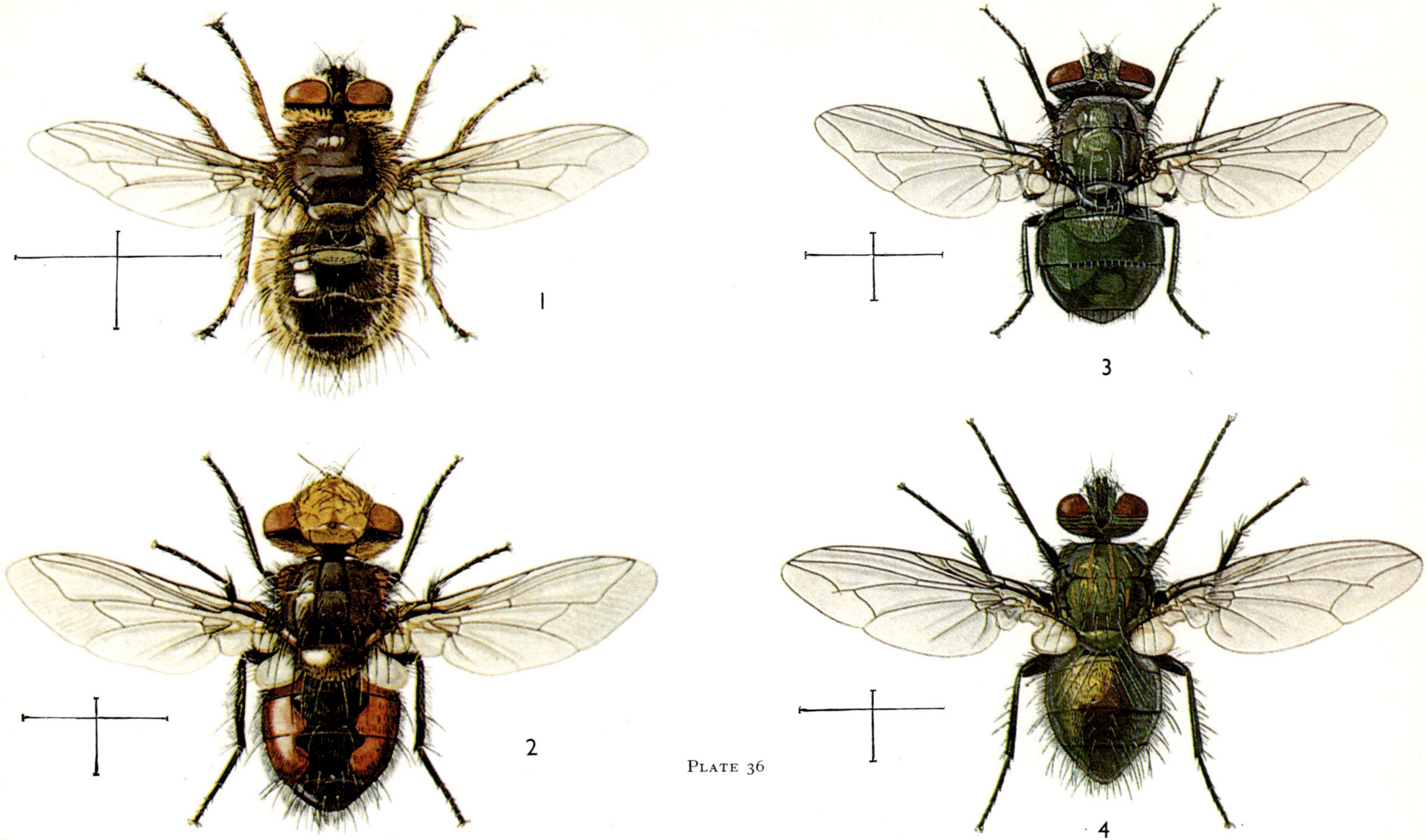

TACHINIDAE-I

1. *Servillia ursina* Mg., ♂, p. 266.

CALLIPHORIDAE-I

3. *Lucilia caesar* L., ♀, p. 283.

TACHINIDAE-II

CHAPTER XXI

TACHINIDAE (=LARVAEVORIDAE) Parasite Flies.

140 genera. Over 250 species. Distinguishable from all other Calypterates, except Calliphoridae, by "fan" of hypopleural bristles; and from them by well-developed postscutellum and the fact that the 2nd abdominal sternite (the apparent 1st) is partly or entirely covered by the edges of the corresponding tergite; arista bare, pubescent or plumose; ♂ eyes usually well separated, ♀ frons in some Phasiinae narrower than ♂; thoracic squama usually much larger than alar squama; Vein 4 more or less sharply bent toward Vein 3 at apex, often with appendix beyond the bend; abdominal bristles mostly well developed; abdominal sternites, in some cases, surrounded by visible membrane. (Pl. **36,** 1, 2, 4: **37,** 1, 2, 3: **40**: **41,** 2: **42,** 7, 8: **43,** 1, 2, 4–6: **46,** 1–6: **47,**4.)

THE larvae of Tachinidae are internal parasites in the living bodies of Arthropods, mainly insects, but in some cases, Spiders, Woodlice and Centipedes; the insect-hosts chosen are predominantly the larvae of Butterflies and Moths, but also include those of Wasps, Bees, Sawflies and Flies; or various stages of Beetles, Bugs, Grasshoppers and Earwigs. A list of the recorded hosts of this family and of Calliphoridae is furnished by Audcent.[1] Although the parasitic larvae derive their sustenance from the body-substances of the hosts, they do not usually cause fatal damage until they, themselves, are fully developed, when they either pupate in the host's skin, or bore their way out and pupate in the ground. They may, therefore, be justly described as the "Ichneumons" of the Fly world (cf. *Anthrax*, p. 111, Cyrtidae, p. 112, Pipunculidae, p. 152, Conopidae, p. 177).

In shape, these parasitic larvae differ somewhat from the generalised Cyclorrhaphous type, the anterior part being tapered very little or not at all, i.e. they are more or less evenly cylindrical; moreover, the segments are not so clearly defined. They are introduced into the bodies of their hosts by various means, all of which exhibit a high degree of special adaptation. From this aspect, Lundbeck,[2] following the work of Pantel, divides the family into ten groups, which for our present purpose, we may conveniently condense into four:

Group I.

Eggs laid on the skin, or attached to the hairs of the host, whence the larvae bore their way into the interior. The eggs may take some time to hatch, or may give rise to larvae immediately.

Group II.

Tiny eggs, often in very large numbers, laid on plants or scattered broadcast over them, where they will probably be eaten by their prospective hosts, usually caterpillars of Lepidoptera. The eggs hatch in the intestines of the hosts, whence the parasite-larvae migrate to the parts where they will feed.

Group III.

Eggs laid in places frequented, or likely to be frequented by suitable hosts, giving rise, almost immediately, to migratory larvae, which bore into any suitable creature which they encounter. These larvae are covered with hard plates, which, when the body is contracted, close tightly and form a case which prevents the tender, moist, larval body from drying up.

Group IV.

Skin of the host punctured by the female Larvaevorid, either by means of a special spine, or by the ovipositor itself, and the immediately-hatching eggs deposited within the host's body. Some species of Larvaevoridae are equipped, in the female, with a special structure which appears to be adapted for holding the host as well as for piercing.

Inside the body of the host, the life of the larvae gives rise to a respiratory problem akin to that of aquatic larvae. Some of them live free in the body-cavity, apparently deriving their oxygen from the blood by absorption, while others exist in this way only in their early stages. Others remain with their rear spiracles connected with the opening in the host's integument, through which their original entry was made, thus breathing outside air; this leaves their anterior part free to move in a prescribed ambit for feeding purposes. Yet others attach themselves, at least in their later stages, to a trachea of the host, again by their rear extremity, so as to be able to breathe some of the air intended for the oxygenation of the host's blood. In the two latter cases, a sheath or funnel is formed round the parasite larva, at least round its hind part, secreted by the host's tissues in response to irritation. Tachinidae, themselves, are sometimes subject to the attentions of hyperparasites, in the form of certain Hymenoptera, including Ichneumons and Chalcids; there are, moreover, flies which, although normally parasitic in the larvae of Lepidoptera, will sometimes adopt the Tachinid parasite itself as a host; so that the principal host, a Lepidopterous caterpillar, may be carrying within its body, simultaneously, a parasite within a parasite (cf. *Anthrax*, p. 111).

Adult Tachinidae are mostly arrayed in various hues of black, grey and brown and are not usually of very striking or handsome appearance, to which there are some notable exceptions. In the field, it is often possible to "place" them by their habits and attitudes; for example, many of them have a characteristic manner of resting, with the wings half-opened, i.e. at roughly a right-angle to one another (Pl. **46,** 5); or they have an equally characteristic, fussy, inquisitive sort of flight, rather like a busy little terrier questing to and fro. This is particularly noticeable in the females when they are seeking hosts for their future progeny. They are much varied in shape, ranging from the broad, almost globose, forms of some species of the sub-family Phasiinae (Pl. **40,** 2) at one extreme, to the narrow, elongated bodies and long legs of some of the Dexiinae (Pl. **37,** 1, 2), or Tachininae (Pl. **40,** 1), at the other.

Tachinidae and Calliphoridae can be separated from the rest of the Calypterates by the " fan " of hypopleural bristles (just in front of the rear thoracic spiracle, Pl. **46,** 6) ; Vein 4 is more or less sharply bent,* near its apex, towards Vein 3, which serves to separate them from all except the sub-family Muscinae (Muscidae) in which Vein 4 is either sharply bent, or curved (Pl. **44,** 3, 4); Calliphoridae may be shut off by the fact that the second abdominal sternite, or ventral plate (the apparent first) appears to be free and more or less overlaps the edges of the corresponding tergite, or dorsal plate, which are extended round to the ventral surface of the abdomen.

Classification within the family is somewhat complicated and not of a nature to be readily appreciated by, or to greatly interest, the field-naturalist; for this reason, the sub-families Tachininae, Goniinae, Maquartiinae† of van Emden[6] will be treated together. Since the family constitutes one of the youngest and most highly-developed of the order, and evolution has not reached the stage where clear-cut definition of limits is always possible, the number of genera is very large, often with only one or two species. It will not be possible, therefore, to do more here than to select a few representatives of the broad groups adopted.

TACHININAE, GONIINAE, MAQUARTIINAE

In this group, the arista is bare, or occasionally plumose; a presutural bristle is present; the abdominal sternites are partly visible, i.e. they are not entirely covered by the edges of the tergites, which do not meet; and the antennae are inserted above or at the middle of the eye.

In the species of *Gonia* (=*Salmacia*), a striking feature is the unusual shape of the head from a dorsal aspect, the frons and face being inflated and noticeably protruding (Pl. **36,** 2). The eyes are bare, and there are stout marginal and lateral bristles on the abdomen, while the arista is short, bare and

* But see Note 1 on p. 311.

† Larvaevorinae (inc. Salmaciinae and Phorocerinae) of first edition and Check List.

three-segmented, the basal segment being a mere button, the second long and thick, and the third a little shorter, coming suddenly to a sharp point at the apex. Two of the British species, *G. divisa* Mg. (Pl. **36,** 2) and *G. sicula* R-D. (=*fasciata* Mg.) (Pl. **46,** 5), are abroad very early in the year, in late March or early April, and may thus be missed by collectors who are not "up and doing" when they *think* the season or the weather may not be propitious. *Gonia* species will usually be seen flying low over the ground in suitable localities, such as heaths or commons, in a characteristic questing fashion, alighting now and then to bask in the early spring sunshine on dried remnants of last year's herbage or on bare patches of earth in sheltered places. *G. sicula* differs from *divisa* principally in having the abdomen completely blackish, very rarely with a reddish tinge at the sides, and with bands of whitish sheen, on the anterior part of the segments, which are most noticeable when viewed from the rear. Larvae are parasitic mainly in the larvae of Agrotid moths, such as the Turnip Moth, but are also recorded as bred from the Earth Humble-Bee (*Bombus*) and Mining Bees (*Anthophora*), which are themselves early spring insects. *Gonia* species belong to Reproductory Group II, the eggs being very small and numerous, and scattered broadcast. The method seems to be a very hazardous way of carrying on the species, but that it is effective is witnessed by the numbers of these flies to be seen.

Nemorilla floralis Fall. (Pl. **42,** 8) is a common fly, usually about 6–7 mm. long, although much smaller, and sometimes larger, specimens occur; it is of a blackish ground-colour, with a yellowish or golden sheen, the abdomen appearing tessellated, particularly as the fly moves about in sunlight. The eyes are covered with pale hairs, the third antennal segment is rather longer than the second, and the bare arista is thicker in the basal half than in the apical; the abdomen, in addition to stout marginals, bears discal bristles, i.e. pairs in the middle of the segments. The species is parasitic in the larvae of a wide range of Butterflies and Moths and belongs to Reproductory Group I.

Those who breed Lepidoptera are almost certain, sooner or later, to encounter the undistinguished-looking *Phryxe vulgaris* Fall., which is extremely catholic in its choice of hosts. It is a shining black fly, varying from about 6 mm. to 9 mm. in length, with a reddish-brown scutellum, and a broad band of whitish sheen on the anterior part of the last three segments of the abdomen, this band narrowing or disappearing altogether at certain angles of view. All bristles are stout; the frontals descend to the middle of the cheeks or below, while crossed apical scutellars and abdominal discals are present. The third segment of the black antennae is about four times as long as the second, the arista is tapered off after the basal part and bare, while the eyes are hairy. This species belongs to Reproductory Group I, the eggs hatching rapidly, while the larvae are of the type which acquire a connection with the air, other than the original entry-hole. Apparently *vulgaris* has two generations per year, and each host may contain one or more parasitic larvae, according to its size. When ready, the larvae escape by boring through the skin of the host and, as a rule, pupate in the earth, but those of the second generation may hibernate within the host, either as larvae or pupae.

Compsilura concinnata Bouché is about 7–8 mm. long, blackish with white and brownish shifting tessellations; the eyes are hairy, the frontal stripe dull black, the antennae black and the arista bare. In the male, the third antennal segment is about six times as long as the second, but in the female only about four times; the females, together with those of two other species in closely related genera, are remarkable in that the ventral surfaces of the abdominal segments are produced downward at the hind margin, so that, in profile, a saw-like effect is produced, which together with the curved, spine-like ovipositor is very distinctive. *C. concinnata* may be distinguished from the other two species referred to, by the absence of ocellar bristles ; it has been recorded from a greater variety of Lepidopterous hosts than any other British Tachinid. With other parasites, it has been imported into America in an attempt to combat destructive defoliating Lepidoptera, which have, themselves, been accidentally introduced, and have often flourished beyond all control in the absence of their own native parasites. Apparently, *concinnata* is now well established in the north-eastern States of the U.S.A. and in eastern Canada, where it is proving itself very useful in keeping down the Gypsy Moth.

The common species of the genus *Tachina* have a rather narrow, elongated body. *T. rustica* Mg. has black antennae, in which the third segment is about one and a half times as long as the second ; a bare arista, with the usual broad-button basal segment, a longer second segment, and the third thickened in approximately the basal third ; bare eyes ; and, in profile, the frons produced forward to the base of the antennae, from which the face recedes slightly, giving the front of the head a somewhat triangular appearance. Like *concinnata*, *Tachina* species, which belong to Reproductory Group I, are catholic in their choice of Lepidopterous hosts. *T. larvarum* L. has also been introduced into America as a biological control; it may be distinguished from *rustica* by a reddish scutellum instead of black, four postsutural dorsocentrals instead of three, and the absence of abdominal discals, these being present in *rustica*.

Species of *Siphona* (=*Bucentes*) (=*Crocuta*) are small brownish-grey flies, about 4–6 mm. long, often to be seen on various flowers of Compositae, such as Dandelions, Hawkweeds, Sow-thistles, etc., from May to October, in the characteristic right-angled wing attitude, and with their long, thin, doubly-bent proboscis probing for nectar (Pl. **43,** 6). *S. geniculata* Deg., the head of which is figured, is exceedingly common. The wings in this genus differ from nearly all other Larvaevoridae and Calliphoridae in that Vein 6 reaches the margin (but cf. *Hypoderma*, Pl. **47,** 3). Larvae of *Siphona* species are parasitic mainly in the larvae of Crane Flies (Tipulidae, p. 46), but are also recorded from Lepidopterous larvae; they belong to Reproductory Group I, with immediately-hatching eggs. A number of authors have recorded *S. cristata* F. from the aquatic larvae of *Tipula maxima* Poda (p. 47, and Pl. **1**). Roubaud [3] suggests that the egg is deposited at the moment when the Tipulid larva protrudes its rear part from the water in order to open the lobes covering the spiracular orifices. Upon submersion, the egg would be enclosed; since the *Siphona* larva is of the type which acquires a connection with the tracheae of the host, it would experience no more difficulty in respiration than the host itself, when submerged. The problem of emergence is solved by the fact that the fully-grown Tipulid larvae, as we have seen earlier, come out of the water to pupate in earth, at which point the *Siphona* larva can safely bore out and enter the earth for its own pupal stage.

Voria ruralis Fall. is a very common species with a most distinctive head (Pl. **43,** 4), the bristles being exceptionally stout and prominent. It is about 6½–10 mm. in length, black, with whitish or pale-bluish sheen, four black stripes on the dorsum, and shifting whitish sheen on the anterior part of the last three abdominal segments; a row of bristles extends the whole length of Vein 1 and about half-way along Vein 3. *Voria trepida* Mg. is very closely similar, but differs in the possession of abdominal discals. Both species are to be found on low vegetation in many kinds of location, particularly in and on the borders of woods; the autumn generation of *ruralis* is usually abundant. They belong to Reproductory Group I, with rapidly-hatching eggs, and are parasitic in a number of hosts, including the Red Admiral Butterfly, the Silver-Y, Burnished Brass, and Cabbage Moths, and the Pine Sawfly. As many as six or seven of the larvae of either species may be present in one host.

The genera *Ernestia* and *Meriania* comprise a series of

medium-sized to fairly large dark-greyish flies of superficially similar appearance, at least sufficiently so as to be confusing in the field. *Ernestia radicum* F. (Pl. **43,** 2) will be encountered, sometimes in considerable numbers on umbels and other flowers such as Ragwort, from July to September; it has a variable band of whitish sheen on the anterior part of the second and third abdominal segments, with, like the other species in this particular group, a tendency toward reddish patches at the sides. The antennae are long and black, the second segment being roughly as long as the third, if anything a little shorter; the arista is faintly pubescent, with a distinctly lengthened second segment and the third thickened for about half its length; bristles are strongly developed, the cheeks bare, eyes hairy and male genitalia fairly prominent. *E. radicum*, like its allies, belong to Reproductory Group III, with armoured, mobile larvae; it is recorded as parasitic upon the Peacock Butterfly, and the Dot, White Ermine and Cabbage Moths; the larvae belong to the respiratory group in which contact is maintained with the original entry-hole.

Ernestia rudis Fall. is about the same size, but the bristles on the ocellar triangle and the vertex are much thinner, almost hair-like; the reddish patches on the sides of the abdomen are, as a rule, a little brighter than in *radicum*, likewise the reddish apex of the scutellum; the antennae which are very similar in shape and segmental proportions are, however, either partly or wholly reddish. *E. rudis* is recorded as bred from the larvae of a number of Moths, including the Pine Beauty and Common Quaker.

Meriania puparum F. is closely similar to both of the preceding species, but has hairy cheeks, hair-like vertical bristles, reddish abdominal side-patches in the male only, the whitish bands of sheen on the abdomen broad and more sharply defined, and the wing-veins, especially the cross-veins, more or less brown-clouded. This is considered to be a rare species, although as we have suggested elsewhere, it is probable that many of these so-called rarities have earned their reputation by appearing very early or very late in the year, when all but the most enthusiastic collectors and observers have retired indoors; or by having a very restricted flight-period, so that week-end collectors, for instance, may visit a suitable locality at roughly the same date for years without, perhaps, being aware of its existence there. *M. puparum* is a case in point; it appears not only to be an early spring species, but to have a fairly short flight-period as well; we have found it in fair numbers in early April, in Surrey not far from London, and specimens have been taken in the precise locality by ourselves and friends, on almost the same dates in different years, but not at any other time. There is no doubt that the species confines its appearance to late March or early April, according to whether or not the season is advanced; males are very much more in evidence than the females and will be found sunning themselves on long grasses and sometimes on tree-trunks. *M. puparum* could easily be passed over for the common *radicum*, *rudis* or allied species, were it not remembered that the latter are hardly likely to be abroad so early. *M. puparum* has been recorded from the Pine Beauty Moth and probably has a very similar life-history to the allied species referred to.

Another early spring species is *Servillia ursina* Mg. (Pl. **36,** 1), which is of bigger build and looks rather like a brown Humble-Bee, on account of its dense hairiness. The antennae are considerably elongated, with the second segment longer than the third, a character shared by only a few British species of the family; the wings are usually somewhat darkened and the veins indistinctly clouded. Through the thick " pile " on the abdomen, bands of pale sheen can be seen on the anterior part of the segments, especially when viewed obliquely from the rear; the male genitalia are exceptionally prominent (Fig. 9D). *S. lurida* F. is similar, but smaller, about 11–12 mm. long, with reddish patches on the sides of the abdomen, usually on the first and second segments, but sometimes on the third. The wings are not darkened to the same extent as in *ursina*, and the veins are narrowly, but perhaps a little more distinctly clouded. *S. lurida* is not so early in appearance and may be encountered as late as June, whereas *ursina* is finished by April. Neither species is common, but both may be found in old forest areas in the south, such as the New Forest. *S. lurida* has been recorded from the Spurge Hawk, Sycamore, Common Quaker and Mullein Moths.

Species of *Echinomya* include some of the largest and most handsome flies in the family. Like the *Servillia* species, they have the second antennal segment longer than the third; their size, rounded abdomens and robust appearance make them not difficult to recognise in the field. *E. fera* L. (Pl. **37,** 3) is a very common fly, sometimes abundant on water-mint in late summer or early autumn around the borders of ponds and streams. It belongs to Reproductory Group III and is parasitic in Lepidopterous larvae; recorded hosts include the Large Cabbage White Butterfly, the Black Arches, Autumnal Rustic, the Broom and the Pine Beauty Moths. Larvae of this and the two succeeding species are of the type which remain attached to the original entry-hole. *E. ferox* Pz. is very similar, but of a darker aspect generally, due to the black abdominal stripe being much more pronounced, and the legs mostly black, instead of reddish as in *fera*. Recorded mainly from Devon and Cornwall, it is generally considered to be rather uncommon, although we have taken it in a Surrey park near London; it has been bred from the Dark Arches Moth. *E. grossa* L. (Pl. **43,** 1) is not likely to be confused with any other British fly on account of its size, build and colouring; it is the largest British Tachinid, and its black, bristly appearance, with bright yellow head, and extremely robust build are quite unique; the third antennal segment is only about half the length of the second. *E. grossa* is not uncommon on the flowers of Umbelliferae and Compositae around woods in the south-west of England, particularly Devon and Cornwall; it is recorded as bred from the Pine Hawk, Brown-tail, Oak Eggar and Fox Moths.

Somewhat resembling *fera*, but smaller, about 9–10 mm. long, with a less rotund abdomen, but similar reddish colour and dark median stripe, is *Linnaemya* (= *Micropalpus*) *vulpina* Fall.; the thorax, however, is greyish with four black stripes and the abdomen has a patchy, shifting golden sheen; the eyes are hairy and the third antennal segment is about one and a half times as long as the second. *L. vulpina* is fairly common in July and August, especially on brambles, and is said to be parasitic in the larvae of the Pine Hawk and Lunar Hornet Clearwing Moths, among others. The larvae are of the armoured mobile type in Reproductory Group III.

Eriothrix rufomaculatus Deg. (Pl. **40,** 1) is another red-and-black species, but of a long, narrow build, and is very common indeed from late June to early September, perhaps one of the most common species in the family; it is, therefore, all the more surprising that very little appears to be known of its life-history and hosts, although it is said to belong to Reproductory Group III.

Thelaira leucozona Pz. (Pl. **42,** 7) is long and narrow with long legs, rather like some of the species of the next sub-family, Dexiinae; the impression is heightened on closer examination by the medium-plumose arista, an exceptional feature in the group now under review. *T. leucozona* is black, with conspicuous, broad shifting grey or whitish anterior bands of sheen on the second, third and fourth abdominal segments; the bands are interrupted in the middle giving the impression of a dark median line; there may also be suggestions of translucent reddish colour on the sides of the abdomen towards the base. The head, in profile, is characteristically semicircular, unlike most Larvaevorids. *T. leucozona* is quite common on low, thick vegetation, especially near water, throughout the summer; it belongs to Reproductory Group I, with rapidly-hatching eggs; recorded hosts include the Poplar Hawk, Elephant Hawk, Small Elephant Hawk, Grass Eggar, Fox, White Ermine, Buff Ermine, Garden Tiger, Dot, and Mullein Moths.

The brilliant green *Gymnocheta* (= *Gymnochaeta*) *viridis* Fall. (Pl. **36,** 4) has hairy eyes, bristly abdomen and legs, a bare arista and an appendix to Vein 4 (i.e. a prolongation beyond the sharp upward bend). It may be distinguished from the other Green Bottles as indicated on p. 283; moreover, its general build is typically Tachinid, rather than Calliphorid or Muscid. We possess a unique male which is deep prussian-blue in colour. *G. viridis* is common in spring, from as early as the latter end of March, when the males are frequently to be seen basking, head downward, on tree-trunks or fences; the females are more likely to be found on low vegetation or early flowers. The species belongs to Reproductory Group III, and recorded hosts include the Black Arches and Small Dotted Buff Moths.

DEXIINAE.

In this sub-family, the species mostly have narrow elongated bodies, long legs, the arista long-plumose * right to the tip, and deep antennal depressions on the face, with a resultant ridge or "keel" between them; the antennae are inserted below the middle of the eye and no presutural bristle is present; the abdominal sternites are concealed by the edges of the tergites, which meet.

Dexiosoma caninum F. (Pl. **37,** 1) is a typical representative, the males of which will often be encountered on bracken or low vegetation in woods from July to September, especially in damp situations. It is recorded as parasitic in the larvae of Cockchafer beetles; the life-history is probably similar to that of a closely-related genus, *Dexia*, which belongs to Reproduc-

* Except in *Trixa*, where it is practically bare.

tory Group III; here the eggs are laid in the ground and the rapidly-hatching armoured larvae migrate in search of a larva of the Cockchafer or related beetles.

Examination of the trunks of pines in sandy areas in July and August may produce males of *Prosena* (=*Calirrhoe*) *siberita* F. (Pl. **37,** 2); these will, when disturbed, fly off with such a rapid dart as to make their disappearance seem almost uncanny, and almost invariably, if the observer stands quite still, they will reappear in almost the same spot, as if by magic. The species, particularly the female, is also to be found on flowers in the same sort of locality, Field Scabious often being favoured; it is rather more robust than *caninum*, the males having a prominent dark stripe on the middle of the abdomen, as figured, while the females are generally darker, with an almost unicolorous grey abdomen. Both have an unmistakably long proboscis, about as long as the body of the fly, which is unlike the form we saw in *Siphona* (cf. Pl. **43,** 6) in that it is not bent, or geniculate, in the middle. Like *Dexia*, *siberita* belongs to Reproductory Group III.

The two British species of the genus *Trixa* * are of very different appearance, rather robust and somewhat reminiscent of the *radicum-rudis-puparum* series, in the preceding section of this chapter, but distinguishable at once, under a lens, by the antennae, which are very short, about half as long as the face, and entirely orange-yellow. *T. caerulescens* Mg. (= *alpina* Mg.) (Pl. **41,** 2) is generally considered somewhat rare; we have, however, taken it fairly frequently in a number of locations in Herts, Surrey and Sussex. The sexes differ a good deal in appearance, the male having the sides of the abdomen reddish and an overall dark-reddish or brownish hue, while the female has no reddish side-patches and an overall pale-greyish or bluish-grey colour. It will be noticed that, in spite of the yellowish tinge of the basal and anterior parts of the wing, the cross-veins are not clouded, whereas in the rather

* See note on p. 269.

more common *T. oestroidea* R-D. they are, thus giving the wing a spotted appearance. Both species differ from the other Dexiinae we have considered in having a bare arista. The life-history is obscure; *caerulescens* is said to deposit living larvae on dung, where they parasitise beetle larvae, while *oestroidea* is recorded as depositing larvae under grass, probably for a similar purpose.

PHASIINAE.

Some of the members of this sub-family may be recognised fairly easily by virtue of their unusual and distinctive shape; most of them have relatively few and weakly-developed bristles; the abdominal sternites are separated from the tergites by a visible membrane, which more or less surrounds the sternites.

Alophora hemiptera F. (Pl. **40,** 3, 4) is a most unusual Dipteron, in shape and coloration and, particularly in the male, is more suggestive of a Bug (Hemiptera) than a fly. The species is liable to extreme variation in size and colour, even in minor details of structure, and provides another example of sexual dimorphism. In contrast to the normal male form illustrated, we possess specimens very little more than 6 mm. in length, with a very sombre appearance and narrower, almost clear wings as in the typical female; the dilated front tarsi of the typical males are also lacking. There is, as might be expected, a range of intermediate forms. The females of this, and closely-related species, are remarkable in having a narrower frons than the males; their wings are narrower, with a straighter costal margin, and almost clear, or at most merely yellow-tinted; the structure of the ovipositor is worthy of notice, consisting of a stout, rearward-pointing instrument, with a sort of sheath of similar shape (Pl. **46,** 3). As regards the wing-venation, the stalked cell formed by the apical sections of Veins 3 and 4 should be noted. *Alophora* is thought to be parasitic upon Hemiptera and Orthoptera.

Somewhat similar, but much smaller and with not so great a tendency to bright colouring, is *Alophora obesa* F. Here also the males are prone to considerable variation, the wings, in extreme forms, being much tinted or clouded with yellowish-brown or brownish, and having the costal margin distinctly curved outwards; the thorax and abdomen do not display any pale colouring, but the rich violet sheen, which is apparent on the dark parts of *hemiptera* is repeated to some extent. A variable whitish sheen occurs on the thorax of some specimens and is broken to leave four dark stripes, the middle pair being abbreviated at, or just beyond the suture (Pl. **46,** 4). Wing-venation is similar to that of the following species, *Alophora pusilla* Mg. (Pl. **46,** 1), except that the stalk of the sub-apical cell is shorter (cf. Pl. **46,** 2). *A. pusilla* is only about 3 mm. long, although larger specimens occur, and one of the smallest species in the family; it does not display the same tendency to variation in colouring or shape, even between the sexes, as in the preceding two species; the wings are clear, the dorsum dull black, and the abdomen dark grey with indistinct darker markings; as before, however, the female frons is narrower than that of the male. All three species belong to Reproductory Group IV.

Gymnosoma (=*Rhodogyne*) *rotundatum* L. (Pl. **40,** 2) is generally considered to be rare, although we have found it in fair numbers in mid-July, on the flowers of Cow-Parsnip, in more than one locality in Surrey, not very far from London. It is said to be parasitic on adult Pentatomid bugs, and belongs to Reproductory Group I, the larvae being of the type which ultimately acquires a connection with the tracheae of the host.

The odd-looking females of the genus *Weberia* have the apical segments of the abdomen curved downwards, then forward, the end-segment being much elongated and bearing a kind of claw-like process at the tip (Pl. **43,** 5). The two British species are said to be uncommon; we have swept *pseudo-funesta* from the grass verge of a Hertfordshire lane. They

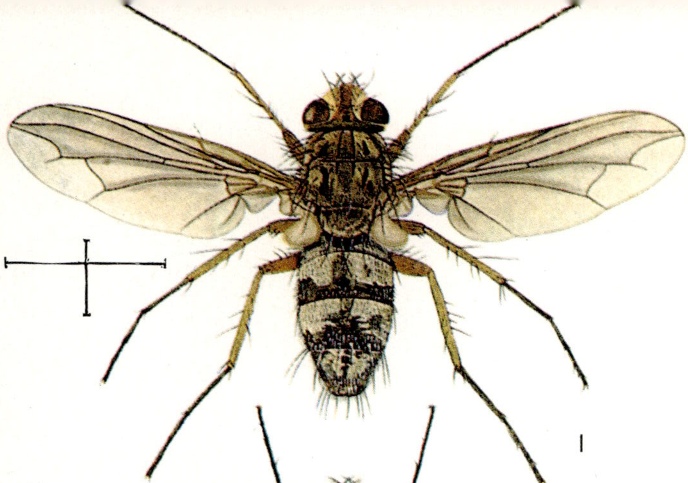

1

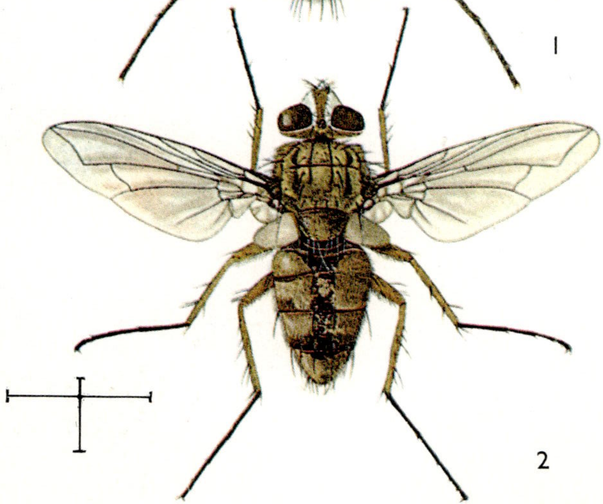

2

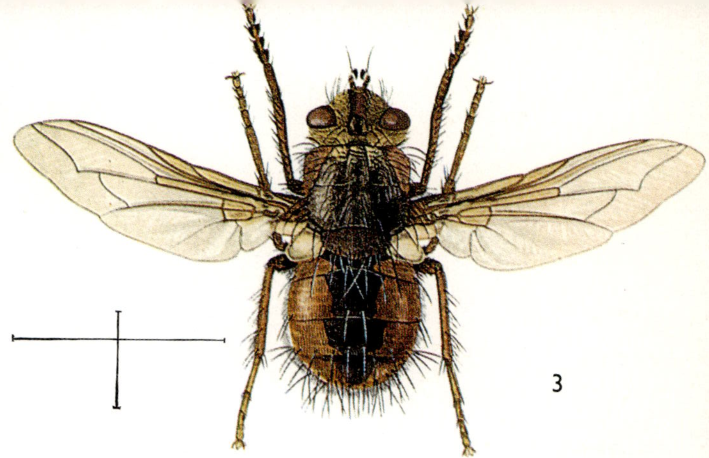

3

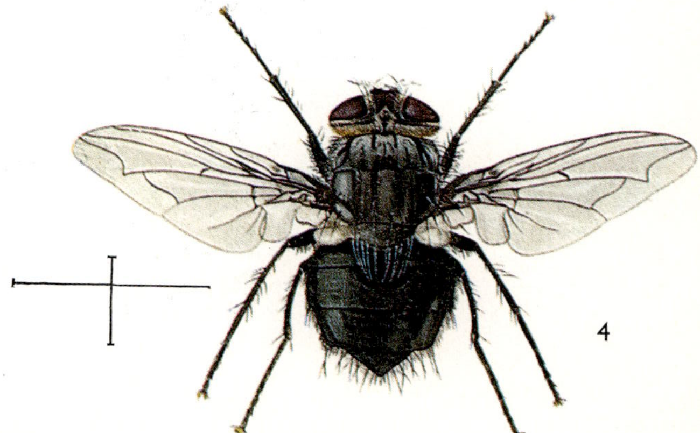

4

PLATE 37

TACHINIDAE-III

1. *Dexiosoma caninum* F., ♀, p. 269.
2. *Prosena* (=*Calirrhoe*) *siberita* F., ♂, p. 270.

TACHINIDAE-IV

3. *Echinomya fera* L., ♀, p. 267.

CALLIPHORIDAE-II

4. *Calliphora vomitoria* L., ♀, p. 280.

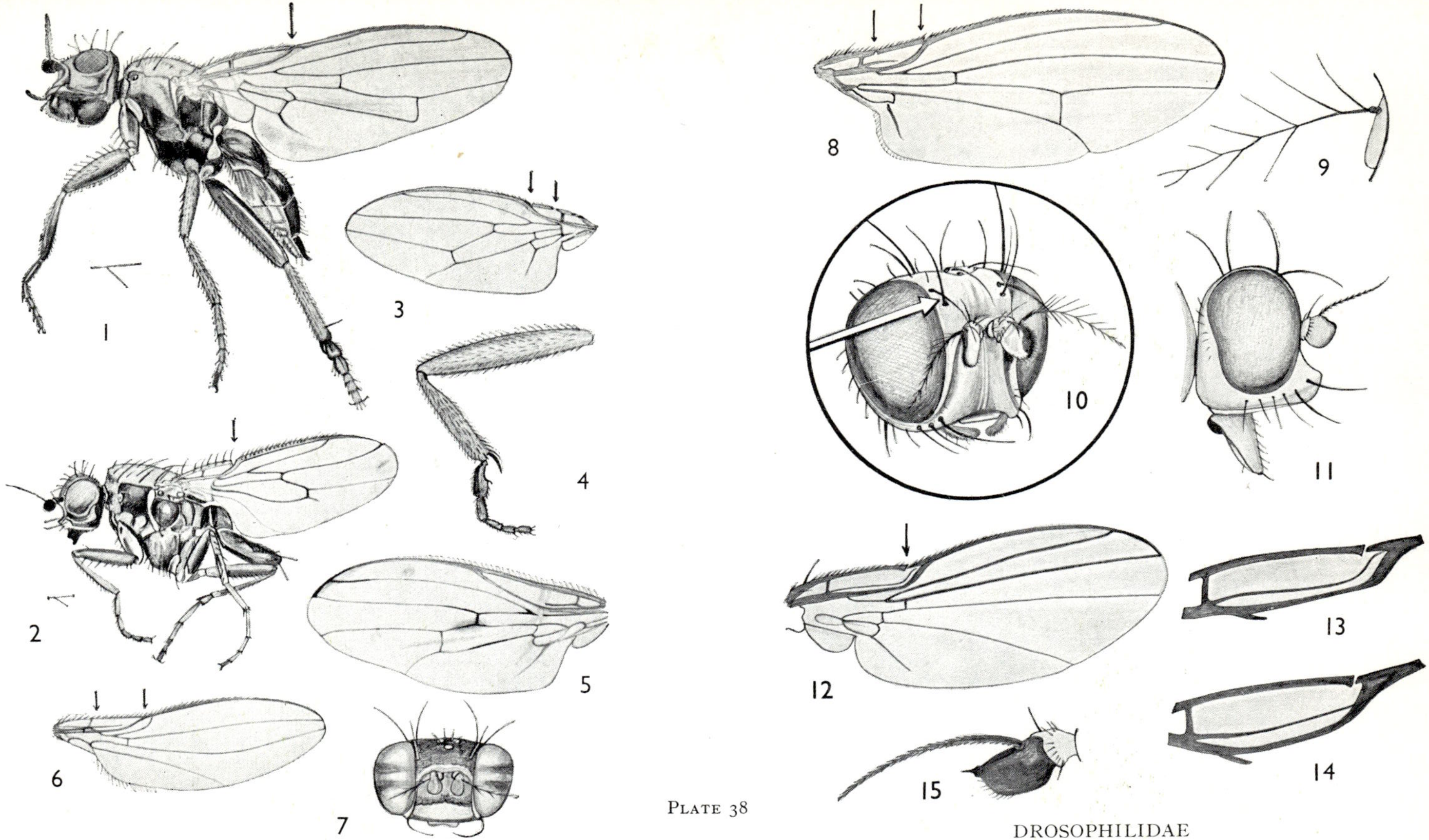

PLATE 38

SPHAEROCERIDAE (=BORBORIDAE)
1. *Trichiaspis* (=*Borborus*) *similis* Collin, ♀, p. 225. 2. *Paracollinella fontinalis* Fall., ♀, p. 225. 3. *Sphaerocera curvipes* Latr., wing, p. 225. 4. *Sphaerocera curvipes* Latr., hind leg, p. 225.

PERISCELIDAE 5. *Periscelis winnertzi* Egg., wing, p. 227.

ASTEIIDAE
6. [illegible] Mg., wing, p. 228. 7. *Asteia amoena* Mg., head, p. 228.

DROSOPHILIDAE
8. *Drosophila*, wing, p. 229. 9. *Drosophila*, arista, p. 229.

DIASTATIDAE
10. *Diastata*, head, p. 232.

AULACIGASTRIDAE
11. *Aulacigaster*, head, p. 233.

AGROMYZIDAE
12. *Phytomyza*, wing, p. 236. 13. *Phytomyza*, sub-costa, p. 236. 14. *Agromyza*, sub-costa, p. 236. 15. *Cerodontha spinicornis* Macq., antenna, p. 237.

are said to be parasitic in the adults of certain Carabid beetles and belong to Reproductory Group IV.

Oestrus, which is included in the Check List as belonging to this sub-family, is referred to Chap. XXIII, p. 290.

Keys to the British species of Tachinidae, together with Calliphoridae, are provided by Emden[6], Wainwright[4] and Day[5]; the last will probably be found easier to use. Lundbeck[2] will also undoubtedly be found useful, a good number of the British species being covered. In addition to Audcent[1], hosts of Tachinidae are given by Herting.[7]

LITERATURE CITED

1. Audcent, H. A Preliminary List of the Hosts of some British Tachinidae (Dipt). *Trans. Soc. Brit. Ent.* **8** (i): 1–42. 1942.
2. Lundbeck, W. Diptera Danica. Vol. VII. Wm. Wesley & Son, London. 1927.
3. Roubaud, E. Biologie larvaire et métamorphoses de *Siphona cristata* F. *Acad. Sci. (Paris) Compt. rend.* **142**: 1438–9. 1906.
4. Wainwright, C. J. The British Tachinidae (Diptera). *Trans. R. ent. Soc. Lond.* **76**: 139–248. 1928. *Ibid.* (Supplement). **80**: 405–24. 1932. *Ibid.* (Supplement). **90**: 411–48. 1940. *Ibid.* (Supplement) (Corrections). *Ibid.* (B) **10**: 70. 1941.
5. Day, C. D. British Tachinid Flies. *North Western Naturalist.* **21** (Supp.): 1–64. 1946. **22** (Supp.): 65–150. 1947.
6. Emden, F. I. van. Handbooks for the Identification of British Insects. Vol. X. Pt. 4(a). Tachinidae and Calliphoridae. Royal Entomological Society of London. 1954.
7. Herting, B. Biologie der westpaläarktischen Raupenfliegen. Dipt., Tachinidae. *Mon. Angew. Ent.*, **16**: 188 pp. 1960.

CHAPTER XXII

CALLIPHORIDAE . Blow-flies, Blue Bottles; Green Bottles; Flesh Flies, etc.

42 genera. Over 90 species. Distinguishable from all Calypterates except Tachinidae by "fan" of hypopleural bristles; and from them by the fact that the second abdominal sternite, or ventral plate (the apparent first) appears to be free and more or less overlaps the edges of the corresponding tergite, and by the absence or weak development of the postscutellum; arista bare, pubescent or plumose; male eyes separated or holoptic. (Pl. **36,** 3: **37,** 4: **41,** 1, 3, 4: **43,** 3, 7: **46,** 7, 8: **47,** 1–3.)

THIS family includes some of the most abundant and familiar of all the British flies, such as the ubiquitous Blow-flies or Blue Bottles (*Calliphora*), the Green Bottles (*Lucilia*) and the Grey Flesh Flies (*Sarcophaga*). Some species bear a superficial resemblance to certain members of the Muscidae, from which, however, they may be readily differentiated by the hypopleural bristles or by the wing-venation.

Life-histories are extremely varied. Some spend their larval phase in decaying matter, especially carrion; the larvae of others prey on the food-stores of burrowing bees and wasps; some infest the living bodies of toads and frogs; others suck the blood of nestling birds, or infest wounds or sores on the bodies of mammals; while occasionally some develop in the bodies of moribund or dead Lepidopterous larvae.

Three sub-families fall for consideration, namely, Rhinophorinae, Sarcophaginae and Calliphorinae; the sub-family Hypodermatinae together with the genus *Cephenomya* (Calliphorinae) being referred to Chapter XXIII, p. 289.

RHINOPHORINAE (now usually given family status).

A characteristic feature of this sub-family is the narrow

thoracic squamae, the inner margins of which curve away from the scutellum ; the fifth sternite of the abdomen in the males is cleft at the hind-margin ; and the posthumeral bristle is as high as, or higher than the presutural, i.e. as far or further from the pleura. None of the species are very distinctive in appearance ; they are mostly small, dark-greyish or black flies, sometimes with the wings heavily suffused or marked with dark colouring ; the males have the eyes somewhat approximated and the arista is pubescent or short-hairy. Rhinophorinae are said to be parasitic in woodlice and beetles ; the larvae of one species are recorded as developing in the egg-masses of a spider.

SARCOPHAGINAE (now usually given family status).

As in Rhinophorinae, the post-humeral bristle is as high as or higher than the presutural ; in this respect, both subfamilies differ from Calliphorinae. The thoracic squamae are large, with the inner margin lying alongside the scutellum ; the fifth sternite of the abdomen is not cleft at the hind-margin ; arista bare, short plumose in basal part or long-plumose ; male eyes always separated ; mostly grey flies with a tessellated or chequered appearance.

Familiarly known as Flesh Flies or Grey Flesh Flies, the species of the genus *Sarcophaga* develop in carrion and decaying animal matter generally ; the adults rarely come indoors. Specimens of a particular species may vary greatly in size according to environmental and nutritional factors affecting the life of the larva. The flies are greyish, with dark longitudinal stripes on the dorsum and a chequered or tessellated pattern on the abdomen, which shifts and changes according to the angle of view ; the arista is short-plumose on the basal part only ; in life, the eyes are bright red ; and the enlarged claws and pulvilli convey a distinctly " big-footed " impression (Pl. **47, 1**). They may quickly be distinguished from certain Muscids which they superficially resemble, especially as regards colour, e.g. *Polietes lardaria* F. (Pl. **48,** 1), by the wing-venation, in which Vein 4 curves sharply upward near its apex, and by their very different build. Determination of species is not easily accomplished without the aid of characters provided by the male genitalia; consequently females will not be satisfactorily named unless taken in copula. The species illustrated is the common *S. carnaria* L.

Larvae of *Sarcophaga* are of the usual Cyclorrhaphous type, elongated-conical, but with the hind extremity rounded, not truncate; the rear spiracles are sunk in a deep pit or crater, the edges of which bear a series of fleshy lobes, capable of being folded over so as to exclude liquids from the spiracles, should the larva become submerged; a little later, we shall see how the necessity for such an arrangement arises.

An interesting account of the life-history of *carnaria* is given by Fabre.[1] The female is viviparous, i.e. she deposits larvae instead of the usual eggs; her uterus is enlarged and adapted to form a kind of incubation chamber, wherein the comparatively large eggs hatch before emission. Larvae thus extruded are in their first stage of development; it is interesting to compare this with the Pupipara (Chap. XXV, p. 313) in which the larvae are retained until they are very near their full development, pupating almost immediately after emission. Having sought out a suitable medium upon which to deposit her young larvae, the female performs the operation with extreme rapidity; a mere touch with the tip of her abdomen is sufficient to release several larvae, which disperse themselves equally rapidly over the food substance. In order to reduce this to a suitable consistency for assimilation, the larvae disgorge a liquefying agent; this is often so effective, that they would ultimately be in danger of drowning in the revolting fluid, were it not for the adaptation of the spiracular cavity already referred to. The rosette of fleshy tubercles, when expanded, serves the purpose of a float, allowing the larva to breathe and simultaneously to feed, head downward in the liquor, when no solid support is available. Fully-developed larvae enter the earth to pupate; the emergence of the adult is a fascinating process. How does the tender crumpled insect, just emerged from the puparium (Pl. **46,** 8), accomplish the seemingly impossible task of penetrating the earth above it to win free and harden into the robust-looking fly (Pl. **47,** 1) we are accustomed to see? In the Introductory Chapter (p. 18), we touched upon the horseshoe-shaped groove or ptilinal suture of the Schizophora, and noticed that it indicates the site of the ptilinum, or inflated frontal sac of the newly-emerged adult; the fly is capable of inflating and deflating this sac by alternate contraction and expansion of the body and consequent expulsion and intake of blood from the body-cavity. During the escape from the puparium and the ascent through the covering earth, this process is going on continuously; the ptilinum forces off the end of the puparium, and the body which has simultaneously become contracted, is rendered free to move, little by little, until the limp, crumpled wings, legs and body are completely withdrawn. Now the ptilinum is used as a sort of pneumatic ram to displace the earth above the insect; the grains as they fall, are passed downwards by the legs which are stretched, like the wings, alongside the body, and furnish support for the next stroke of the ram. The process is quite rapid in dry soil of a loose texture; according to Fabre, a depth of six inches is covered in less than fifteen minutes. On arrival in the open air, the fly cleans itself with its legs, brushes the ptilinum with its front tarsi and permanently withdraws it; the wings, which are still crumpled, with a sort of notch in the middle, are cleaned and rapidly expand to their proper shape and harden. Pl. **46,** 7 shows the appearance of the ptilinum when expanded; it will be seen that the whole of the face falls forward as a sort of flap, hinged, as it were, near the mouth-edge; the frontal stripe is also expanded; when the ptilinum is withdrawn, the face is drawn back into the normal position, and the frontal stripe contracts; all that remains to denote the erstwhile location of the ptilinum is the horseshoe-shaped suture. Once this process is understood, there should be no difficulty in recognising Schizophora, by failure to appreciate the presence of the suture. It is, perhaps, worthy of mention that the actual subjects of the drawings referred to were bred from pupae found in the debris at the bottom of a Black Redstart's nest, on a bombed site in the heart of the City of London. The larvae had evidently developed in the bodies of two nestlings, the remains of which were found trampled into the matted mass of plant stems, feathers and excrement forming the floor of the nest.

A much smaller fly, about 6–8 mm. long, of very similar general appearance, with the exception of a pubescent, not a plumose, arista, is *Brachicoma devia* Fall. This species is associated with Humble-Bees, particularly the Common Carder Bee, *Bombus agrorum* F. According to Cumber,[2] the larviparous females enter the nest and deposit their larvae, which are found wandering over the eggs and succeeding stages of the host up to the early pupal stage; they attack and enter the pre-pupa, emerging while themselves still in the larval state. There are two generations, the early one pupating in the nest-material, and the late one overwintering as a pupa, outside.

The larvae of *Miltogramma* species also live on stored honey and pollen, in this case in the nests of Mining Bees; or on the paralysed caterpillars, flies, beetles and other insects brought home by fossorial wasps and buried by them in a little earthen cell for the purpose of feeding their own larvae. There are two British species, *M. punctatum* Mg. and *M. germari* Mg., both of which are obscure little insects, about 5–9 mm. long, and extremely agile on the wing, or in running over the sandy ground near the homes of their hosts. When this happens to be a Sand Wasp or Digger Wasp, the female fly " shadows " it as it carries its prey back to its cell and, seizing a suitable moment, such as when the wasp turns to enter, darts in, lays an egg or two rapidly on the prey and retires in good order. The wasp, all unsuspecting, solemnly packs the prey in the cell, deposits its own egg upon it, and seals up the cell. When the host is a bee, the fly enters the nest to lay its eggs, and, as in the case of the wasp, the fly-larvae live on the food intended for the host-larva ; since they eat more and grow faster, the host-larva ultimately dies of starvation, an ironical twist of the usual relations between wasp and fly. Fabre [1] gives an account of a *Miltogramma* species in France, which seems to overawe and intimidate a large wasp whose normal prey consists of very large and active flies, such as Tabanidae, and might therefore reasonably be expected to make short work of the *Miltogramma*. This wasp, which does not sting and paralyse its victims, but brings fresh prey periodically back to the nest, actually feeds the fly-larvae, at the expense of its own.

Closely allied to *Miltogramma*, with very similar habits and locations, but having a rather more striking appearance, are the species of *Metopia*. There are two on the British list ; *M. argyrocephala* Mg. and *M. campestris* Fall., both about 5–8 mm. long. The male of *argyrocephala* makes a fine sight in bright sunlight, either in its fussy, gyratory flight, or running on broad leaves or on the ground, when it progresses by little jerky leaps, or in rapid circles, its silvery orbits winking and gleaming in the sun's rays like a heliograph. Sometimes the males congregate in small swarms, which make a striking spectacle. The female, which lacks the silvery orbits, also has a fussy, agitated sort of flight, progressing rapidly in tight circles, close to the ground, over sandy banks and similar places in quest of the burrows of her hosts. On close examination the frons of the male is seen to be almost obliterated by the orbits, which are very broad and appear to be partly fused together near the bases of the antennae (Pl. **43,** 3) ; it is the front part which gives the shining, silvery reflections, while the posterior half of the orbits is greyish-black. Both sexes of *campestris* are not unlike the female of *argyrocephala* and lack the silvery orbits. The females of both species are thought to be viviparous and associated with Mining Bees (*Andrena*).

Macronichia ungulans Pand. (Pl. **47,** 2) is worthy of note, on account of its unusual appearance, particularly as regards the female genitalia. It is considered rare, but we have taken it in a small park in a built-up North London suburb. *M. ungulans* has been recorded as bred from logs, together with a large species of Digger Wasp, which affects stumps and posts sufficiently decayed or soft to be bored for cells, instead of constructing the more normal earthen cell. This suggests the possibility that its habits may be similar to those of *Metopia* and *Miltogramma*.

For the identification of the British species of Sarcophaginae, see Patton and Wainwright,[3] van Emden,[12] Day and Fonseca.[13]

CALLIPHORINAE.

Posthumeral bristle lower (nearer the pleura) than the presutural; arista long-plumose; male eyes holoptic or very closely approximated; bristles on dorsum of abdomen generally weak; often metallic blue or green flies; larvae of the generalised Cyclorrhaphous type.

The females of *Calliphora erythrocephala* Mg. and its "double", *C. vomitoria* L. (Pl. **37,** 4), are the common Blue Bottles or Blow-flies so detested by the housewife and the cook; the heavy buzzing of the females as they career clumsily round the room portends the "blowing" or depositing of eggs upon any flesh-food left unprotected from their attentions. Unlike most flies, they are not deterred by the cool darkness of pantries or cellars, and their uncanny instinct enables them to locate meat in spite of extraordinary obstacles. Outdoors, they chiefly frequent decaying animal matter and excrement. Each female is said to be capable of laying up to 600 eggs;

these are creamy-white, about $1\frac{1}{2}$ mm. long, and are laid in clusters in the folds and creases of the meat, indoors; or outdoors, often on the exposed flesh of dead animals or their orifices, such as the eye-sockets, nostrils and genital or excretory openings. Sometimes *Calliphora* species, like those of *Lucilia* (p. 283), lay their eggs on the wool of sheep near sores, or open wounds, or the genital openings of ewes, which the larvae subsequently enter, causing great irritation and lost of condition. Such infestations, however, are always of a secondary nature, the initial attraction apparently being pus or offensive mucous secretions. Under normal conditions, the eggs take about a day to hatch; the resulting larvae are the "gentles", much prized by anglers as bait. The larval stage lasts, in favourable circumstances, about a week, by which time the larva may be as much as 18 mm. long. Pupation normally takes place in the ground, under or near the larval habitat; the pupae are dull red-brown, and short-cylindrical in shape with rounded ends, after the usual Cyclorrhaphous type (cf. Fig. 11F). Adult flies emerge about a fortnight later. In the males, the eyes are closely approximated, the orbits almost touching. The abdomen has whitish dusting upon it, and shows shifting reflections of lighter blue; there is a complete absence of strong abdominal bristles, but there are some weak ones at the sides of the first and second segments, and marginals on the third and fourth; the lower (thoracic) squamae are dark with light edges and have hairs on the disc.

Males are not interested, in the same way as the females, in the attractions of flesh; they are much more likely to be found on flowers, especially umbels, sipping nectar; or basking on broad leaves; or on bare earth or paths in the sun; their occasional presence on carrion is probably due to their seeking the females for mating in places where these are likely to come for oviposition (cf. *Scatophaga*, p. 252).

C. vomitoria is less common than *erythrocephala*, but sufficiently so to be a nuisance in places. It may be known by the fact that it has black jowls with red hairs, whereas *erythrocephala* has reddish jowls with black hairs. The life-history and structure of Blow-flies has been exhaustively studied by Lowne [4] and the mouthparts, in particular, by Graham-Smith.[5]

Another interesting " Blue Bottle " is *Protocalliphora azurea* Fall. (Pl. **41,** 3), a rather smaller fly of a lighter, more brilliant, metallic blue, and lacking the white dusting and somewhat tessellated appearance of the *Calliphora* abdomen ; it differs, furthermore, in having a row of hairs on the posterior side of the thickened common stem of Veins 1, 2 and 3 ; even weaker abdominal bristles, and no hairs on the disc of the thoracic squamae. The pupae, along with those of certain other Diptera, are found in the nests of swallows, sparrows and other birds ; the larvae are said to attach themselves to the nestlings by means of a kind of suction-pad on the first segment, and to suck their blood, often causing their death.

Melinda caerulea Mg. is superficially somewhat similar to *azurea*, but rather smaller and darker ; the abdomen has white dusting and a dark middle line ; there are no hairs on the rear side of the common stem of Veins 1, 2 and 3 ; and the squamae are bare on the disc. The species is closely allied to the genus *Onesia* (p. 285), which are greenish-black, with lighter green reflections ; the larvae of *caerulea* are said to be internal parasites of snails.

As regards the Green Bottles, there are several common metallic-green flies, in more than one family, to which this descriptive name might well be applied ; their superficial resemblance to one another may well cause the collector, at first, to overlook some of them, by assuming that they are the same species. The following rough-and-ready table may be of assistance in sorting out the commoner species which are likely to be encountered :

1. Frons and jowls metallic-green ....*Orthellia*, Muscidae (Pl. **44**, 3)
   Frons and jowls not metallic-green........................2
2. Vein 4 bowed or curved toward Vein 3, not sharply bent....3
   Vein 4 sharply bent toward Vein 3 4
3. Vein 1 hairy at base; eyes hairy...*Dasyphora*, Muscidae (Pl. **44**, 4)
   Vein 1 bare; eyes bare ....*Pyrellia*, Muscidae
4. Eyes hairy; discals present on abdomen; abdominal bristles stout ....*Gymnocheta*, Tachinidae (Pl. **36**, 4)
   Eyes bare; discals absent; abdominal bristles generally weak..5
5. Face and jowls yellow ....*Cynomya*, Calliphoridae (Pl. **41**, 1)
   Face and jowls not yellow ....*Lucilia*, Calliphoridae (Pl. **36**, 3)

The Green Bottles belonging to the Calliphorinae are mainly the species of *Lucilia*, which are all very similar to one another, and distinguishable on rather critical characters; the females are most satisfactorily named when taken in copula. One of the most frequently seen is *L. caesar* L. (Pl. **36**, 3). *Lucilia* species are extremely variable in size, according to the conditions and degree of nutrition experienced by the larvae. Colour also seems to vary considerably with age; newly-emerged specimens are of a beautiful deep bluish-green colour with violet reflections; later this develops into emerald-green with darker green reflections; and finally into a dull coppery-green with red reflections. In this last state, it will nearly always be seen that the wings are worn and frayed. There is a faint whitish dusting over the base-colour of the body, which is more clearly seen from an oblique angle, especially at the front of the thorax; the abdomen is covered with numerous minute black hairs, and there are the usual Calliphorine weak lateral and marginal bristles.

Unlike *Calliphora*, *Lucilia* is seldom attracted indoors, preferring to work in the sunlight, and is always numerous on dead animals, birds and fish, excrement or any decomposing animal substance. An account of the life-history of *caesar* is given by Fabre.[1] The larvae require a moist environment, so the female endeavours to find a part of the carcass which is sheltered from the sun's rays, on which to deposit her eggs. To reduce the flesh to a consistency suitable for assimilation, the larvae emit liquids which have the effect of liquefying the tissues, in the same fashion as we saw in *Sarcophaga* (p. 276). The eggs hatch rapidly, usually in about twenty-four hours, and the larvae mature in two or three days under normal conditions. They are then about 11 mm. long and closely resemble those of *Calliphora* ; often they migrate considerable distances before burrowing in the ground to pupate, the adults emerging, as a rule, after about eight days. These flies are said sometimes to be a great source of trouble to sheep, the larvae entering open wounds, sores or occasionally the body-cavities ; *caesar* and *sericata* appear to be the chief culprits in this country, but the extent of their depredations is not so marked, or such a serious economic factor as it is in the great sheep-raising countries, such as Australia, where the damage attributed to them amounts to millions of pounds annually. Offensive humours or pus seem to be the initial attraction ; the larvae, once established, not only damage the tissues themselves but prepare the ground for other invaders, including other Diptera (*vide* Imms [6]). On the other hand, according to Steinhaus,[7] the healing propensities of some maggots infesting wounds has long been known ; during the war of 1914–18, it was found that soldiers' wounds, which had been unattended for as long as seven days and were infested with maggots, healed up rapidly without infection. Surgeons subsequently discovered that, when reared under sterile conditions, the larvae of *caesar, sericata* and allied flies could be put to very effective practical use in clearing away suppurating tissue and promoting the growth of healthy flesh. It seems, then, that where damage from infestations is imputed to Calliphorine larvae, it is probable that they may only form part of a chain of biological sequences.

L. bufonivora Moniez deposits its eggs on the eye-rims and nostrils of the Common Toad, and sometimes of Frogs, whilst alive; the larvae penetrate the body and death usually follows, though whether or not this is partly due to secondary causes is open to question. Keys to the British species of *Lucilia* are provided by Richards.[8]

Another brilliant, deep metallic blue-green fly, usually rather larger than the average *Lucilia*, is *Cynomya mortuorum* L. (Pl. **41,** 1); it differs, moreover, in having the face and jowls yellowish, and there is no white dusting on the body. Habits are similar, the larvae living in carrion and decaying animal matter generally. The species can hardly be called very common, especially in the south, but may be seen from time to time in situations similar to those affected by *Lucilia* and *Calliphora*, the males, particularly, being fond of sunning themselves on bare patches of earth, footpaths and on walls with a southern aspect.

Species of *Onesia* are somewhat smaller than the average *Lucilia* and of a blackish ground-colour with greenish or greenish-blue reflections and very slight white dusting. The cheeks and disc of the thoracic squama are hairy, while not very strong marginals are present on the abdomen. *O. aculeata* Pand. and *O. agilis* Mg. may be encountered not uncommonly from May to September on flowers, especially umbels, or on over-ripe blackberries. The larvae are similar to most of the Calliphorines in development-habits, but one species has been recorded as parasitic in earthworms; the females are said to be viviparous, incubation having already taken place within the parental body (cf. *Sarcophaga*, p. 276).

Pollenia rudis F. (Pl. **41,** 4) is superficially quite unlike the preceding metallic-coloured species, although on closer examination, it will be seen to be of very similar build. The thorax is clothed with dense, matted, crinkly golden, or occasionally silvery, hairs, particularly on the pleura, a characteristic, in varying degree, of the genus; the abdomen has shifting, irregular tessellate reflections and a dark middle line. Hibernating in the adult stage, it may be seen abroad very early in spring, sunning itself on tree-trunks, fences, telegraph-poles, etc. It is in evidence throughout the summer and until very late in the year, if the weather is not severe; and is often referred to as the " Cluster Fly " owing to its habit of sometimes congregating in clusters in crevices or corners of attics and box-rooms (cf. *Asteia*, p. 228, and Chloropidae, p. 241), in late autumn. At such times it is extremely sluggish and shows a great disinclination to escape from the room when disturbed. The larvae are said to be parasitic in earthworms; the eggs are laid in the earth in autumn and the young larva enters the host and hibernates. In the spring, it proceeds to move along the host's body, and, at one stage, perforates the body-wall so as to expose its rear spiracles to the air. Subsequently it eats its way back again, until very little of the host remains, when it pupates. *P. vespillo* F. is another very common species, with the abdomen shining black, but otherwise very similar to *rudis*.

Before leaving the Calliphorinae, passing reference should perhaps be made to *Stomorhina lunata* F., first recorded in England in 1901, when five specimens were taken; apart from two or three isolated records, it was not seen again until 1947, when there was a considerable number of captures, the last specimen in that year, to the best of our knowledge, being taken by us on October 19th in a park on the northern outskirts of London. Associated abroad with the Migratory and Desert Locusts, *lunata* is said to lay its eggs in their egg-pods, where the larvae feed and pupate. Observation has shown that wherever and whenever a swarm of the locusts appears, the fly is practically certain to turn up simultaneously, and it is probable that it moves automatically with the swarm. There may be some significance in the fact that the two years quoted above were notable for insect immigration in this country, 1947 yielding many records of the Migratory Locust.

It may not however be beyond the bounds of possibility that the fly has adapted itself to a British Locustid. *S. lunata* is about 7 mm. long, and may be distinguished from all other British Calliphorines by the produced, snout-like mouth-edge, the arista plumed on the upper side only (Pl. **43,** 7) and by the indistinct reddish-brown lunules or bands on the second, third and fourth abdominal segments.

Keys to the British species of Calliphoridae, together with Tachinidae, are provided by Emden[12], Wainwright[9] and Day[10]; Lundbeck[11] will also be found useful. Day and Fonseca[13] key females of *Sarcophaga*; Spence[14] keys females of *Lucilia*.

LITERATURE CITED

1. Fabre, J. H. The Life of the Fly. Hodder & Stoughton, London. 1927.
2. Cumber, R. A. Humble-bee parasites and commensals found within a thirty mile radius of London. *Proc. R. ent. Soc. Lond.* (A) **24**: 123–4. 1949.
3. Patton, W. S., and Wainwright, C. J. The British Species of the sub-family Sarcophaginae, with illustrations of the male and female terminalia. *Ann. trop. Med. Parasit.* **29**: 73–90, 517–32. 1935. **30**: 187–201, 337–50. 1936. **31**: 303–17. 1937.
4. Lowne, B. T. The Anatomy, Physiology, Morphology and Development of the Blow-fly (2 vols.). R. H. Porter, London. 1890–5.
5. Graham-Smith, G. S. Further observations on the anatomy and function of the proboscis of the Blow-fly. *Parasitology.* **22**: 47–115. 1930.
6. Imms, A. D. A General Textbook of Entomology. 676 Methuen, London. 1925.
7. Steinhaus, E. A. Insect Microbiology. 34–6. Comstock Publishing Co., New York. 1946.
8. Richards, O. W. Notes on the British Species of *Lucilia* (Dipt.). *Trans. ent. Soc. Lond.* **74**: 255–60. 1926.
9. Wainwright, C. J. The British Tachinidae (Diptera) (includes Calliphoridae). *Trans. R. ent. Soc. Lond.* **76**: 139–248. 1928. *Ibid.* (Supplement). **80**: 405–24. 1932. *Ibid.* (Supplement). **90**: 411–48. 1940. *Ibid.* (Supplement) (Corrections). *Ibid.* (B) **10**: 70. 1941.

10. Day, C. D. British Tachinid Flies (includes Calliphoridae). *North Western Naturalist.* **21** (Supp.): 1–64. 1946. **22** (Supp.): 65–150. 1947.

11. Lundbeck, W. Diptera Danica. Vol. VII. Wm. Wesley & Son, London. 1927.

12. Emden, F. I. van. Handbooks for the Identification of British Insects. Vol. X. Pt. 4(a). Tachinidae and Calliphoridae. Royal Entomological Society of London. 1954.

13. Day, C. D. and Fonseca, E. C. M. d'A. A key to the females of the British species of *Sarcophaga* (Dipt.,Calliphoridae). *J. Soc. Brit. Ent.*, **4**: 36–7. 1951.

14. Spence, T. A taxonomic study of the females of the British *Lucilia* species (Diptera, Calliphoridae). *Proc. R. ent. Soc. Lond.* (B) **23**: 29–35. 1954.

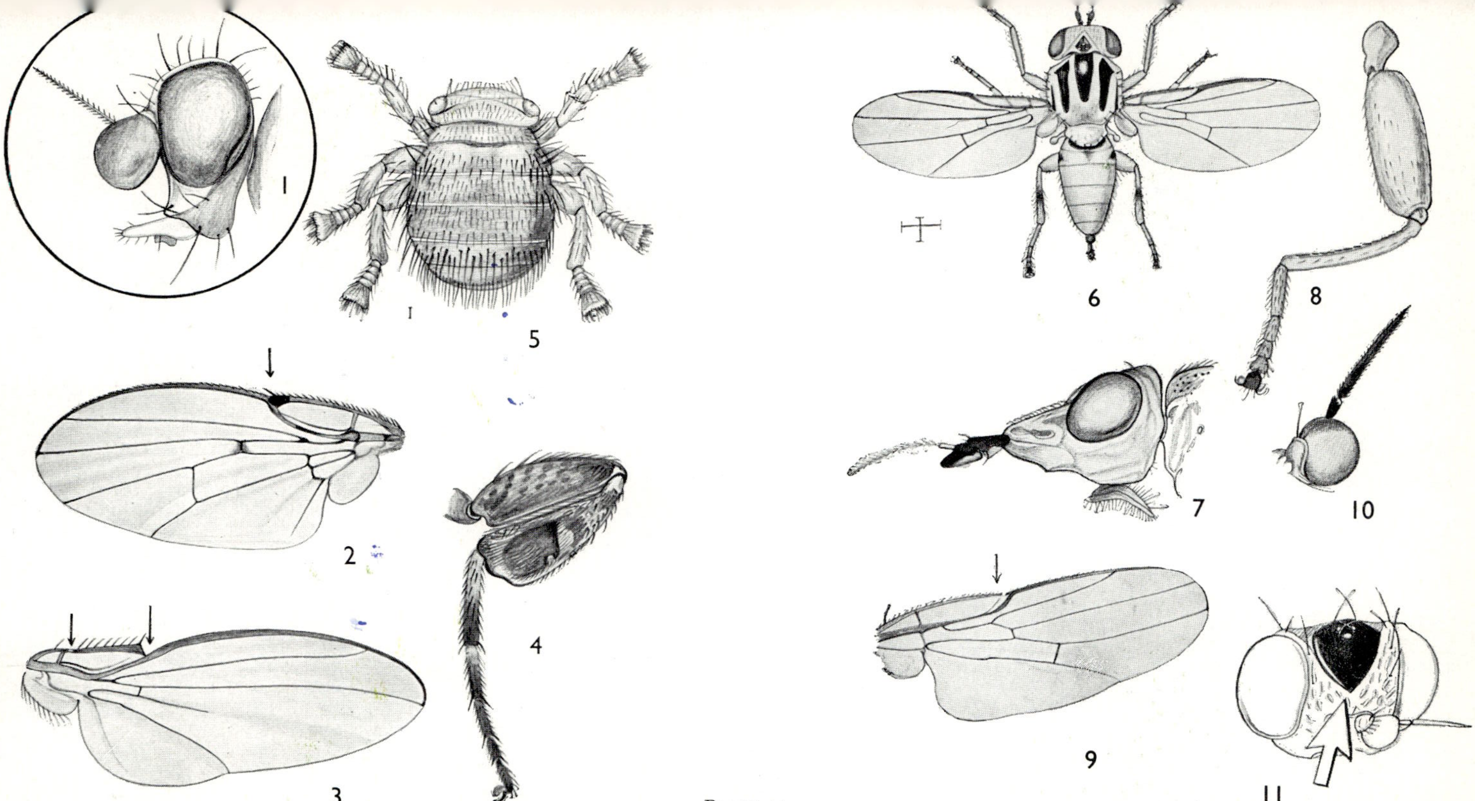

PLATE 39

MILICHIIDAE

1. *Phyllomyza securicornis* Fall., ♂ head, p. 237. 3. *Carnus*, wing, p. 238. 4. *Leptometopa latipes* Mg., hind leg, p. 237.

ODINIIDAE 2. *Odinia*, wing, p. 239.

BRAULIDAE 5. *Braula coeca* Nitzsch, p. 242.

CHLOROPIDAE

6. *Chlorops hypostigma* Mg., ♀, p. 240. 7. *Platycephala planifrons* F., head, p. 241. 8. *Meromyza*, hind leg, p. 241. 9. *Oscinella frit* L., wing, p. 240. 10. *Elachiptera cornuta* Fall., antenna, p. 240. 11. Chloropid ocellar triangle, p. 240.

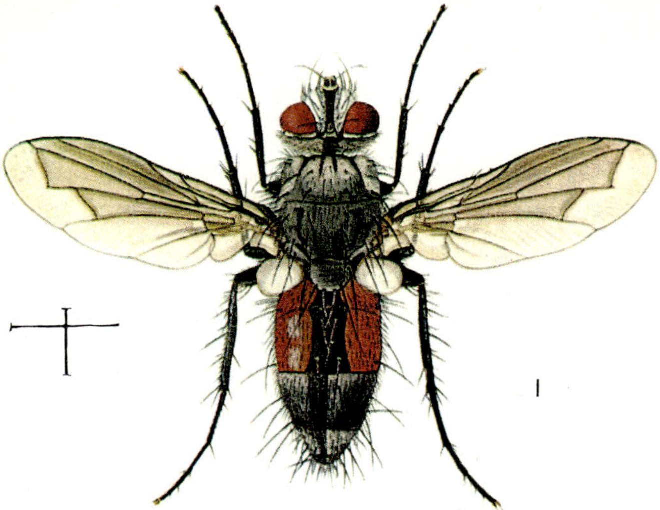

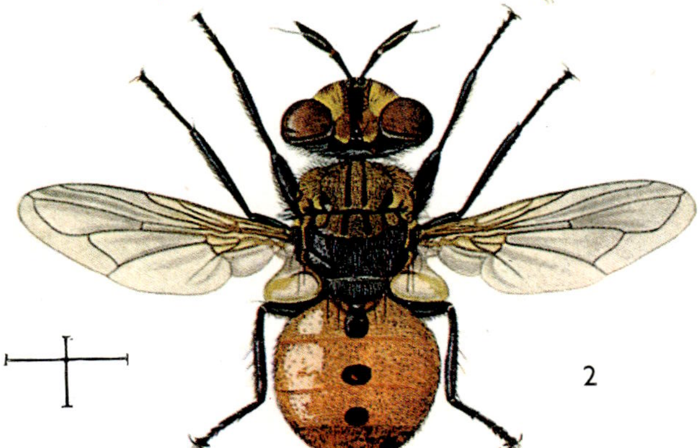

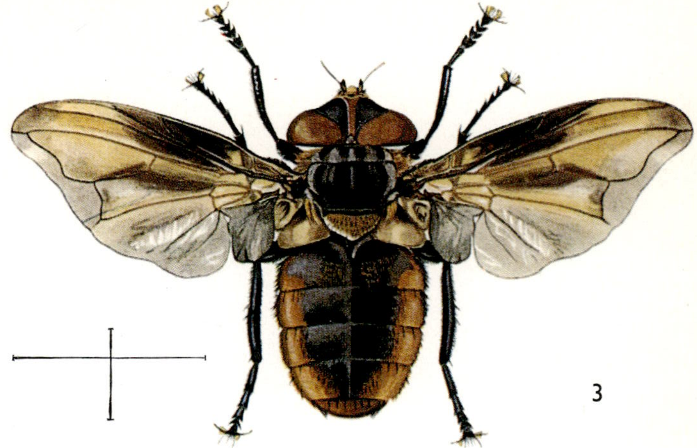

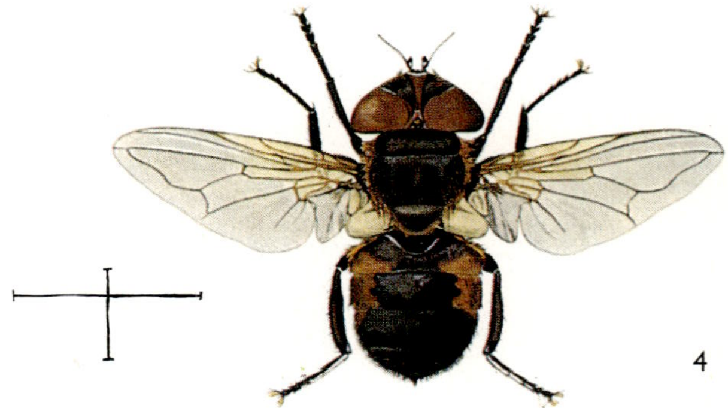

Plate 40

TACHINIDAE-V

1. *Eriothrix rufomaculatus* Deg., ♀, p. 268.

3. *Alophora hemiptera* F., ♂, p. 271.

CHAPTER XXIII

Gen.	*Hypoderma*, Calliphoridae .	Ox-warble Flies.
,,	*Cephenomya*, ,, .	Deer-bot Flies.
,,	*Oestrus*, Oestridae .	Sheep-bot Flies, Sheep Nostril Flies.
,,	*Gasterophilus*, Gasterophilidae	Horse-bot Flies.

FROM the biological standpoint, the life-histories of the flies covered by this chapter are of considerable interest. Since they are all, in the larval stage, dependent for their existence upon large mammalian hosts, they have been grouped together here, despite the fact that they belong to more than one family in the Check List; in fact, they are often classified as one family, the Oestridae.

Hypoderma females deposit their eggs, as a rule, on the lower parts of the legs of cattle, although sometimes the upper parts or the flanks are chosen; the eggs, usually about a dozen, are attached firmly to hairs by means of a clamp-like modification of the egg-covering, and of an adhesive substance. After about three days, they hatch and the resultant larvae pass down the outside of the hair and penetrate the skin at the base; for the next two months, they work their way as far as the chest, or the abdominal cavity; occasionally they appear in the tissues of the gullet. Thereafter, they proceed quickly to a position just beneath the skin of the back, sometimes by way of the spinal canal. Here begins a change of habit which lasts up to two months; the larvae remain stationary after making a perforation in the skin of the host for a breathing-hole and become surrounded by pus, secreted by the host's tissues in response to the irritation; the swelling which results is familiarly known as a "warble". When fully developed, the larvae emerge through the breathing-aperture by alternate expansion and contraction of their bodies, and drop from the animal to pupate in the ground, a process which occupies about a month.

Hypoderma larvae are short, fat, corrugated grubs armed with spines. The adults are fairly large, hairy, bee-like flies; their association with the Tachinidae-Calliphoridae group is marked by a fan of yellow hypopleural hairs (not bristles), and by the sharp bend of Vein 4 toward Vein 3; it will be noticed that the posterior cross-vein is so placed as to be almost in line with the apical section of Vein 3 (Pl. **47,** 3). Vein 6 reaches the margin, an unusual feature in the group (but cf. *Siphona*, p. 263); the front view of the head is very distinctive (Pl. **47,** 3-inset); and the mouthparts are vestigial and non-functional, no feeding taking place in the adult stage. An affinity with the Tachinidae rather than with Calliphoridae is exhibited in the presence of a strongly developed, convex postscutellum. Pairing takes place soon after emergence, the span of adult life occupying only a very small proportion of the life-cycle; the greater part is taken up in the larval stage in the host and the pupal stage in the earth.

There are three British species, *H. lineatum* de Villers (Pl. **47,** 3), *H. bovis* L. and *H. diana* Brauer, the first two affecting cattle and sufficiently common to be of economic significance, while the third affects deer and is therefore not so well known, in this country. Hides are ruined by the "warble" perforations and the cattle lose condition through the irritation caused by the larvae, both during their travels in the body and during the "warble" phase. In England, the larvae of *lineatum* appear about February, and are ready to pupate about April or May. Those of *bovis* are generally a few weeks later, the adult flies appearing in May or June. *H. lineatum* appears to be prepared to lay its eggs under any reasonable circumstances, even in cool weather, but *bovis* apparently prefers hot, sunny conditions.

The life-histories of *Cephenomya auribarbis* Mg., the Deer-bot Fly, and *Oestrus ovis* L., the Sheep-bot Fly (Pl. **47,** 4) are very similar; we will consider the latter, since it is much more common and an economic factor of some significance. *Oestrus* females are viviparous, like some of the species of Calliphoridae (cf. *Sarcophaga*, p. 276, and *Onesia*, p. 285), i.e. living larvae are deposited instead of eggs. The place chosen is the nostrils of sheep, as a rule in hot, sunny weather; thence the young larvae find their way to the frontal sinuses. They attach themselves by their mouth-hooks to the mucous membrane and feed, remaining here for about nine months and undergoing changes of form meanwhile. By spring or early summer, the third larval stage is reached, and this causes, in the host, fits or a dazed condition known as "false gid"; the term is used in contradistinction to "true gid", which is caused by the infestation of the sheep's brain by a tapeworm. In this stage, the larva is a fat, corrugated maggot, somewhat narrowed at the anterior end, with large mouth-hooks and armed with transverse rows of spines on the ventral surface. When fully-fed, the larvae relinquish their hold on the membrane and are sneezed out by the sheep on to the ground, where they pupate under stones or grass-tufts. Adult life is short and no feeding takes place, the mouthparts being vestigial; the flies are to be found on walls and in shady places in sheep-raising localities. The animals try to evade the attentions of the *Oestrus* females by keeping their noses to the ground on bare patches of earth, where the flies can be easily seen. Generally, only a few larvae are present in any one animal, but they may cause a serious loss of condition.

As regards the structure of the adult, it will be seen that the frons is considerably produced; this, together with the dorsum and scutellum, bears tiny, wart-like excrescences. The eyes are small; the abdomen is brownish-black, marbled with white, and, in life, has a silvery sheen; the wing-venation has a feature in common with *Hypoderma*, i.e. the posterior cross-vein and the apical section of Vein 3, although more sinuous, are more or less in line; a convex postscutellum is also present. Unlike *Hypoderma*, however, Vein 4 joins Vein 3 before the margin and the face is divided by a median channel; Vein 6 does not reach the margin. Yellow hypopleural hairs are present, but are somewhat more bristle-like than in *Hypoderma*.

The larvae of *Cephenomya* lead a somewhat similar existence in deer, but apparently reach their full development in the throat or windpipe of the host. The adults are hairy flies, about 15–16 mm. long, with a prominent mouth-beard of bright reddish hairs; reddish-brown hairs on the thorax, with an indistinct darkish transverse band; yellowish or yellowish-brown hairs on the anterior part of the abdomen; blackish in the middle; and yellowish posteriorly. The prosternum is prominently developed, and the postscutellum very slightly convex. Females are viviparous.

The species of *Gasterophilus*, or Horse-bot Flies, constituted the sub-family Gasterophilinae (Muscidae), of the Check List; nowadays they are treated as a separate family, Gasterophilidae. *G. intestinalis* Deg. (= *equi* F.) (Pl. **44,** 1) is the Common Horse-bot Fly and will be seen to differ very much from the flies previously discussed in this chapter. The ovipositor is strongly developed; Vein 4 is not bent toward Vein 3; no hypopleural bristles or hairs are present; and the squamae are very small. Mouthparts are again vestigial.

In addition to horses, which are the usual hosts, donkeys and mules are liable to receive the attentions of these flies. The eggs are usually laid on the anterior parts of the front legs and are attached by two long flange-like processes to the hairs; sometimes they are deposited on the shoulders, belly or the upper parts of the hind legs. Whilst laying, the female hovers with her body in almost a vertical position as each egg is attached to a hair. Hatching is accomplished, in about a week, as a result of friction caused by the rubbing of the horses' lips; this action no doubt provoked by the irritation of the eggs on the hairs. It would appear that the larvae gain entry to the body of the host by means of the lips; by a route apparently unknown at present, they make their way to the stomach-lining, where, in due course, they are to be found attached by their mouth-hooks. Here they remain for some nine or ten months; when fully developed, they release their hold and are passed out with the host's excretions. A period of three to ten weeks is passed as a pupa in the ground, the duration of the pupal phase being apparently regulated by ground temperatures. Fully-grown larvae are barrel-shaped, not corrugated as in *Hypoderma* or *Oestrus*, and with double transverse rows of spines on the segments.

There are three other British species of *Gasterophilus*, viz., *G. haemorrhoidalis* L., *G. nasalis* L., and *G. pecorum* F.; the life-histories are similar to that of *intestinalis*, except that the eggs are able to hatch without friction and, in the case *haemorrhoidalis*, the larvae re-attach themselves, in subsequent stages, to the rectum and opening of the anus, where they protrude and drop to the ground at any time.

It is interesting to recall that, whereas the "Gad" Flies or "Stouts" (Tabanidae, p. 104), whose bite is usually severe and painful to human beings, do not cause any undue panic in horses and cattle, the flies of this group, which do no more, in the first instance, than to deposit eggs or larvae, seem to excite much fear and annoyance in their potential hosts. Piffard, quoted in Edwards, Oldroyd and Smart,[1] states that stark terror was induced in peacefully grazing cattle by a French peasant-girl, merely by making, some distance away, a faint hissing sound, imitative of the Warble Fly. Smith [2] states that *Gasterophilus* and *Oestrus* cause fear in horses and sheep respectively. It does not seem reasonable to suppose that the hosts have come instinctively to associate the ovipositing of the flies with the much later irritation caused by the larvae; it is also difficult to understand why the immediate wounds inflicted by Tabanidae do not occasion any pronounced exhibition of feeling.

Whilst bots do not appear to be the cause of permanent damage to their hosts, merely by the fact of their presence in the stomach or other situations, the fear of the adults and the gastric inflammation caused by the larvae, with attendant indigestion and other secondary complaints, may produce a general loss of condition which constitutes an economic factor of no little importance.

In addition to the key works cited under the families to which the species of this group belong, reference may be made to Seguy,[3] and Zumpt,[4] the latter dealing with the immature stages.

LITERATURE CITED

1. Edwards, F. W., Oldroyd H., and Smart, J. British Blood-sucking Flies. 149–52. British Museum, London. 1939.
2. Smith, K. M. A Text Book of Agricultural Entomology. University Press, Cambridge. 1931.
3. Séguy, E. Études sur les Mouches Parasites, 1. Conopides, Oestrides et Calliphorines, etc. Lechevalier, Paris. 1928.
4. Zumpt, F. Myiasis in animals and man. Butterworths. London. 1965.

CHAPTER XXIV

MUSCIDAE*
(= ANTHOMYIIDAE) — House Flies, Biting "House Flies" and Stable Flies, Cabbage Root Fly, Onion Fly, Wheat Bulb Fly, Beet or Mangold Fly, etc.

84 genera. Over 450 species. Distinguishable from Tachinidae-Calliphoridae group by absence of hypopleural bristles and (with exception of sub-family Fucellinae) from Scatophagidae by fact that thoracic squama is as long as, or longer than, alar squama (see p. 249). Fucellinae may be differentiated from Scatophagidae by the presence of a costal spine, near apex of *sc*, which is absent in Scatophagidae; the females have crossed bristles on frontal stripe, absent in Scatophagidae; abdomen with four to five visible segments, against five or six in Scatophagidae. Arista bare, pubescent or plumose; ♂ eyes holoptic (e.g. Muscinae, Phaoniinae) or separated (e.g. Fucellinae, Coenosiinae). (Pl. **42,** 6: **44,** 1, 3, 4: **45**: **48**: **49**.)

In this family a wide variety of forms exists; habits, likewise, are extremely diverse. Some species subsist, in the adult stage, on the blood of mammal hosts; some adults are predatory upon other insects and the larvae of others prey on insect larvae; many species are attracted, as adults, to excrement and decaying matter of various kinds, in which their larvae develop; certain larvae are pests of vegetables and cereals; others are leaf-miners; and some species frequent the seashore, where the larvae develop in rotting seaweed and other decomposing matter. We must content ourselves here, in view of the size of the family, with selecting from the

* Some workers now consider that the broad group covered in this chapter constitutes three separate families: viz Muscidae, Fanniidae and Anthomyiidae, the sub-family Fucellinae being part of the Family Anthomyiidae.

various sub-families one or two representative species for brief review.

MUSCINAE.

Species characterised by shape of Vein 4, which is either sharply bent as in *Musca domestica* L. (Pl. **49,** 1) or curved in a long bow, with the end converging toward Vein 3, as in *Dasyphora* (Pl. **44,** 4) or *Mesembrina* (Pl. **45,** 3); arista always plumose; male eyes holoptic; pteropleuran setulose.

If more than casual attention be given to the flies which occur in dwellings, it will be noticed that some of them appear to be perfectly at home and quite content to stay indoors, while others display a desire to depart which frequently leads them to a window, upon which they wander, in varying degree of agitation, seeking egress. Examination of the latter kind, over a period, will reveal that they consist of many species and will lead to the conclusion that they are accidental visitors; whereas the domesticated kind will be seen mainly to comprise two principal groups, whose occurrence indoors, by virtue of its frequency and regularity, is obviously not fortuitous; it is with this latter kind that we are now concerned.

One of the two groups consists of both larger and smaller flies, which, if allowed, will wander over food, especially sweet substances, investigating it with the proboscis, or will rest upon walls, lamp-pendants and other places in the room where they leave traces of their stay in the form of "fly-spots" (see p. 298). The other group comprises smaller flies, which fly in a characteristic manner for long continuous periods beneath lamp-pendants, in shafts of sunlight from windows and similar places. This flight is made on a series of irregular triangular or quadrilateral courses, an almost imperceptible hovering taking place at the corners and the sides being covered in a rapid dart. When undisturbed and alone, these flies maintain a more or less constant height and regular course, but when more than one decides to patrol the same "beat", it usually happens that one darts rapidly toward the other, a sharp flurry and prompt dispersal ensues and eventually one of them recommences the patrolling. If one of these flies be taken, it will be seen, against the light, to have pale yellow, semi-transparent patches at the base of the abdomen. This is the male of the Lesser House Fly, *Fannia canicularis* L. (Pl. **49**, 2), a species belonging to another sub-family, which we shall consider a little later; its flight-habit is characteristic of the males in the genus. The larger flies of the first group will probably be found to have an appearance similar to that depicted on Pl. **49**, 1, which is the female of the House Fly, *Musca domestica* L.; the male has a narrower frons; in both sexes, the ground-colour of the abdomen is more or less yellowish or buff-tinted. The smaller flies which occur with them will probably prove to be the females of *canicularis*, which are dull-coloured and lack the semi-transparent patches at the base of the abdomen which are present in the male. It will be noticed that *domestica* may be distinguished at once by the sharp upward bend of Vein 4 toward Vein 3, whereas in *canicularis* Vein 4 is more or less straight.

The life-history, structure and medical significance of *domestica* have been very fully studied and there is a wealth of literature which may be consulted; the works of Graham-Smith [1] and Hewitt [2] cover the subject in detail, while a condensed account is given by Austen.[3] Briefly, the eggs are laid in many kinds of refuse, but it is probable that the original development-medium of the species was the excrement of horses and other animals, in which situations it is frequently found today. This fly has accompanied man all over the globe and has adapted itself to breeding in various kinds of kitchen rubbish, provided it is moist, and in human excreta where this is accessible; even today, there are plenty of country districts where sanitation is of a primitive order. The eggs are about 1 mm. long and glistening-white; they are deposited in masses in crevices of the development-medium ; each female fly may, it seems, lay up to as many as 900 eggs, in batches of about 150. Temperature of the medium regulates the time taken for the eggs to hatch, which may vary from about eight hours to three days. The larvae feed rapidly under favourable circumstances, and in warm weather may reach their full development, when they are about 10–12 mm. long, in just under two days. This process may be delayed in colder temperature up to as much as eight weeks. In manure-heaps, where great heat is often generated, they are found near the surface, since they could not exist at a deeper level. The larvae are of the general Cyclorrhaphous form, long-cylindrical, tapered anteriorly and truncate posteriorly ; the pupa, likewise, is of the usual barrel shape, pale-coloured at first, becoming successively reddish, brown and finally very dark brown. Escape of the adult fly is accomplished in a manner similar to that already described for *Sarcophaga* (p. 277). Under favourable conditions, the whole life-cycle may be passed in as little as eight or nine days. It can readily be seen, therefore, how large numbers of these flies can appear in a very short time where conditions of primitive sanitation exist. Whilst individual flies do appear in an active condition throughout the year in places where the temperature permits, such as kitchens, bakeries, restaurants and boiler-rooms, the main broods of the year are not generally in evidence before June or July, and continue until about September.

Structure of the mouthparts is similar to that of *Calliphora*. The pseudotracheae or collecting-channels are extensively developed and the prestomal teeth very little ; feeding may take place on liquids or solids. In the case of the latter, a drop of liquid is expelled from the mouth and the solid food-substance, either in solution or in the form of minute suspended particles dislodged by the prestomal teeth, is re-imbibed with it. This process can be watched very easily when, for instance, *domestica* is feeding on sugar. Sometimes, when the fly is resting, as if its crop were too full, a drop of liquid is regurgitated and deposited on the surface upon which it stands; these drops, together with excretions, dry to form the familiar "fly-spots".

Another common species in the genus *Musca*, normally occurring outdoors, but occasionally finding its way indoors in country districts, is *M. autumnalis* Deg. (= *corvina* F.). It may easily be recognised, in the male, by the orange-yellow abdomen with black median stripe. The female is very similar to *domestica*, but distinguishable by the breadth of the frontal stripe, which is not twice that of one of the orbits, whereas in *domestica* it is three or four times as wide (Pl. **49,** 1); there is, moreover, a trace of buff colour at the base of the abdomen in *domestica* which is not present in *autumnalis*. The flight-time of *autumnalis* is normally from very early in the spring until very late in the autumn. Males may often be seen in large numbers, congregating in the sun on fences or tree-trunks; they seem to be especially fond of white surfaces, such as the white-painted rails and posts which are frequently erected round ponds or on the sides of little country bridges over streams. Almost invariably, they seem to select the point which receives the maximum amount of the sun's rays; thus they may be found, in the morning, in a tight swarm on the eastern surface of a post, on the upper surface at midday, and on the western surface after the sun has passed its zenith. The larvae develop in various kinds of excrement.

In *Graphomya maculata* Scop. (Pl. **45,** 1, 2) there is a striking example of sexual dimorphism; the colour difference between the sexes will be seen to be most marked. *G. maculata* is a bigger and broader insect than the two preceding *Musca* species and moreover has hairy eyes. The larvae live in mud at the bottom of puddles or in decaying vegetable matter, and are carnivorous (vide Séguy [4] and Keilin [5]); they are furnished with transverse rows of little hooks on the ventral face of some of the segments.

Reference has already been made to the Green Bottles of

the genus *Orthellia* (p. 283). These will be known by their brilliantly-metallic green jowls and frons, the latter, of course, being more noticeable in the females, in which the eyes are widely separated. *O. cornicina* F. (Pl. **44,** 3) is the more common of the two British species and may be distinguished by the fact that there are no presutural acrostichal bristles and that four pairs of postsutural dorsocentrals are present ; the other species, *O. caesarion* Mg., has a strong pair of presutural acrostichals and only three pairs of postsutural dorsocentrals. The larvae live in excrement and decaying matter generally ; the adults are likely to occur in much the same sort of situations as *Lucilia* (p. 283).

All the foregoing have Vein 4 sharply bent, whereas in the following group Vein 4 is bowed or curved ; two more Green Bottles are here included. *Dasyphora cyanella* Mg. (Pl. **44,** 4) is metallic-green in newly-emerged specimens, but is always darker and less brilliant than *Orthellia*, and has whitish dusting on the front part of the dorsum ; as in other Green Bottles, older specimens tend to turn reddish. *Pyrellia cyanicolor* Zett., and the other species of the genus, may be distinguished from *Dasyphora* by the fact that the eyes and Vein 1 are bare ; both are hairy in *Dasyphora*. The larvae of both genera develop in decaying matter.

Muscina species have bare eyes, two distinct rows of strong acrostichals and the end of the curved Vein 4 rather widely separated from the apex of Vein 3. They are greyish-flies of undistinguished appearance, with tessellated abdomens and the common characteristic of a reddish tip to the scutellum. *M. stabulans* Fall. is a cosmopolitan species, which may be known from the others by its yellow or orange tibiae ; the remainder have black legs. *M. assimilis* Fall. may be picked out by its black palpi, and is not uncommonly seen basking on tree-trunks or bare patches of earth in early spring. The two remaining species have reddish or yellowish palpi. *M. pabulorum* Fall. has no trace of red on the sides of the basal part of the abdomen and is common in similar locations to those affected by *assimilis*. Larvae in this genus develop in decaying animal or vegetable matter; those of *stabulans* and *pabulorum* are said to prey on other larvae.

Morellia species are somewhat similar to those of *Muscina*, but have the ends of Veins 3 and 4 more approximated; the scutellum is never reddish at the apex; and acrostichals are very weak or absent. The four British species are fairly common but not easy to differentiate, the females, in particular, having no pronounced distinctive characters. *M. aenescens* R-D. is, in the male, perhaps more easily recognised than the others by virtue of its curved hind tibiae and brush of hairs on the posterior surface of the hind metatarsus. *Morellia* larvae are said to develop in similar media to those of *Muscina* and *Graphomya* and to be carnivorous.

Mesembrina meridiana L. (Pl. **45**, 3), a large black fly with conspicuously yellow wing-bases, is very common from early until late in the year; it may be seen basking on tree-trunks or the ground; or, in the season, sipping nectar from umbels and other flowers. Séguy [4] states that the female lays a single large egg, about 4½ mm. long, containing a larva which hatches almost immediately; the medium chosen is horse- or cattle-dung.

In this group, with a curved Vein 4, there are three British species which differ strikingly from the rest of the group, and for that matter, from the rest of the family; they are sometimes referred to as a separate sub-family, the Stomoxyinae. These are the Biting Muscids, *Stomoxys calcitrans* L., *Haematobia stimulans* Mg., and *Lyperosia irritans* L., all of which, in both sexes, suck the blood of mammals. Here we find that the retractile proboscis of most of the Muscidae is replaced by an elongated, rigid, non-retractile organ (Pl. **49**, 3, 4), capable of piercing the skin of the host-animal. *S. calcitrans*, otherwise known as the Stable Fly or Biting House Fly, is not dissimilar in general appearance to *domestica*, but the abdomen has no buff-colour ; it is dark-greyish with a number of indistinct dark brown spots. The projecting proboscis, moreover, is sufficiently prominent to reveal its identity, either when viewed from above or from one side. It sometimes occurs indoors in country districts, particularly in farmhouses close to stables. *S. calcitrans* sucks the blood of warm-blooded animals, such as horses, cattle, dogs, cats, chickens and, occasionally, human beings ; on horses and cattle, the lower parts of the legs and under the belly seem to be favourite feeding-places. The larvae are of the general Cyclorrhaphous type and develop in straw, hay and other debris impregnated with animal excretions, such as stable litter. Sometimes old decaying hay or straw alone is chosen, or horse-manure, but never cattle-dung. Pupation takes place in drier parts of the habitat or in the ground. Adult flies are in evidence from May to October and are often so numerous near farms as to constitute a serious nuisance. *S. calcitrans* may be distinguished from the other two species by the fact that the palpi are small and thin and only about a quarter of the length of the proboscis (Pl. **49**, 3, 4). In *stimulans*, which is a little smaller, the palpi are much longer, relative to the proboscis, and club-shaped, i.e. dilated apically ; moreover, they are yellowish in colour which serves to differentiate the species from the even smaller *irritans*, in which the palpi, although similar in shape and relative length, are dark greyish. *H. stimulans* has similar habits to those of *calcitrans*, but *irritans*, which is sometimes known as the Horn Fly, from its habit of congregating on the heads and particularly the horn-bases of cattle, is not known to bite human beings. The larvae of both *stimulans* and *irritans* usually develop in cattle-dung, pupating in the earth beneath ; they will however, on occasions, select horse-droppings.

The British Stomoxyinae are covered by Smart.[6]

GASTEROPHILINAE.

See Chap. XXIII, p. 292.

PHAONIINAE.

Fourth vein more or less straight, i.e. Veins 3 and 4 do not converge apically (Pl. **45,** 4); Vein 6 does not reach the margin; two or three sternopleurals, if the latter, never arranged as a symmetrical equilateral or isosceles triangle; male eyes holoptic; hind tibia bears a characteristic dorsal bristle or bristles, apart from apical or sub-apical spurs or bristles, a character sometimes not easily appreciated (cf. Intro. Chap., p. 30, and Fig. 8D; and compare Pl. **48,** 4, Phaoniinae, with Pl. **48,** 3, Mydaeinae, in which the dorsal bristle is absent). Doubt may usually be resolved by considering the tibia as if it were stretched out straight from the body and locating the various sectors of the surface with the aid of the longitudinal rows of tiny bristles. The sub-family contains about 130 species in 22 genera, of which the largest are *Phaonia,* with 29 species, *Hydrotaea,* with 22, and *Fannia,* with 45.

Within the genus *Phaonia,* there is a small group of species, in which the abdomen is yellowish; *P. populi* Mg. (Pl. **48,** 4) is an example. *P. pallida* F. is pale all over. *P. erratica* Fall. (Pl. **45,** 4) is representative of the remainder of the genus, which consists of somewhat undistinguished-looking grey or brownish-grey flies, the abdomen often with shifting tessellations. They are frequently to be seen basking on tree-trunks, palings and fences from very early until very late in the year, in company with other superficially similar Muscids (e.g. *Muscina* species) and *Pollenia* (Calliphoridae, p. 285); they also occur on flowers, sap flowing from tree-wounds, decomposing fungi and excrement. In some cases the larvae are said to be carnivorous (*vide* Keilin [5]); in others to develop in rotten wood and other decaying matter; we have bred *P. gobertii* Mik from larvae found beneath the bark of a rotten oak stump, in company with other Dipterous larvae.

Males of *Hydrotaea* species have oddly deformed or modified fore femora and sometimes fore tibiae (Pl. **49,** 6); the females have crossed bristles on the frontal stripe; only two sternopleurals are present. *H. irritans* Fall. is very common indeed and frequently congregates in large numbers round the heads or bodies of horses and cattle; it is especially attracted to mucous secretions, suppurations from sores or blood oozing from wounds. Equally, it is attracted to human perspiration and can be an almost intolerable nuisance to the outdoor man on warm days in the woods, when the forehead and neck are moist; the apparent purpose of the flies is to sip the sweat, no biting taking place, but the sensation caused by the alighting and application of the mouthparts is vastly irritating. Other species in the genus have similar habits; some have a particular liking for damp situations; males occasionally fly in a manner not unlike that of *Fannia* males (p. 296), and usually high up. The larvae are either excrement-feeders or prey on other insect larvae (*vide* Keilin [5]).

A paper by Grimshaw [7] on the British species of the genus, though now rather old, may prove useful if more up-to-date monographs on the family are not accessible.

Mention of the Lesser House Fly, *Fannia canicularis* L. (Pl. **49,** 2), has already been made; this is the commonest species of the genus, in which, as in *Hydrotaea,* the males often have curious deformities or ornamentations of the legs, but in this case, the middle ones (Pl. **49,** 7). Sometimes there are one or more bent thorns projecting downward from the mid-coxae; in profile, the heads are characteristically hemispherical; the arista is bare; and the arrangement of Veins 6 and 7 is usually distinctive, i.e. Vein 6 is short and Vein 7 sharply curved (Pl. **49,** 2). Females may be known by the arrangement of the bristles on the frons, i.e. there are two strong upper fronto-orbitals lying outside the usual row of frontals (Pl. **48,** 2). The trace of yellow at the base of the abdomen which we noticed in *canicularis* (p. 297) is extended in some species to a much larger area, giving them a sharply-contrasted black-and-yellow appearance, as in *F. hamata* Macq. Pl. **48,** 2).

F. scalaris F., sometimes called the Latrine Fly, is a greyish

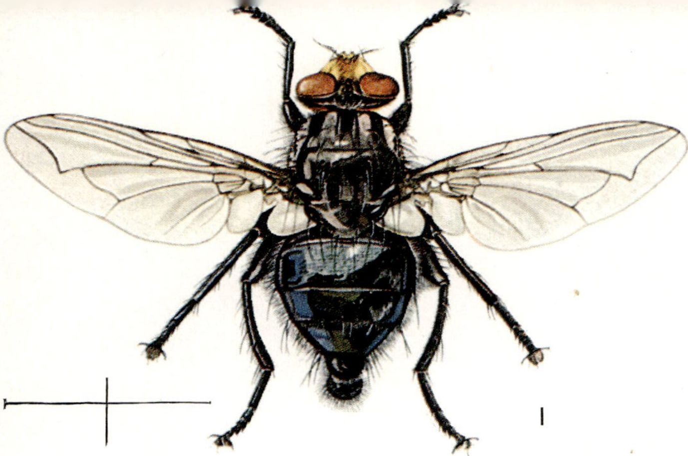

1

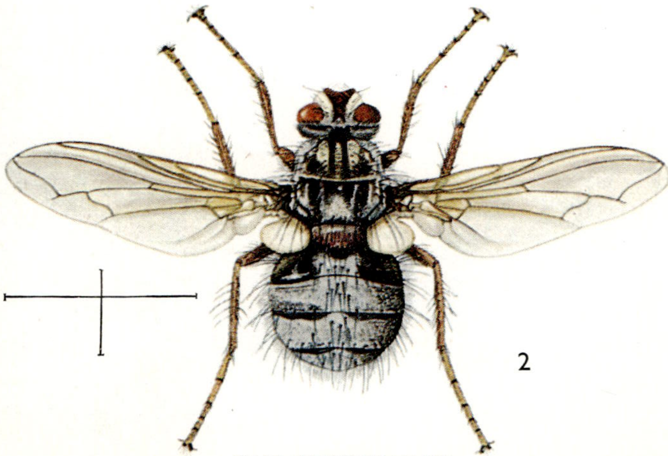

2

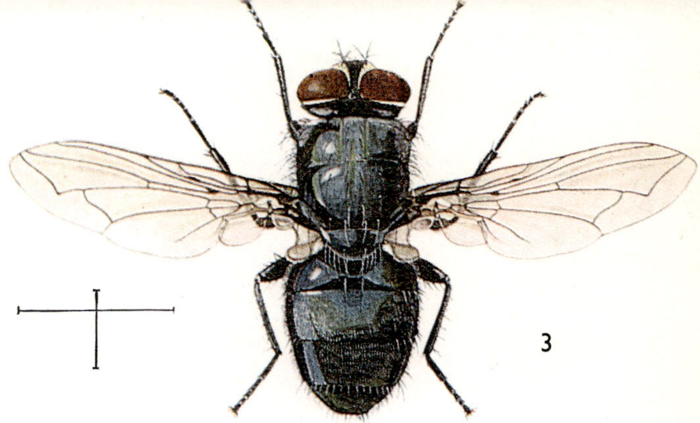

3

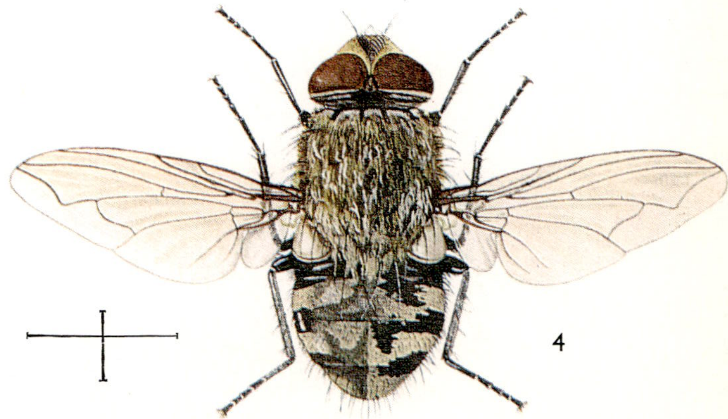

4

PLATE 41

CALLIPHORIDAE-III

1. *Cynomya mortuorum* L., ♂, p. 285.

TACHINIDAE-VI

2. *Trixa caerulescens* Mg. (=*alpina* Mg.), ♀, p. 270.

CALLIPHORIDAE-IV

3. *Protocalliphora azurea* Fall., ♂, p. 282.
4. *Pollenia rudis* F., ♂, p. 285.

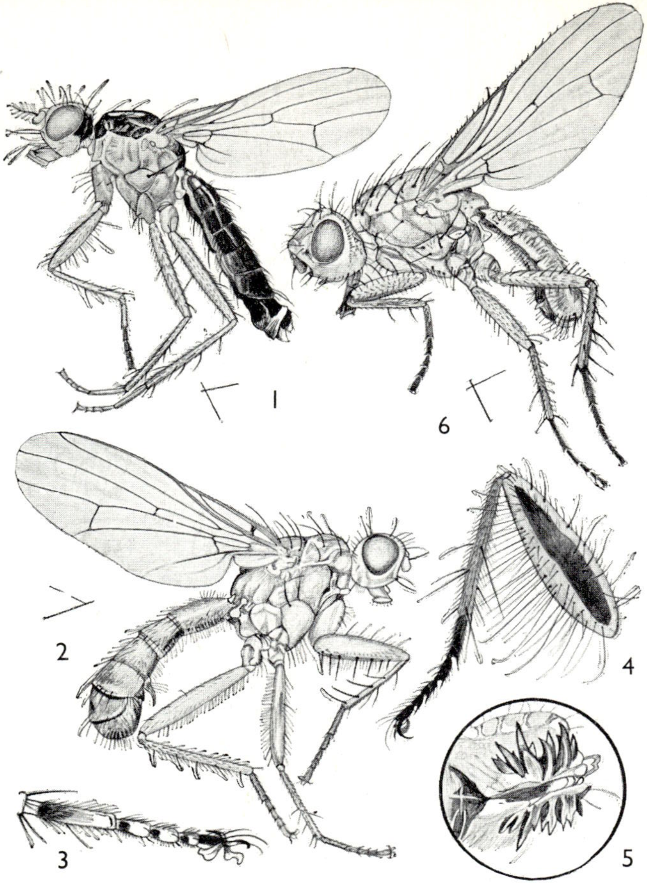

SCATOPHAGIDAE-II

1. *Parallelomma albipes* Fall., ♀, p. 251. 2. *Norellisoma spinimanum* Fall., ♂, p. 255. 3. *Cordilura pubera* L., underside of tarsus, p. 250. 4. *Cnemopogon apicalis* Wied., fore leg, p. 251. 5. *Scatophaga stercorarium* L., prestomal teeth, p. 254.

PLATE 42

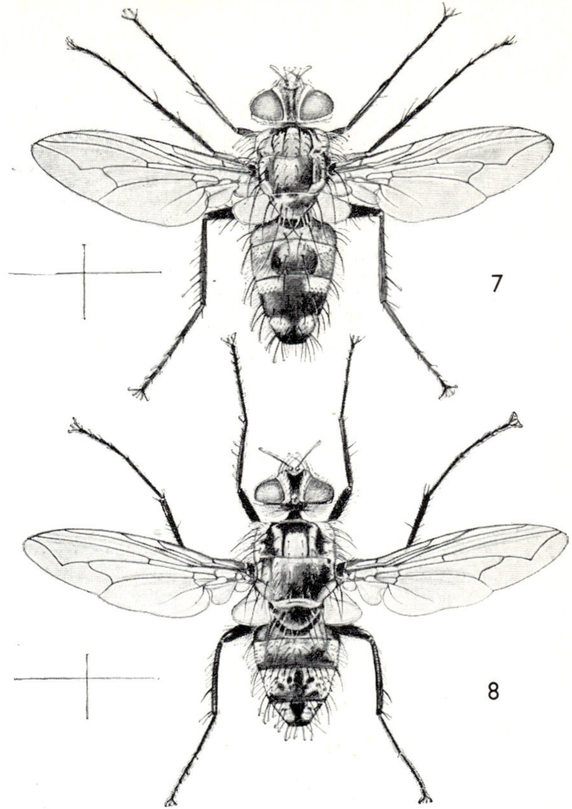

TACHINIDAE-VII

7. *Thelaira leucozona* Pz., ♂, p. 268.
8. *Nemorilla floralis* Fall., ♂, p. 261.

species which is not uncommonly found indoors in country districts where sanitation is not of a very high order; the larvae will readily develop in human excrement when and where opportunity permits.

Reference has been made to the peculiar flight-trajectory of *canicularis* males (p. 296); this is very noticeable also in males of other species outdoors. They may often be seen thus indulging in sheltered spots under trees, and seem particularly fond of flying in and out of shafts of sunlight striking through the foliage; at one moment they are brilliantly illuminated and the next, they have apparently vanished.

Fannia larvae are generally associated with decaying matter or excrement, often of a semi-liquid nature. Those of *canicularis* are of unusual structure, very unlike the usual Muscid form; the segments are equipped with appendages or protuberances bearing cilia, which serve as a means of progression or floating; similar processes occur in the larvae of other species. The eggs, also, have two float-like modifications of the covering, which probably enable them to remain suspended upon the surface of liquid matter in the habitat.

Polietes lardaria F. (Pl. **48,** 1), a fairly large fly of light bluish-grey and dark grey colouring, is very common and, in life, somewhat reminiscent of *Sarcophaga* (p. 275) by virtue of its striped thorax and tessellated abdomen; the wing-venation and general build remove any doubt, of course. The genus is characterised by a number of short, fine bristles at the base of Vein 3, two complete rows of acrostichals extending back to the scutellum and facial bristles which extend well beyond half-way from the mouth-edge to the bases of the antennae. *P. lardaria* larvae live in animal dung, where they prey upon other Dipterous larvae.

LISPINAE.

One genus, with nine species, characters as for Mydaeinae (p. 306) except for a stout proboscis and very dilated, spoon-shaped palpi (Pl. **49,** 8); in other respects, rather undistinguished-looking flies, of greyish-brown or brownish-black colouring; male eyes well separated. They are predatory in both adult and larval stages, the adults running about actively on waterside plants, while the larvae either live in the water or develop in the flowers of aquatic plants. Several species, including the common *Lispe tentaculata* Deg., are to be found on the seashore.

MYDAEINAE.

Over 90 species in 6 genera. Fourth vein straight; distinguishing characters as for Phaoniinae, except that the dorsal bristle on hind tibia is wanting (Pl. **48,** 3); in *Mydaea* and *Helina*, close superficial resemblance to *Phaonia* species.

In *Mydaea*, there are a few bristles at the base of Vein 3; *M. scutellaris* R-D. (= *pagana* F.) (Pl. **48,** 3) is one of the common yellowish-grey or greyish species of the group, which may be seen basking on foliage, feeding on umbels or at over-ripe blackberries, and sometimes on decomposing fungi. According to Keilin,[5] the larvae of some *Mydaea* species are carnivorous; those of a certain South American species are said to live as parasites beneath the skin of birds.

Helina species are very similar, but lack the bristles at the base of Vein 3; in both genera, where longitudinal stripes are present on the dorsum, they are of even number.

The species of *Hebecnema*, are dark, sombre-looking flies, much like the foregoing two genera in structure, but without distinct stripes on the dorsum and with the wings more or less tinged with smoky-black or brownish-black. In certain lights, as for instance when basking on light-coloured foliage or umbels, the wings appear quite black, which gives the flies a distinctive appearance. *H. affinis* Malloch is a fairly common example, which may sometimes be seen in some abundance on bracken.

Greyish or yellowish-grey species superficially similar to

those of *Mydaea* and *Helina* are included in the complex genus *Limnophora*; longitudinal stripes on the dorsum, when present, are of odd number; and the abdomen may have a number of indefinite, shifting spots. Habitats near water are mostly favoured; the larvae, in fact, are said to be aquatic and of carnivorous habit. A key to the British species of the genus is furnished by Collin.[8]

Azelia species are active little greyish flies, about 2–4 mm. long, to be found on umbels, foliage or excrement in summer. They are characterised by the shape of Vein 6, which is short, thick and blunt-ended (Pl. **49,** 9).

FUCELLINAE.

Fourth vein straight; Vein 6 reaches the margin; sternopleurals not arranged as equilateral or isosceles triangle; male eyes widely separated; females very much like males; distinguished from Anthomyiinae by the fact that thoracic squamae are much shorter than alar squamae, and from Scatophagidae (p. 249) by presence of costal spine; also in the females by presence of crossed bristles on frontal stripe.

Mycophaga fungarum Deg. has a greyish thorax and reddish-yellow abdomen; the carnivorous larvae develop in fungi. *Chirosia* species are little greyish flies whose larvae mine fern leaves. *Fucellia maritima* Hal. (Pl. **42,** 6) and the closely-similar *F. fucorum* Fall. frequent heaps of rotting seaweed thrown up by the tide on the seashore.

ANTHOMYIINAE (now usually given family status).

Fourth vein straight; Vein 6 reaches margin; sternopleurals not symmetrically arranged; male eyes holoptic; squamae comparatively large; nearly 150 species in 23 genera.

This is a very difficult and complex group, since many characters used to differentiate genera and species are inconstant; this is particularly true of *Hylemya, Delia, Phorbia* and closely-related genera (*vide* Collin[9]); consequently the use of keys, even when a large material is available, can be very trying. There are, however, certain common and fairly easily-recognisable species which will serve as representatives for our present purpose.

Anthomyia pluvialis L. and the very closely similar *A. procellaris* Rond. are common enough on umbels, brambles and foliage generally and are sufficiently striking, in their dress of dove-grey with deep black markings (Pl. **49,** 11), to catch the eye, among their drab-coloured relatives. Males are said to dance in the air at the approach of rain, a performance we have not, personally, witnessed; the larvae have been bred from fungi.

Some of the species of *Pegomya* are fairly distinctive in appearance, with greyish thorax and reddish or yellowish abdomen; fungi and rotten wood are the usual habitats for the larvae, although some are leaf-miners; we have bred *P. iniqua* Stein from cultivated mushrooms.

Egle muscaria Mg. is generally abundant, in early spring, on Sallow catkins, where the larvae develop; the long spatulate palpi and protruding mouth-edge are particularly noticeable (Pl. **49,** 10).

A number of species of Anthomyiinae find their place in textbooks on agricultural entomology, as pests of economic significance. The larvae of *Hylemya coarctata* Fall., the Wheat Bulb Fly, *Delia* (=*Hylemya*) *cilicrura* Rond. and *Phorbia genitalis* Schnabl & Dziedzicki, attack cereals (*vide* Smith [10]); *Erioischia* (=*Hylemya*) *brassicae* Bouche, the Cabbage Root Fly, in the larval stage, causes damage to the roots of cabbages, radishes, turnips and swedes; *Delia cepetorum* Meade (=*Hylemya antiqua* Mg.) is the destructive Onion Fly; *Pegomya hyoscyami* Pz., and its var. *betae* Curt., is a pest of beet and mangolds; while various other species are an occasional source of annoyance to the farmer and horticulturist, as for example, *Hylemya variata* Fall., which sometimes mines the leaves of pinks and carnations. Most of these species seem to have alternative wild hosts.

Paraprosalpia (= *Prosalpia*) species are recorded by Hobby [11] as predatory, in the adult stage, upon other flies. The British species of this genus, also of *Leucophora* (= *Hammomyia*) and *Hylephila*, are dealt with by Collin.[12]

COENOSIINAE.

Fourth vein straight; sternopleurals arranged in the form of a symmetrical equilateral or isosceles triangle (Pl. **49,** 5); eyes well separated in the male; over 40 species in 14 genera.

Caricea tigrina F. is a fairly regular member of the "small fry" which turn up in the sweep-net from grasses and low herbage generally; it is a dark greyish or greyish-brown fly, about 4–7 mm. long, with all the tibiae reddish. Some of the species of *Coenosia*, although smaller, are particularly noticeable in the sweep-net, by virtue of their light bluish or silvery-grey body-colour and reddish or yellowish legs, which usually show up in sharp contrast to the remainder of the "catch". Hobby [11] gives records of these and other Muscid flies as predators and has studied the mouthparts, in which the prestomal teeth are very strongly developed (cf. Scatophagidae, p. 254, and *Paraprosalpia*, above).

In addition to the special publications already cited, Hennig[15] monographs the palaearctic Muscidae *s. s.* but the earlier works of Seguy[4] and Karl[13] are now of limited use. Works by Ringdahl[14], especially those translated by Fonseca[16], the keys to genera by Fonseca[17] and papers by Collin[18] treat British species.

LITERATURE CITED

1. Graham-Smith, G. S. Flies in Relation to Disease: Non-blood-sucking Flies. University Press, Cambridge. 1913.
2. Hewitt, C. G. The House Fly (*Musca domestica* Linn.), etc. University Press, Cambridge. 1914.
3. Austen, E. E. The House Fly, etc. Econ. Series No. 1 (Revised). British Museum, London. 1939.
4. Séguy, E. Faune de France. **6.** Diptères Anthomyides. Lechevalier, Paris. 1923.

5. Keilin, D. Recherches sur les Anthomyides à larves carnivores. *Parasitology*. **9**: 325–450. 1917.
6. Smart, J., in Edwards, Oldroyd and Smart. British Blood-sucking Flies. 115–18. British Museum, London. 1939.
7. Grimshaw, P. H. On the British Species of *Hydrotaea* Dsv. *Ent. mon. Mag.* **41**: 239–47. 1905. **42**: 8–11, 72–7. 1906.
8. Collin, J. E. British Species of the Anthomyiid Genus *Limnophora* Dsv. *Ent. mon. Mag.* **57**: 94–100, 162–75, 238–48. 1921.
9. Collin, J. E. On some characters of possible generic importance in the *Hylemyia-Chortophila* group of the Anthomyidae. *Ent. mon. Mag.* **63**: 129–35. 1927.
10. Smith, K. M. A Text Book of Agricultural Entomology. University Press, Cambridge. 1931.
11. Hobby, B. M. Notes on Predacious Anthomyiidae and Cordyluridae (Dipt.). *Ent. mon. Mag.* **70**: 185–90. 1934.
    Hobby, B. M. Predacious Diptera and their Prey. *J. Soc. Brit. Ent.* **1**: 35–9. 1934.
12. Collin, J. E. The British Species of *Prosalpia* Pok. *Ent. mon. Mag.* **79**: 83–6. 1943.
    Collin, J. E. A Contribution towards the knowledge of the Anthomyid genera *Hammomyia* and *Hylephila* Rond. *Trans. ent. Soc. Lond.* **1920**: 305–26. 1920.
13. Karl, O. Die Tierwelt Deutschlands. Zweiflügler oder Diptera. **3.** Muscidae. Gustav Fischer, Jena. 1928.
14. Ringdahl, O. Svensk Insect Fauna. Vol. I. Muscidae. 1:334 pp.
15. Hennig, W., in Lindner, E. Die Fliegen der Palaearktischen Region. 63b. Muscidae. Schweizerbart, Stuttgart. 1955–64.
16. Fonseca, E. C. M. d'A. Translations of Ringdahl's Muscidae tables. *J. Soc. Brit. Ent.* as follows: Coenosiinae, **4**: 47–59; *Hydrophoria* and *Acroptena*, **4**: 75–83; *Mydaea* and *Helina*, **4**: 95–111; *Lasiops*, **4**: 161–7; *Hydrotaea*, **4**: 193–203; *Phaonia*, **5**: 1–17. 1952–4. Additions and corrections, **5**: 17–22. 1954.
17. Fonseca, E. C. M. d'A. A review of the British subfamilies and genera of the family Muscidae (Diptera). *Trans. Soc. Brit. Ent.*, **12**: 113–28. 1956.
18. Collin, J. E. Some additional British Anthomyiidae. *J. Soc. Brit. Ent.*, **4**: 169–77. 1953.
    Collin, J. E. The genus *Chiastochaeta* Pokorny. *Proc. R. ent. Soc. Lond.* (B) **23**: 95–102. 1954.
    Collin, J. E. Genera and species of Anthomyiidae allied to *Chirosia*. *J. Soc. Brit. Ent.* **5**: 94–100. 1955.

KEY TO THE PRINCIPAL GROUPS OF THE BRITISH CALYPTERATES

(Partly based on Lindner, *Die Fliegen der Palaearktischen Region*, Part 1, 1925)

1. Proboscis rudimentary; palpi rudimentary or absent 2
   Proboscis distinct...4
2. Vein 4 straight (Pl. **84,** 1)....GASTEROPHILIDAE p. 292)
   Vein 4 bent, converging toward Vein 3 apically...........3
3. Face broad and flat (Pl. **47,** 3 and Ins.)
   HYPODERMATINAE
   (Calliphoridae, p. 289)
   Face very narrow, with deep median channel (Pl. **47,** 4 and Ins.)
   OESTRUS (Oestridae, p. 291)
4. Hypopleural bristles present; Vein 4 bent [1] toward Vein 3 apically, often with appendix 5
   Hypopleural bristles absent; if Vein 4 bent, never with appendix 10
5. Abdominal tergites (dorsal plates) and sternites (ventral plates) separated by visible conjunctiva (outstretched membrane) which more or less surrounds the sternites (Pl. **40,** 2, 3, 4)
   PHASIINAE
   (Tachinidae, p. 271)
   Abdominal sternites not surrounded by visible conjunctiva ..6
6. Second abdominal sternite (the apparent first ventral plate) appears free and extends over edges of corresponding tergite (dorsal plate); postscutellum [2] absent or weakly developed.7
   Second sternite partly or entirely covered by edges of corresponding tergite; postscutellum present, strongly convex..9
7. Posthumeral bristle lower (nearer the pleura) than the presutural; arista long plumose; bristles on dorsum of abdomen generally weak (Pl. **36,** 3: **37,** 4)
   CALLIPHORINAE
   (Calliphoridae, p. 280)
   Posthumeral bristle as high or higher (as far from or further from the pleura) than the presutural 8
8. Thoracic squamae large, inner margin close to scutellum; fifth abdominal sternite not cleft at hind margin; grey, tessellated flies (Pl. **47,** 1, 2).........SARCOPHAGINAE
   (Calliphoridae, p. 275)

[1] Except *Cinochira* (Vein 4 complete and almost straight); in *Racodineura*, *Phytomyptera* and some species of *Actia*, apical or upturned portion of Vein 4 is missing altogether.

[2] See Glossary, p. 362; and *Cephenomya*, p. 292.

Thoracic squamae small, inner margin curving away from the scutellum ; fifth abdominal sternite, in males, cleft at hind-margin ..................RHINOPHORINAE
(Calliphoridae, p. 274)

9. Arista plumose ;[1] antennae inserted below middle of eye ; no presutural bristle present ; mostly slender flies [1] (Pl. **37**, 1, 2)
DEXIINAE
(Tachinidae, p. 269)

Arista bare or pubescent ;[2] presutural present ; if not, abdominal sternites partly visible ; antennae inserted above or at middle of eye (Pl. **36**, 1, 2, 4: **37**, 3: **40**, 1 ; **42**, 7, 8) **43**, 1)
TACHININAE, GONIINAE, and MAQUARTIINAE
(Tachinidae, p. 260)

10. Thoracic squama smaller than alar squama, vestigial or reduced to a narrow strip 11

Thoracic squama as long as, or longer than, alar squama, never reduced to a narrow strip ; lower, outer corner of anal cell, formed by cross-vein and Vein 6, never sharp-pointed ; abdomen with not more than five visible segments 12

11. Wings with costal spine near apex of *sc* ; abdomen with four or five visible segments ; female with crossed bristles on frontal stripe ; Vein 6 reaches margin (Pl. **42**, 6)
FUCELLINAE (Muscidae, p. 307)

Wings without costal spine ; abdomen with five or six visible segments ; female without crossed bristles on frontal stripe (Pl. **33**, 3, 4: **42**, 1, 2)......Scatophagidae (p. 250)

12. Vein 4 bent sharply or in a long curve, converging toward Vein 3 apically ; arista long-plumose (Pl. **44**, 3, 4: **49**, 1)
MUSCINAE (Muscidae, p. 296)

Vein 4 more or less straight, not converging toward Vein 3 apically ; if bent, then arista not long-plumose........13

13. Sternopleurals forming inverted isosceles or equilateral triangle (Pl. **49**, 5) ....COENOSIINAE (Muscidae, p. 309)

Sternopleurals otherwise disposed ; lower sometimes absent..14

14. Vein 6 reaches margin ......ANTHOMYIINAE
(Muscidae, p. 307)

Vein 6 not reaching margin 15

15. Hind tibia with dorsal bristle (Pl. **48**, 4)
PHAONIINAE (Muscidae, p. 303)

Hind tibia without dorsal bristle (Pl. **48**, 3)
MYDAEINAE
(Muscidae), including
LISPINAE, pp. 305 and 306

[1] Except *Trixa*, p. 270, and Pl. **41**, **2**.
[2] Except *Thelaira*, p. 268, and Pl. **42**, 7.

CHAPTER XXV

SECTION VI. PUPIPARA

Our last Section, a small one, comprises a group of insects which are hardly likely to be confused with any others in the order Diptera, except perhaps *Braula* (p. 242) and certain aberrant, wingless species of families such as Phoridae, Sphaeroceridae, etc., or *Carnus* (p. 238). Some of them are so different (Pl. **50**, 3, 4) that they hardly correspond to the usual conception of a fly, at all. They are generally considered to be aberrant descendants from the Muscid type and therefore really to belong to the Cyclorrhapha. Their present form seems to have evolved as a result of their specialised habits as external, bloodsucking parasites of warm-blooded animals, on the bodies of which they spend all their lengthy adult life. Man seldom becomes a victim to their attentions, and then only under fortuitous circumstances. Structures unnecessary to their way of life have, in a number of cases been shed, and others, more useful, developed to an extraordinary degree. Thus we find wings and wing-venation in various stages of atrophy (Pl. **44**, 2: **50**, 1, 2) or completely absent (Pl. **50**, 3, 4); where atrophy of the venation is in progress, it follows the usual tendency toward strengthening and crowding of the anterior, and reduction of the posterior veins (cf. Pl. **7**, 4: **10**, 1, 4: **22**, 1, 2: **38**, 2). Although the fully-winged species are capable of flight, they seldom indulge, except when disturbed or seeking a fresh host. The head and its appendages, particularly the eyes, antennae and mouthparts, also show various degrees of degeneration; sometimes the head is capable of being folded back into a sort of sheath or groove in the dorsum (Nycteribiidae).

The free larval stage is dispensed with, the female depositing a single fully-developed larva, which immediately pupates; hence the name Pupipara, which means, literally, "pupa-bearing". Usually the pupa is of a yellowish, reddish or brownish colour, more or less spherical in shape, with a dark patch at one pole, and looks rather like a small berry or seed.

On the other hand, the structure exhibits certain specialised adaptive developments related to existence as an external parasite; the integument is tough and leathery and the body depressed, i.e. flattened in profile. The legs and claws, especially the latter, are extraordinarily developed and make it possible for these flies to cling with great tenacity to the hairs or plumage of their hosts. Although the mouthparts have certain features in common with those of the Biting Muscids (p. 254), they appear as slender curved tubes, with a dilated or bulbous basal part; here, also, both sexes are bloodsuckers.

There are two British families for review, viz., Hippoboscidae and Nycteribiidae. The Palaearctic species of both families have been monographed by Theodor and Oldroyd[1] and Theodor.[2]

HIPPOBOSCIDAE.

9 genera. 13 species. Head fairly large, not folded back on to dorsum, but closely fitting into an accommodating excavation of the thorax; claws long, curved, toothed at base; wings sometimes reduced or absent; parasitic on birds and large mammals. (Pl. **44,** 2: **50,** 1–3.)

Hippobosca equina L. (Pl. **50,** 1) is the common "Forest Fly" of the New Forest and may be found there on wild forest ponies and domestic animals, such as horses and cattle; it has been recorded occasionally from other parts of the country, as far north as Scotland. The adults are in evidence from May to October and are winged in both sexes; overwintering probably takes place in the pupal state. Sometimes on cattle, where the species is easier to observe than upon the shy and elusive forest ponies, large clusters may be seen under the tail, on the base of the udder, or on the inner surface of the upper parts of the hind legs. When disturbed, these flies move rapidly sideways, rather like a crab, and are not easily dislodged, since their long, curved claws enable them to maintain a firm hold on the hairs. Their bite seems to cause little inconvenience or irritation to the native beasts of the forest, but their clinging movements usually cause an animal, not acquainted with them, to become panic-stricken; this will be better understood if one of the flies be allowed to crawl on, say, the hairy part of the forearm. We have often been told by elderly foresters that, in the days of horse-drawn vehicles, when forest folk went, on market days, to the neighbouring large town of "Bourne" (Bournemouth), much excitement would ensue when a "Forest Fly" deserted its host for the more tender hide of a "town" relative—resulting in wildly-careering milk-carts and tradesmen's vans ; cursing, perspiring drivers, and sudden scrambles to safety by the pedestrian population. Other tales told to us include variations on the theme of " Forest Flies " being gathered by youthful, mischievous hands, imprisoned in a match-box and surreptitiously released at home, thereby causing long-haired young ladies, suddenly feeling the clinging movements of the flies, to become much put out, and subsequently to inflict well-merited chastisement upon younger brothers.

Ornithomya, Lynchia, Stenepteryx and *Crataerina* are bird-parasites, which are winged in both sexes ; the first two possess moderately-developed wings, but in the latter two cases, these organs are so much atrophied that their owners are incapable of flight. *Ornithomya avicularia* L. (Pl. **44,** 2) is commonly to be found on a variety of birds, especially nestlings, including hawks, owls, pigeons, jays and finches. When disturbed, it moves rapidly sideways amidst the plumage of the host ; occasionally it will transfer itself from a shot bird to one's person, where it produces little else than an uncomfortable sensation. *O. chloropus* Bergr. (=*lagopodis* Sharp) is smaller and darker, with dark palpi (in *avicularia* they are pale), and dark markings on the underside of the head and thorax; it seems to occur mainly on red grouse, but has been recorded from a number of other birds. *O. fringillina* Curt. is smaller still, pale green in colour, also with dark palpi, but without dark markings below; in England it has been recorded mainly from robin, dunnock, warblers and finches. *Ornithomya* species appear generally not to occur on aquatic birds, except waders, but Smart[3] records *chloropus* from the Arctic skua. Papers by Hill[4] treat the British species of *Ornithomya.*

Stenepteryx hirundinis L. (Pl. **50,** 2) occurs on martins and swallows, and the similar *Crataerina pallida* Latr. on swifts. Overwintering pupae may be found in the nests while the birds are away from this country. In *hirundinis,* it will be seen that minute ocelli are present and that the strap-like wings are nearly as long as the legs; in *pallida*, ocelli are absent and the wings are broader, about half the length of the legs.

The Deer-Ked, or Deer Fly, *Lipoptena cervi* L., is interesting on account of its habit of shedding its wings; the newly-emerged adults possess these appendages in both sexes, but, it seems, dispense with them when they have taken up their abode upon a host (cf. *Carnus*, p. 238), usually the roe, red or fallow deer. Winged males are occasionally encountered in the free, but, according to Smart[1], the females lose their wings on being killed. The wingless form is not unlike the following species, except for the remains of the basal part of the wings and the presence of halteres. Haarlov[5] has studied the biology.

Melophagus ovinus L. (Pl. **50,** 3) is variously known by country folk as the Sheep-Ked, Sheep-Tick or Sheep-Louse, and is completely wingless in both sexes; even halteres are absent. The whole life-history is passed within the fleece of the host, where the adults look like small spiders crawling about. Whilst very little suffering seems to be occasioned by the bites, the attentions of *ovinus* may sometimes cause the sheep to rub themselves against fences, posts or other places and the resulting sore patches may present an opportunity to *Lucilia* or *Calliphora* species (p. 284) to infest the part. The life-history of *ovinus* has been studied by Evans.[6]

NYCTERIBIIDAE.

3 genera. 3 species. Head very small, capable of being folded back into a groove on the dorsum; legs long, claws well-developed; wings absent; parasitic on bats. (Pl. **50,** 4.)

Nycteribiidae are tiny, spider-like, wingless insects, parasitic upon bats. *Stylidia biarticulata* Hermann (Pl. **50,** 4), which is a little over 3 mm. long, occurs on several species of host, including the greater and lesser horseshoe bats and the common pipistrelle. *Nycteribia kolenatii* Theo. is slightly smaller; it is usually recorded from Daubenton's bat. *Basilia nana* Theo. has been recorded once, from Bechstein's bat.

Adult Nycteribiidae leave their hosts to deposit their larvae on beams and walls, where they immediately pupate. The structure of the adults is somewhat similar to that of *ovinus*, but even more degenerate; the head is much reduced and, when at rest, is folded back in an accommodating groove or sheath on the upper surface of the thorax; the femora and tibiae are thickened and the metatarsi long, slender and curved; at the hind extremity of the abdomen there are two cerci with apical bristles.

Further reading may be sought in Smart[3], Theodor and Oldroyd[1], Theodor[2], and Falcoz.[7]

This brings us to the end of the Order. To the reader who is unfamiliar with Diptera, it will, no doubt, seem a far cry from the large, long-winged, long-legged Tipulids and the robust, fast-flying Tabanids to the tiny, wingless Nycteribiidae, which only take to the air when transported by their hosts. Doubtless, mental comparison will unconsciously be made between the beautiful Stratiomyids and, say, some of the drab Calypterates, or between the pleasant, flowery haunts of some of the Syrphidae and the often revolting habitats of the Scatophagids and Calliphorids. Whatever the reactions may be in other respects, we feel that the non-entomological reader will agree with us, that here, indeed, is another " world ", of which so very much may still be learned by those who are fortunate in possessing a patient and enquiring mind and who are prepared to observe.

LITERATURE CITED

1. Theodor, O. and Oldroyd, H., in Lindner, E. Die Fliegen der Palaearktischen Region. VIII(65). Hippoboscidae. Schweizerbart, Stuttgart. 1964.
2. Theodor, O. *Ibid.* VIII(66A). Nycteribiidae. 1954.
   Theodor, O. An illustrated Catalogue of Nycteribiidae in the Rothschild Collection and in the British Museum (Nat. Hist.). Cambridge. 1967.

3. Smart, J. in Edwards, Oldroyd and Smart. British Blood-sucking Flies. 118–27. British Museum, London. 1939.
4. Hill, D. S. Revision of the British species of *Ornithomyia* Latreille (Dipt., Hippoboscidae). *Proc. R. ent. Soc. Lond.* (B) **31**: 11–8. 1962.
Hill, D. S. A study of the distribution and host preferences of three species of *Ornithomyia* (Dipt., Hippoboscidae) in the British Isles. *Proc. R. ent. Soc. Lond.* (A) **37**: 37–48. 1962.
Hill, D. S. The life-history of the British species of *Ornithomya* (Dipt. Hippoboscidae). *Trans. R. ent. Soc. Lond.* **115**: 391–407. 1963.
5. Haarlov, N. Life Cycle and Distribution Pattern of *Lipoptena cervi* (L.) (Dipt. Hippoboscidae) on Danish Deer. *Oikos*, **15** (1): 93–129. 1964.
6. Evans, G. O. Studies on the bionomics of the sheep ked, *Melophagus ovinus* in West Wales. *Bull. Ent. Res.*, **40**: 459–78. 1950.
7. Falcoz, L. Faune de France. 14. Diptères Pupipares. Lechevalier. Paris. 1926.

APPENDIX

COLLECTION, PRESERVATION AND EXAMINATION OF FLIES

What to Collect.

This will depend on the spare time available to the collector and the kinds of locality which are accessible. At first, it would probably be easier and more satisfactory to collect the largest and commonest flies seen on each expedition ; only to take home such a number of specimens as can comfortably be dealt with in the time available ; and to become as familiar as possible, by carefully recorded observation, with their habits in the field. Concurrently with this, as an aid to further observation, the foregoing chapters and any works available on the families to which they belong, should be consulted.

Later, it is probable that interest will tend to centre on one or two particular groups. In order to make a more thorough study of these, it will be necessary to identify species accurately ; this may be essayed with the aid of keys such as those cited in the bibliographies and by reading very carefully special works on the groups chosen. Identifications should be checked by close comparison with reliably-named specimens in private or museum collections, if accessible, or by submitting specimens to specialists for confirmation. Most specialists are usually pleased to assist and expect very little for their services except, occasionally, permission to retain a specimen or two of particular interest to them. It is of course advisable to write to specialists first as they are very busy people.

It is felt that to collect flies for the sake of collecting is a mistake ; to observe and learn something of their way of life will be found most satisfying to the collector and of real scientific value.

If specialisation be undertaken, there is plenty of scope for original observation, the present state of knowledge of some families being very restricted. However, it would perhaps be advisable for the beginner to study very carefully, in the first instance, a family which has been well monographed and keyed, and of which specimens are not so uncommon as to be difficult to obtain in reasonable numbers, e.g. Stratiomyidae, Tabanidae, Tipulidae, Syrphidae and Tachinidae; thus gaining valuable practice in determining specimens and learning something of the methods and terminology adopted by experienced dipterists.

For the reasons stated under "Genitalia" in the Introductory Chapter (p. 32), pairs in copula should always be taken. Immature stages, empty larval skins or puparia, plant-galls, etc. should be preserved, where possible, together with the corresponding adults.

WHERE AND WHEN TO COLLECT.

There can be few places where flies of one kind or another cannot be found. Whilst the greater number are in evidence during spring, summer and autumn, others may be found by the patient seeker in the winter months, flying in mild, damp weather, or hibernating in various situations. Most flies delight in the sun, but some may normally be found in shady places, under foliage, in caves, animal burrows, or the nests of bees, wasps and ants. On the whole, damp situations are favoured; but the dry, sandy commons and heaths have their own special fauna. If collecting adults should prove unprofitable on any occasion, recourse may always be had to searching for immature stages; the water-net can be used in ditches, ponds and streams. Earth, mud, flood-refuse, dead leaves, rotten wood, fungi (especially decomposing ones), excrement, carrion and plant-galls may all be investigated and attempts made, where feasible, to breed out the adults.

PLATE 43

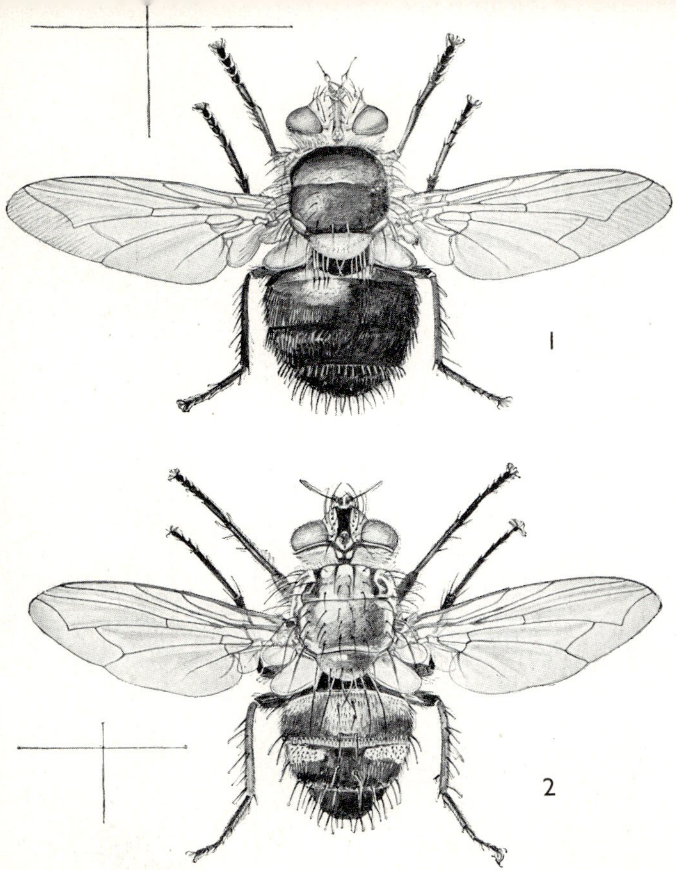

TACHINIDAE-VIII

1. *Echinomya grossa* L., ♀, p. 267.
2. *Ernestia radicum* F., ♀, p. 265.

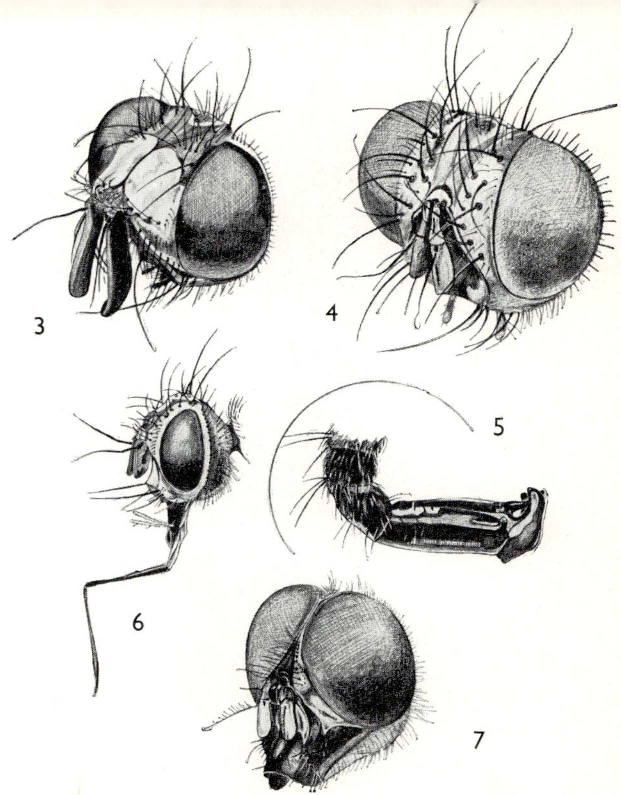

CALLIPHORIDAE-V

3. *Metopia argyrocephala* Mg., ♂ head, p. 279. 7. *Stomorhina lunata* F., ♂ head, p. 286.

TACHINIDAE-IX

4. *Voria ruralis* Fall., head, p. 264. 5. *Weberia pseudofunesta* Vill., ovipositor, p. 272. 6. *Siphona geniculata* Deg., head, p. 263.

Plate 44

GASTEROPHILIDAE
1. *Gasterophilus intestinalis* Deg., ♀, p. 292.
HIPPOBOSCIDAE-I
2. *Ornithomya avicularia* L., ♀, p. 315.

MUSCIDAE-II
3. *Orthellia cornicina* F., ♀, p. 300.
4. *Dasyphora cyanella* Mg., ♀, p. 300.

BREEDING.

This is a very interesting and highly instructive way of collecting and studying flies ; in fact, it is sometimes the only practical way of obtaining various stages of certain species. For instance, where immature stages are little-known or difficult to come by, breeding out from a gravid female (another reason why pairs in cop. are important) may be possible. Conversely, other species, of which the larvae are common but the adults little in evidence, may be bred out to the adult stage and, after pairing, round to full-cycle again.

Not a great deal of assistance will be obtained from the literature, but much may be accomplished by patience, experiment, and ingenuity in simulating, as far as possible, the natural conditions experienced by the species which is being bred. It is sometimes possible to breed " out of season " with species which have a short life-cycle. Whilst the breeding of some flies, such as Simuliidae (p. 79), the immature stages of which require fast-flowing, highly-oxygenated water for maintaining existence, may necessitate laboratory conditions and complicated equipment, much may be done in the home with simple gear, such as jam-jars, paste pots, test-tubes, etc. Larvae which normally pupate in the earth may be provided with clean sand, fine earth, leaf-mould, or similar material, according to judgment or experience, in the jars ; humidity may be provided by not-too-liberal applications of mist from an ordinary scent-spray or atomiser (care being taken, of course, not to use water from the domestic supply which may be chlorinated, but from, say, a water-butt, or unpolluted stream or pond). Where the life-cycle is short, often a glass slip will suffice as a cover for the breeding receptacle, for the purpose of preventing the escape of the emerging adults ; in most cases, however, so as to avoid excessive condensation on the sides of the receptacle, which would probably result in damage to specimens when they emerge, or encourage the growth of moulds, it is better to use some such material as cheese-muslin, which will permit egress of moisture, which can be renewed, and likewise permit the passage of air.

Where a temperature higher than the normal for the room is required, it should not be too difficult for the handyman to devise some form of continuously-heated breeding receptacle by means of electric lamps and suitable flues for conducting warm air therefrom round the outside of the chamber, or through it; humidity could be provided by allowing the heated air to pass over a container of water, to enter the chamber through, say, muslin at a low point and escape from a high one, likewise. In the case of aquatic immature stages, the usual types of aquarium immersion-heaters could be utilised, and so on.

In order to avoid arriving at erroneous and misleading conclusions, care should be taken to ensure that adults bred out do, in fact, correspond with what the collector *thinks* are the immature stages, i.e. he should be satisfied, where possible, that such immature stages (where they are not already identifiable) are the only ones present in the development-medium used.

Examples of ecological studies which indicate what species may be found in a specialised habitat are given by Buxton, Hackman, Laurence, Smith and Woodroffe.

Collecting Apparatus and Collecting Techniques.

In our experience, the simpler the gear used the better; we have tried many special contrivances devised by our friends and ourselves (some of them highly ingenious in theory but mostly a great nuisance in practice); whatever is carried should be light, easily stowed and of such a nature as not to impede the free movement of the collector, who may require sometimes to move quickly and easily over difficult terrain. These things are, however, very much a matter of personal taste; a series of collecting expeditions, particularly in the company of other dipterists, will soon make the advantages or disadvantages of particular pieces of equipment quite evident.

An ordinary butterfly-net with a fine-mesh bag alone will suffice for the capture of larger flies; most needs will be served with the aid of such a net, together with a sweep-net, small water-net, suction-bottle, a supply of assorted collecting tubes (empty aspirin tubes are useful), some rubber bands, a wide-necked jar with bung, a tin or two, and a stout knife or trowel, most of which items can be carried in, say, a rucksack or haversack and the pockets. We consider a haversack more convenient, since it can be taken off the shoulder rapidly and set down if the need arises.

As regards the butterfly-net, we have found that, for most occasions, one of the small, inexpensive, collapsible nets sold by most entomological dealers, with a spring-steel frame, which, when folded, will fit easily into the pocket, is a most useful weapon, used with or without a short handle. Of course, when out for particularly fast or high-flying "game", e.g. *Criorrhina* (p. 169), a long-handled, wide-aperture net will increase the chances of success, but much may be done with ordinary appliances and a reasonable amount of patience. In any case, interesting specimens almost invariably (and most perversely) appear when the special apparatus is not being carried, or equally perversely disappear as it is being hurriedly assembled. There are occasions when a net with a very small aperture, like a boy's fishing-net (Pl. **51**), may be distinctly advantageous, as for instance, when working broken foliage and being desirous of not disturbing the quarry by prior contact with adjacent leaves, which would probably result from the use of a net with a larger aperture. Moreover such a net, in ordinary circumstances, often makes up for the smallness of its aperture by its greatly increased manoeuvrability. A long bag, as depicted, also makes it more difficult for an insect, once within it, to escape, since after a stroke, a quick twist of the wrist will cause the bag to hang in such a way as to seal off the aperture, a procedure which is necessary with any kind of net unless the quarry is on the ground, when the net is usually placed over it quickly and the frame kept flush with the ground.

Another piece of equipment which is useful under certain circumstances is the " forceps ", a scissor-like contrivance in which the " blades " take the form of small frames over which is stretched some fine muslin or similar soft material. The edges of the frames are thickly bound or padded, so that when they are tightly closed a space is left between the two pieces of material, thus forming a trap from which the captured insect cannot escape. This instrument can often be used under conditions where a net-bag is liable to be torn, e.g. on thorn or bramble. Extraction of the insect is not so simple, however; it is best accomplished by inserting the closed forceps into the ordinary net and opening it within, trapping the escaping insect in a fold. There is a modification of this appliance, known as the net-forceps, in which one of the frames bears a miniature bag; this we consider is unsatisfactory in practice. With both instruments, in making a stroke, care is necessary to avoid closing the frames on, say, a thick stem which will keep them apart and allow the quarry to escape.

Strokes should always be decisive; if the quarry is missed, then the collector is unlucky; a hesitating stroke will usually achieve little, in any case; the time for caution is before the stroke is made and here experience will play a large part. Precaution should be taken not to allow one's shadow to fall upon the quarry (many flies are extraordinarily sensitive to changes in light-intensity); for the same reason, sudden movements near a fly may cause it to start off. On the other hand, certain flies are very sluggish and may often be captured merely by placing a collecting-tube over them and corking it as they fly up into it.

All bags should be furnished with a strong hem through which the frame is passed, consisting of some such material as canvas, unbleached calico, drill or soft leather which will

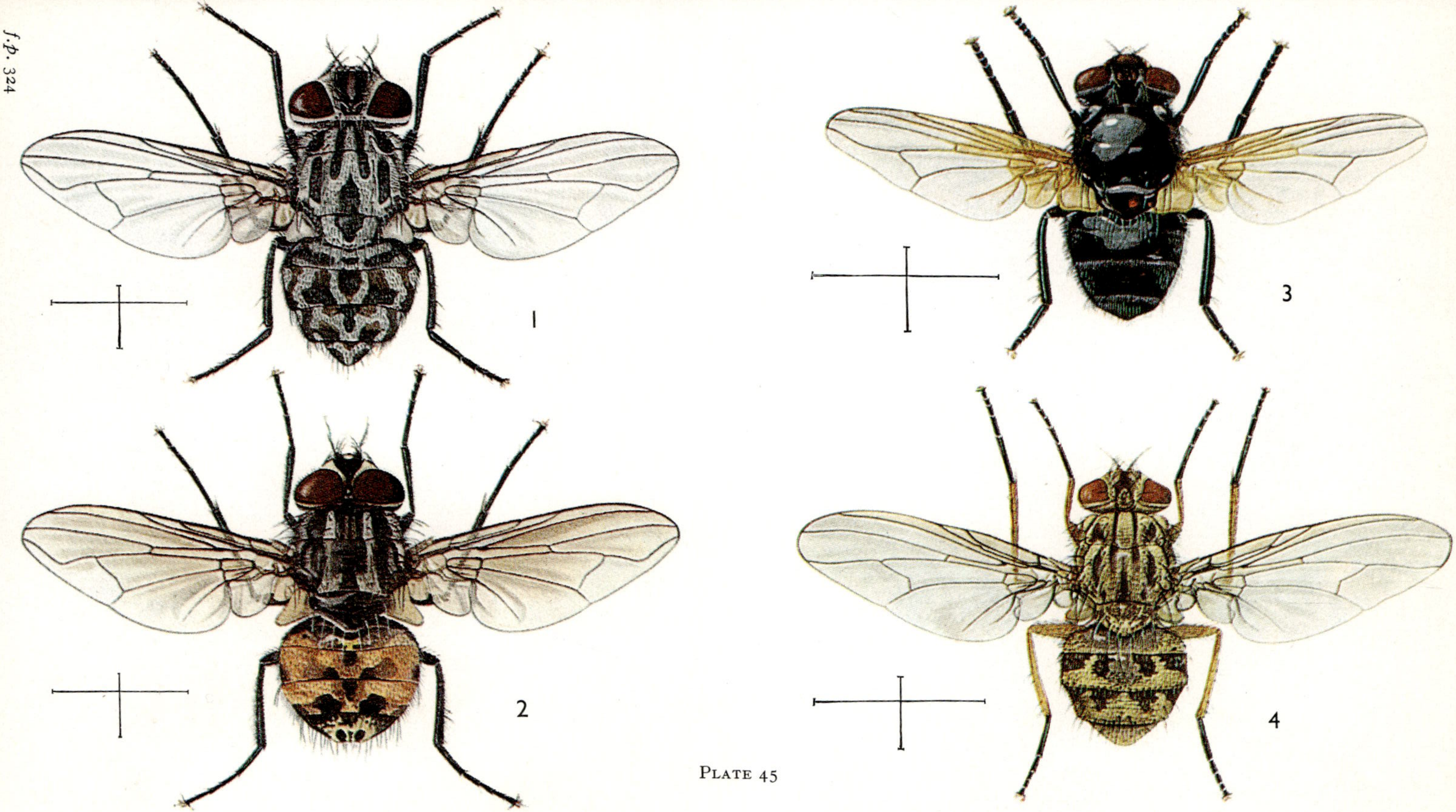

PLATE 45

MUSCIDAE-III

1. *Graphomya maculata* Scop., ♀, p. 299.
2. *Graphomya maculata* Scop., ♂.
3. *Mesembrina meridiana* L., ♀, p. 301.
4. *Phaonia erratica* Fall., ♀, p. 303.

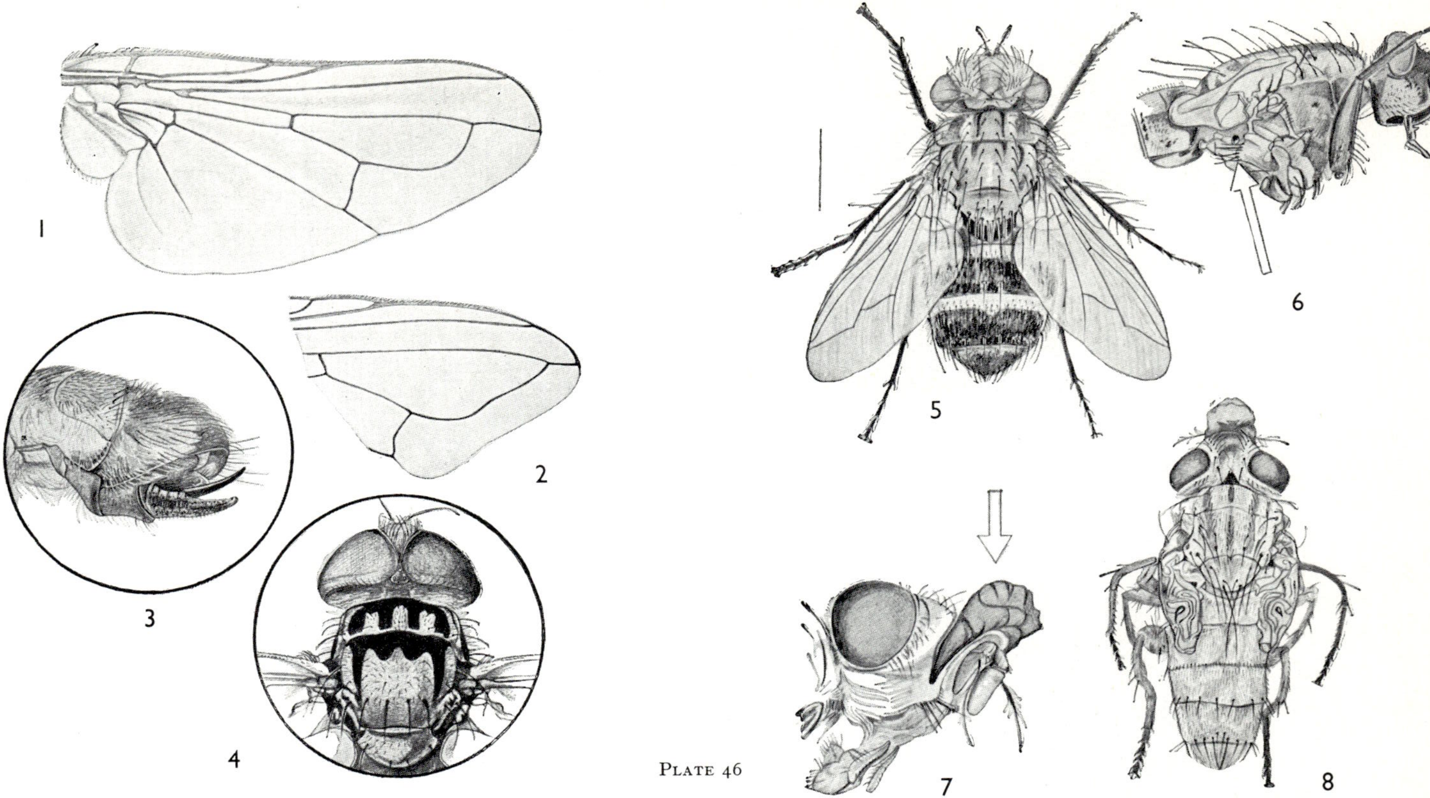

PLATE 46

TACHINIDAE-X

1. *Alophora pusilla* Mg., wing, p. 272. 2. *Alophora hemiptera* F., apex of wing, p. 272. 3. *Alophora hemiptera* F., ovipositor, p. 271. 4. *Alophora obesa* F., dorsum, p. 272.

5. *Gonia* (=*Salmacia*) *sicula* R–D., ♀, in resting attitude, p. 261.

6. "Fan" of hypopleural bristles (Tachinidae and Calliphoridae), p. 260.

CALLIPHORIDAE-VI

7. *Sarcophaga carnaria* L., newly emerged, with ptilinum inflated, p. 277.

8. *Sarcophaga carnaria* L., newly emerged, with ptilinum inflated, dorsal aspect, p. 277.

withstand the hard wear this part receives in contact with grasses, twigs, plant-stems, etc.

Removal of specimens from the net-bag is easily accomplished with the collecting-tubes; the specimen is trapped in a fold, the uncorked tube cautiously inserted and the fly induced to enter; the net material may then be held with the finger over the mouth of the tube while the cork is introduced into the net with the other hand and inserted; or, the finger placed over the aperture of the tube, the tube withdrawn and the cork inserted outside the net. The former procedure is advisable since, if the fly is very active, it merely escapes into the net and may again be trapped in a fold; nothing is more disappointing than losing a coveted specimen after it has been captured! Removal may also be accomplished, in the case of small or very small flies, by means of the suction-bottle (p. 327).

So far we have considered the hunting of individual flies, which is often an exhilarating and absorbing occupation in fruitful territory; there are, however, times when such sport is found to produce poor results, e.g. on dull, cool, or excessively windy days. The collector may then be compelled to resort to collecting, or if specialising may specially desire to collect, small or very small flies which do not freely expose themselves and therefore are not so readily taken with the ordinary net; moreover they may, in some cases, be so small as to be able to escape through the mesh of such a net. Here then the sweep-net comes into action. This should be of strong construction, so far as the frame and handle are concerned, but the bag-material needs to be fairly soft and as transparent as possible so as to permit the location of enclosed insects from outside (at least when the bag is held up to the light). White material will show up small insects more clearly when the inside of the bag is being examined. We have found that organdie is very suitable for the purpose, since it has a sufficiently fine mesh to prevent the escape of the smallest flies and yet is sufficiently transparent, when clean, to allow insects to be located from outside, in fact for some kinds to be recognised by general shape or gait, e.g. Phoridae, Lonchopteridae, Opomyzidae, etc.; such can be usually picked out and tubed in much the same way as with the ordinary net, even though a large number of other "small fry" be present.

In using the sweep-net, the usual method is to sweep it rapidly to and fro amongst grasses, low herbage, waterside plants, the foliage of bushes and trees, and over mud or water, twisting the wrist at the end of each stroke so as to keep a continuous flow of air into the net and to prevent the egress of any captured insects. At the end of the desired number of strokes, the wrist is turned as with the ordinary net, so as to let the bag fall over and seal off the aperture; or the "tail" of the net, in which most of the occupants will be concentrated immediately after the net has ceased moving, may be sealed off by gathering in the middle of the bag with the hand, thus restricting the movement of the enclosed insects. Removal may then be effected with tubes or the suction-bottle (p. 327). Some collectors employ a net with a very wide aperture and very large bag into which the head is inserted after sweeping, the tail of the net being held up with one hand so as to induce the occupants upward (flies in captivity usually gravitate towards the strongest light, therefore upward in this case) and the specimens taken as desired with a suction-bottle.

The hem of the sweep-net, especially, should be of stout material to resist wear; this will be greatly assisted by attaching the bag to the frame with metal curtain-rings, so that the metal frame takes the brunt of impact at each stroke. A folding or collapsible frame is of great advantage for storing the net in the pocket or haversack, providing that, when assembled, it is sufficiently strong and rigid. Such a net is illustrated on Pl. **51**; it will be noticed that it is being used without a handle, merely the ferrule being grasped; this facilitates twisting with the wrist at the end of each stroke. If, however, sweeping is being done at ground level, this is tiring, owing to the back being bent for protracted periods; a short handle, which can be slipped into the ferrule, is then useful.

A sweep-net should not be used when grasses or herbage are wet from recent rain or dew; any specimens thus taken will almost inevitably be in a bedraggled condition and ruined for the purpose of critical examination or retention in the collection.

The suction-bottle or "pooter" is a simple piece of apparatus obtainable from the entomological dealers or easily made for one's self. A variety of forms may be found in use but the operating principle is the same, i.e. the quarry is sucked into a container through a narrow tube and is prevented from being drawn therefrom by means of a piece of muslin or similar material over the aperture through which the suction is applied. A typical "pooter" constructed from an ordinary glass bottle is shown in action on Pl. **52**; the detail in Fig. 13 should be self-explanatory. As an alternative to suction by the mouth, rubber bulbs with valves, as used in scent-sprays or atomisers, may be used. We find it a good plan to have the small cork which fits into the inlet tube (to prevent the escape of captured flies) secured to the bottle by a thin cord or piece of twine, since it is one of the things which may easily be lost in the excitement of a sudden chase just after using the bottle. A little experience with the "pooter" will bring the realisation that it is wise, when bringing it into action, to give a sharp intake as the small cork is removed from the inlet tube, in order to clear any small flies which may have wandered up the inlet tube and thus be liable to escape. Secondly, having placed the inlet tube as near the quarry as possible, a sudden and decisive intake is necessary, since some flies are able to cling very firmly to the surface upon which they are resting and to resist moderate, prolonged suction. One objection to the use of the suction-bottle is the difficulty of recording the exact locality in which each of the contained specimens is taken if more than one locality is worked on one expedition. This, of course, is an additional argument for not taking too many specimens at a time; in any case, overcrowding, especially if any robust and active flies are included, will increase the chances of damage through collision and tangling, and of condensation within the bottle, which will probably render most, if not all, the

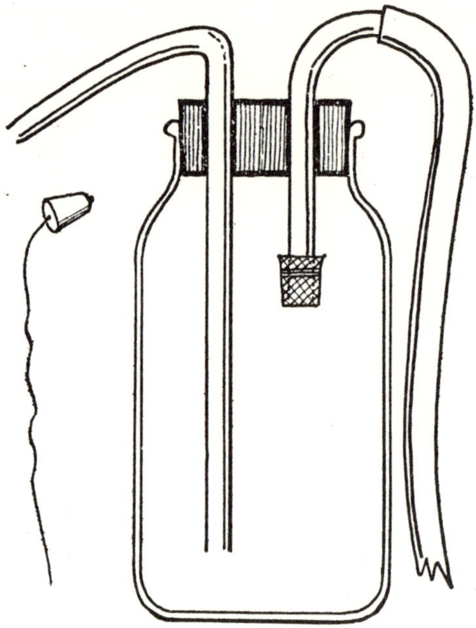

FIG. 13.—THE SUCTION-BOTTLE OR " POOTER ".

specimens useless. Predators should not be taken in the " pooter " since some of them, e.g. *Scatphaga* (p. 252), will, on occasions, take prey even when thus closely confined.

Corked collecting tubes are very useful; it is a good plan to stick on one side of each tube a piece of plain label-paper on which lightly pencilled data can be recorded and removed with an eraser for re-use on the next expedition. The white background thus provided is often very helpful when examining captures in the field with a pocket lens. Tubes should also never be overcrowded; we find it convenient to use a separate tube for every capture, so that there is no risk of damage or confusion of data. It is surprising how many tubes can be carried in the pockets of an old jacket; they can, of course, be carried thus in tins to lessen the risk of breakage, but we prefer to carry more tubes and dispense with the tins. The arrangement should be systematically worked; for instance, we carry unused tubes in a left-hand pocket and transfer them, as they are used, to a right-hand pocket; tubes containing captures from a particular locality are secured together with the rubber bands, which is a convenience later when killing, mounting and recording. Various sizes of tubes should be carried, experience demonstrating the appropriate sizes of tubes which are convenient to use for various sizes of flies. It is a good fault to err on the side of too large a tube rather than too small, as thus the risk of condensation is lessened. Tubes should never be left lying in the sun, as this will also cause condensation and death of the flies, in all probability. The latter may not matter a great deal if mounting is to be done shortly afterward, but as a rule flies stiffen fairly quickly and it is therefore better to keep them alive as long as possible before mounting.

It is a good practice to wash out all tubes fairly regularly with hot water and thoroughly dry them before replacing the corks; some flies, such as Syrphidae and Sphaeroceridae, regularly foul the inside of the tubes with excretions which may spoil later captures and make observation difficult.

As regards the water-net, the usual small funnel type, with a glass tube at the "tail", is adequate. The wide-necked jar with bung will serve as the receptacle for captures, or as a general receptacle for transporting home fungi, rotten wood fragments, galled stems, etc., containing immature stages for breeding out. In this connection, the tins and trowel or knife will be found to have many uses; a white waterproof sheet, or failing this, sheets of newspaper will be found useful for spreading out debris for examination.

Killing.

At home, the equipment required will include killing agents, setting-needles, a pair of large curved entomological forceps, a pair of fine-pointed forceps, some artist's brushes (preferably camel-hair), some pieces of thick sheet cork to one side of which white paper has been pasted, two or three small pipettes, a watch-maker's eye-glass (or large lens on an adjustable stand, so as to leave both hands free to work), some sheet celluloid, amyl acetate, alcohol, blotting paper, entomological pins, and storeboxes.

The large, curved forceps and fine-pointed forceps may be obtained from entomological dealers and the latter also from suppliers of microscopical requisites; as a makeshift, ordinary "tweezers" can be made to serve for many operations. Setting-needles may be easily made by inserting fine sewing-needles into soft wooden handles, or into sealing-wax melted into the end of a piece of thin glass tubing; the eye-end of the needle may be heated in a flame with the aid of the forceps and thus easily fixed. Fountain-pen fillers will serve quite well as pipettes.

In our experience, the most generally useful killing agent is ethyl acetate (acetic ether) which may be obtained from the usual entomological suppliers; other agents may be employed, such as carbon tetrachloride, ether, chloroform, or various highly-volatile spirits including benzene, trichlorethylene, petrol, etc., but these all tend to stiffen the specimens unduly and make them difficult to arrange when mounting. Ethyl acetate, however, leaves them limp for a while and, moreover, usually causes them to die with the mouthparts and genitalia extruded, which is a very great advantage. Fresh young laurel leaves, chopped or crushed, are satisfactory as a killing agent, but a little tedious to prepare and have no particular advantage over ethyl acetate, properly used.

Certain kinds of flies should, if they are to be preserved in good condition, be killed at once in the field, and forthwith pinned or mounted, e.g. Mosquitoes, the scales of which are soon detached, or Bombyliidae, in which the dense pubescence is very easily and quickly damaged. The type of killing-bottle normally used by the lepidopterist or coleopterist is not very convenient for the dipterist, except perhaps for use with very large flies. Large collecting-tubes containing a small, tight wad of cellulose wool or blotting paper, *lightly* damped with ethyl acetate, at the end, will serve admirably for the purpose; they should not be left in the sun or kept in a pocket where they will become warm, otherwise the sides will be covered with ethyl acetate " sweat ". The specimen should be removed shortly after death, i.e. before it has had time to be damaged by condensation or movement in the tube. As an alternative to the cellulose wool or blotting paper " pad " in the tube, a few small pieces of celluloid sheet wetted with ethyl acetate may be inserted and retained in position with a disc of stiff paper or wad of cellulose wool; the celluloid swells and absorbs the acetate, releasing it slowly; such a tube should last out a day of collecting, but in any case may easily be recharged in the field.

Tipulids, Chironomids, etc. are liable to lose legs if kept alive in captivity for any length of time; one way of overcoming this is to kill them in a tube such as that described and remove them immediately after death into a larger storage-tube or wide-necked bottle or jar, with bung or screw-top. A good number of specimens may thus be kept, separately, between thin layers of cellulose wool or blotting paper, lightly damped with ethyl acetate and will be in good condition for mounting at the end of the day.

Mass collections of specimens in the sweep-net, which it

may be desired to retain and examine later, at leisure, may be quietened by inserting the sealed-off tail of the net into the wide-necked jar, which should contain for this purpose a wad of cellulose wool or blotting paper moistened with a few drops of ethyl acetate. The bung should be replaced in position to wedge the net and seal off the jar. After two or three minutes the insects will have quietened off sufficiently to be tipped out of the net into the jar or into a tin containing blotting paper damped with acetate, where they should remain fairly well relaxed for the rest of the day. Again it is important to ensure that not too much acetate is used; if considered necessary, a very small piece of leaf or one or two freshly killed large flies will provide all the moisture required, provided the receptacle is airtight.

On return home, live specimens in the collecting-tubes may be killed by introducing a tiny slip of blotting paper, lightly damped with ethyl acetate, carefully into the tubes and replacing the corks. They should be removed fairly quickly for mounting; if a large amount of material awaits attention, it is better to deal with a few tubes at a time.

Specimens in the "pooter" may also be killed by introducing small pieces of rolled blotting paper, damped with acetate, into the inlet tube, replacing the cork, and jerking them gently down into the bottle; care should be taken, however, after removing the dead flies, to *blow* any traces of acetate out of the bottle and allow any moisture from the breath to dry out before replacing the corks. In this way, inhalation of ethyl acetate vapour on the first use of the "pooter" on the next expedition will be avoided; it is most disconcerting and unpleasant for this to happen accidentally.

Relaxing.

It is possible for large and medium-sized flies to be relaxed by enclosing in a moist atmosphere for about twenty-four hours or such time as may be required to soften them suffici-

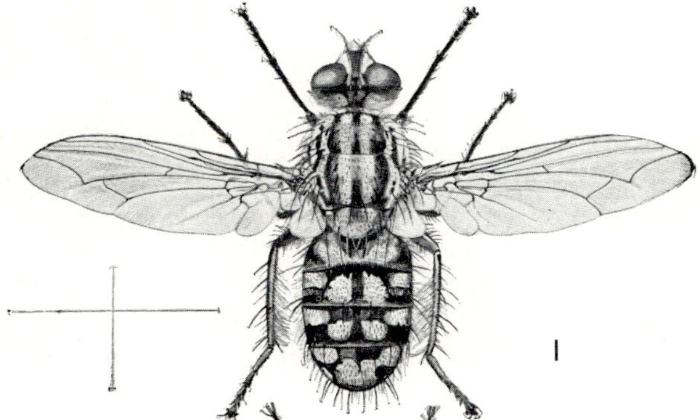

1

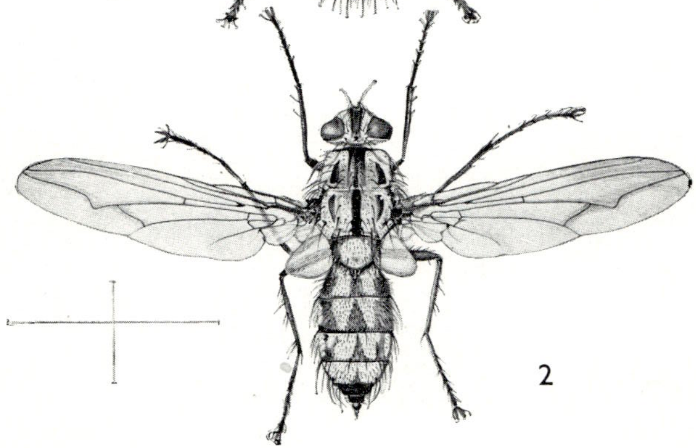

2

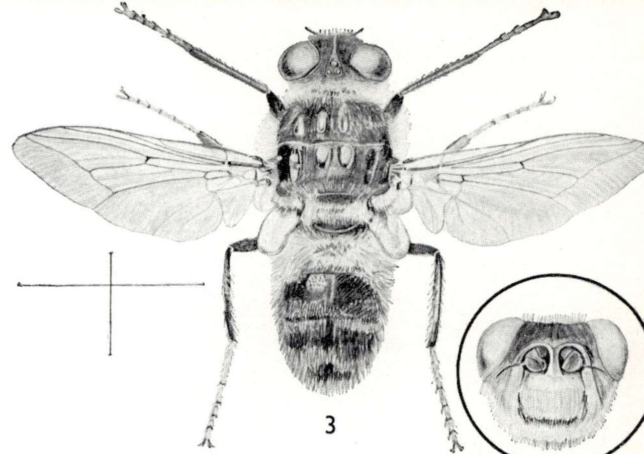

3

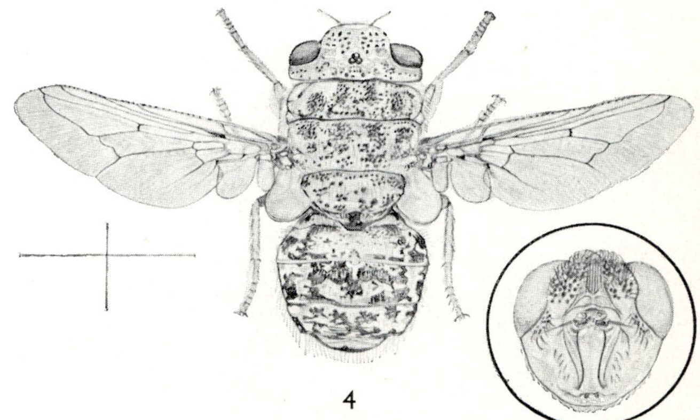

4

PLATE 47

CALLIPHORIDAE-VII

1. *Sarcophaga carnaria* L., ♂, p. 275.
2. *Macronichia ungulans* Pand., ♀, p. 280.
3. *Hypoderma lineatum* de Villers, ♂, p. 290.

Inset 3. *Hypoderma lineatum* de Villers, front of head, p. 290.

TACHINIDAE-XI

4. *Oestrus ovis* L., ♀, p. 290.

Inset 4. *Oestrus ovis* L., front of head, p. 291.

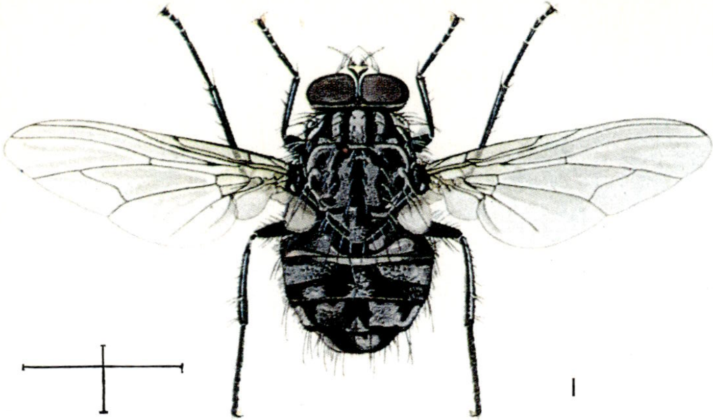
1

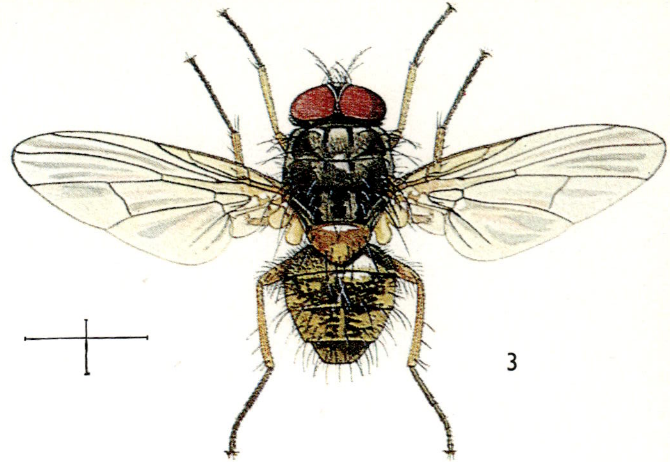
3

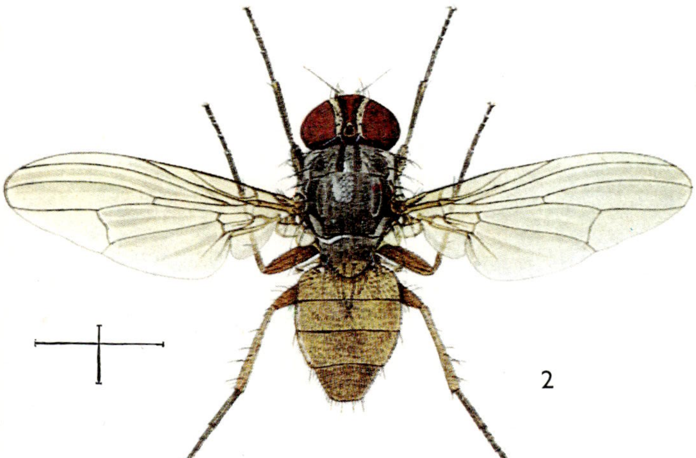
2

PLATE 48

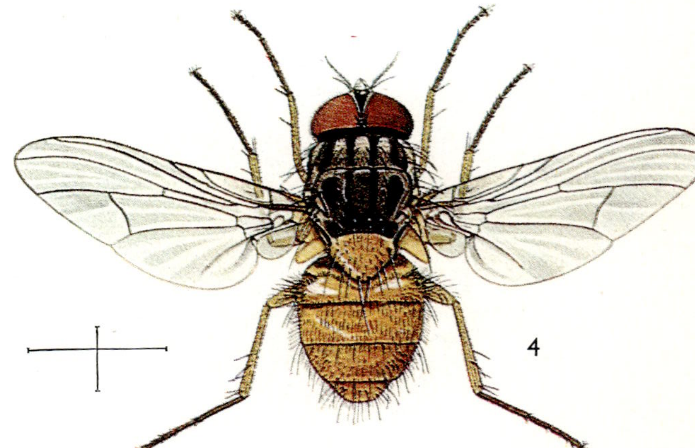
4

MUSCIDAE-IV

1. *Polietes lardaria* F., ♂, p. 305.
2. *Fannia hamata* Macq., ♀, p. 304.
3. *Mydaea scutellaris* R–D. (=*pagana* F.), ♂, p. 306.
4. *Phaonia populi* Mg., ♂, p. 303.

ently to arrange wings and legs in a desired position. The usual type of relaxing-tin or jar containing sand or blotting paper moistened with water (to which should be added a little carbolic acid to inhibit the growth of moulds) will answer quite well. In any attempt to relax dried flies, care must be taken to see that they do not come into actual contact with the water, or the sides of the container on which condensation will almost certainly be present. Specimens should be laid on a dry piece of blotting paper or cork sheet, well above the wet sand or blotting paper, on, say, a piece of lead or in a small saucer or watch-glass.

Specimens so relaxed, however, are seldom as satisfactory as those mounted immediately after death. With smaller flies relaxation is impracticable; the bodies, wings and legs are apt to break up or tear when being pinned or arranged.

MOUNTING.

All except minute adult flies may be pinned either in the orthodox manner adopted by lepidopterists or in various other ways; the larger flies may also be set in much the same fashion as Lepidoptera; in fact, most of the flies illustrated in this volume were thus set. With ingenuity and patience, even quite small flies may be artistically arranged with the aid of fine pins, bristles, and tiny braces.

Nevertheless, it is not at all necessary, and in some cases definitely inadvisable, to set Diptera; particularly if the objective is critical study of external structure and positive identification of species. For example, where the examination of characters provided by the pleura is important, the orthodox method of setting may result in the pleura being obscured by the legs or wings. Generally it is important to see that the legs are not tangled, or folded up under the body, but extended sufficiently for their characters to be seen, and in a downward direction, so as not to obscure the pleura or genitalia; for a similar reason, the wings should be slightly raised and never left folded over the abdomen, since thus the venation will be difficult to appreciate.

Obviously the ideal method of mounting would be that in which it would be possible to examine all the parts of the body and which would simultaneously cause the minimum of damage to the specimens. Pinning is not usually prejudicial to the condition of large and medium-sized flies, especially if an appropriate size of pin is used. Too large a pin will destroy too much of the body-surface ; too small a pin will make the specimen, when dried, liable to move upon it. Long Continental pins, black-lacquered or of stainless steel, preferably the latter, should be used for direct pinning. If it is desired to pin medium-sized or small flies with short pins, they can be subsequently " staged " on polyporus or slips of celluloid. In the latter case, the pin should be " pricked " through the celluloid carefully on the pinning-stage, an initial tiny hole being made, if necessary, with the extreme point of a fine needle ; the stage is then mounted on a long Continental pin. The specimen or stage should be about three-quarters of the way up the long pin, so that a firm grip may be obtained on the upper part of the pin with the large entomological forceps when fixing in storeboxes or removing, etc. ; and so that both ends of the pin may, if necessary, be grasped with the fingers when manipulating for examination under the microscope.

It will be realised that, in the case of very small or minute flies, even the finest pins obtainable are likely to damage a relatively large area of the body-surface (unless, of course, the point of the pin be inserted from below or from one side, and not allowed to go right through ; small specimens, when dry, however, are very liable to " spring off " such points at the slightest jarring or vibration). Therefore, where only one side of the thorax need be seen, the celluloid cement method (p. 335) could be employed.

Where pinning is adopted, experience will show that certain flies should be pinned in somewhat different fashion from others. Large Brachycera, such as Asilidae, Tabanidae and Stratiomyidae, may be pinned on the centre line of the dorsum, but where acrostichals or dorsocentrals are important, e.g. Calypterates, Empididae, and Dolichopodidae, the pin should be inserted on one side of the centre line, so as to leave one half of the dorsum unobscured; or if a sufficient number of specimens is available, some of them may be pinned through the dorsum and some through the pleura, i.e. mounted in profile.

When mounting with an adhesive, it is obvious that the specimen must be attached to the mount in such a way as to expose all the principal features of the structure; since the pleura are usually alike, attachment by one side of the thorax offers a solution to the problem. Thus the dorsum, one side of the thorax and the remainder of the body will not be obscured. This may be accomplished satisfactorily with the aid of a solution of celluloid in amyl acetate, of a fairly thick, cream-like consistency. The solution is easily prepared by cutting up some sheet-celluloid into very small pieces, putting them into a bottle and adding a little amyl acetate gradually; the celluloid will swell and absorb it, when more acetate may be added and the mixture stirred with a glass rod, and so on until the desired consistency has been attained. A balsam-bottle with a dipping-rod as used by microtomists will serve the purpose admirably. The fly may now be attached with this solution by one side of the thorax to the point of a small celluloid triangle, which serves as a stage and should be impaled on a long pin; an advantage of the method being that a data label impaled on the same pin, below, may be read through the celluloid mount, without removing the specimen from the storebox (Fig. 14A). Only the tiniest spot of the solution should be used, to avoid embedding or smearing the specimen and obscuring the characters; it may be applied to the point of the triangle, already impaled on the long pin, with the dipping-rod; some difficulty may be experienced at first in negotiating the treacle-like cement, but a little practice in drawing up the rod firmly but decisively in a vertical direction from the bottle and rapidly twirling it between forefinger and thumb, so as to " wind in " the hanging drip, will soon obviate the trouble. Alternatively the rod with some cement adhering to it may be drawn up vertically and held

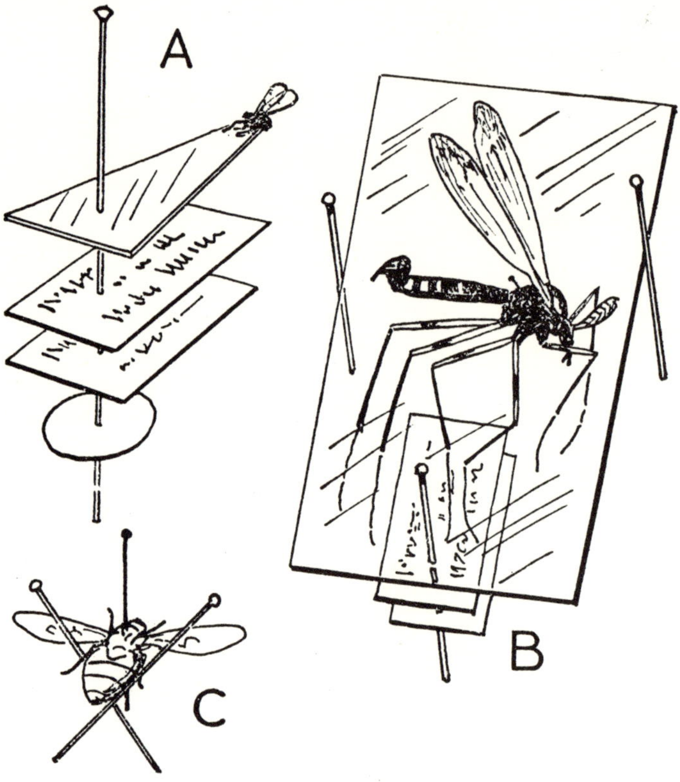

FIG. 14.—MOUNTING, LABELLING AND SECURING FOR TRANSIT.
A. Triangular celluloid stage.
B. Large rectangular celluloid stage, secured.
C. Direct-pinned specimen, secured.

over the mouth of the bottle, and the point of the triangle just touched to it. If the solution is of exactly the right consistency, the tendency for hanging drip will disappear ; adjustments to the consistency may be effected by allowing the acetate to evaporate by leaving the cap of the bottle off for a time, or by adding more acetate very gradually. The specimen may be applied to the point of the triangle with fine forceps or picked up by just touching the point of the triangle itself to one side of the thorax. As an alternative to the dipping-rod, a small camel-hair brush, in a metal ferrule, with the handle piercing the cork of a bottle, may be kept immersed in the solution, when not in use.

In place of the celluloid cement method, card triangles may be used and the specimens affixed with gum applied with a fine brush. The disadvantages of this system include a greater tendency for specimens to " spring off " their mounts, and the fact that data labels cannot easily be seen without removing specimens from their storeboxes.

It is often desirable to provide a more protective mount than a triangle for flies with long, delicate legs or fragile bodies, e.g. Tipulidae, Chironomidae, Mosquitoes, etc. A rectangular celluloid slip of suitable size may be employed ; the specimen is attached by placing a tiny blob of cement at the appropriate point on the slip and causing one side of the thorax to adhere to this. After allowing a minute or two for the cement to set and firmly secure the specimen, the lower wing is stuck flat on the slip, by lightly painting the slip with a fine brush charged with amyl acetate, and lightly dabbing, not brushing the wing with a *dry* brush, or manipulating it with the side, not the point, of a setting-needle, so as to cause it to adhere to the softened celluloid ; similarly with the legs, antennae and abdomen (Fig. 14B). The wings, legs, etc., should not be " brushed on " with a wet brush in the manner adopted by coleopterists when " carding " beetles, since scales, hairs and bristles, etc., may be removed, damaged or obscured by so doing.

Pairs taken in cop. should always be mounted together,

i.e. on the same mount; likewise predators and prey.

Mounted specimens should be allowed to dry thoroughly before being placed in the permanent collection and during this period should not be allowed to stand anywhere where they will accumulate dust or become liable to accidental damage. To achieve this, the setting-boards or cork boards may be placed in a large box, suitably perforated with large holes covered with muslin or gauze. Such a box should be treated with disinfectants or contain some substance which will deter pests or prevent the growth of moulds (p. 345). The time taken for flies to dry varies from about a day for very small species to two or three weeks in the case of larger flies; if uncertain, it is better to err on the side of drying for too long a period.

Empty larval skins or puparia may be attached to small celluloid slips with cement and the slips impaled on the same pin below the principal mount. Eggs and larvae, or pupae which fail to mature, may be preserved in an 80 per cent. solution of alcohol; for this purpose industrial methylated spirit is most suitable and economical. This is colourless and contains about 95 per cent. ethyl alcohol; an 80 per cent. solution (approximately) may be made by mixing eight parts of the 95 per cent. spirit, by volume, with one part of water. The violet-coloured methylated spirit sold for household use is not suitable, since the addition of water causes the formation of a milky precipitate; the introduction of specimens will have a similar effect, any water present in the tissues being extracted by the spirit. Industrial spirit may be obtained from certain entomological dealers or suppliers of microscopical requisites, but first it is necessary to make application for permission to purchase, to the local Customs and Excise officer, stating the purpose for which it is required, i.e. preservation of entomological specimens, or for microscopical preparations. If the Customs officer is satisfied with the application, a permit will be issued and this must be presented to the supplier. There are other small formalities, including the periodical inspection of the spirit and its storage by the officer, and the making of occasional returns as to amounts used; in our experience, the authorities have always been very courteous and helpful in these matters.

Small collecting-tubes may be used for the reception of individual specimens, or groups of specimens, stored in alcohol; a small paper label, with data written thereon in waterproof Indian ink, inserted, and after filling with the 80 per cent.

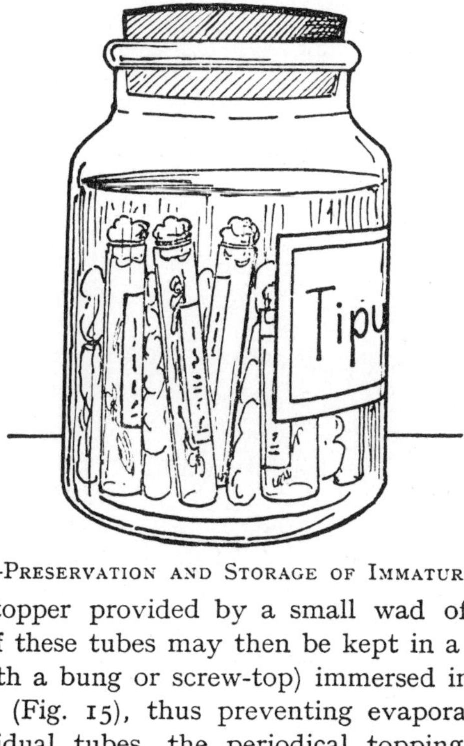

FIG. 15.—PRESERVATION AND STORAGE OF IMMATURE STAGES.

spirit, a stopper provided by a small wad of cotton-wool. Numbers of these tubes may then be kept in a tightly-sealed jar (e.g. with a bung or screw-top) immersed in more 80 per cent. spirit (Fig. 15), thus preventing evaporation of spirit from individual tubes, the periodical topping-up of which would be tedious. Until the large jar is full of small tubes, the unoccupied space may be packed with cotton wool.

Male genitalia of many flies are sufficiently visible without preparation, e.g. Tipulidae, Asilidae, Empididae, Dolichopodidae, etc.; or may be extended, after killing and pinning, with the aid of a fine needle or pin, the end of which has been heated and bent into a tiny hook with forceps or pliers. They may be retained in position while drying, by hooking round a fine pin, or by means of small pins and braces, if necessary—a simple matter if side-pinning has been adopted. This is often satisfactory with certain fairly large or medium-sized flies, e.g. Calypterates, Syrphidae, Sciomyzidae, etc. In the case of smaller flies, however, it is usually necessary to remove the genitalia and make a microscopic preparation.

For permanent preparations a simple method used by Edwards [1] will be found useful. The tip of the abdomen is removed and placed in potash solution, where it is left for twenty-four hours or gently heated in a water-bath. Afterwards the specimen is removed to glacial acetic acid to which has been added, if necessary (it seldom is), a drop or two of Acid Fuchsin stain. If the stain is used, the specimen is removed to pure glacial acetic acid after a minute or two. The specimen is allowed to remain in the acetic acid for about five minutes (or only a few seconds if stained) and is then removed to clove oil. Thence it is mounted in balsam on a small celluloid slip, only a little balsam being used at first until the specimen is arranged as desired with needles under the microscope; subsequently, when dry and set, more balsam is added until the specimen is completely covered; the mount is then put away in a drying-box finally to harden, after which the slip may be impaled on the same pin as the fly from which the genitalia were removed. Care must obviously be taken, if more than one preparation is being made to ensure that mounts ultimately find their way back to the correct principal mounts.

One or two hints in connection with the foregoing procedure may not be out of place. In removing the tip of the abdomen (or the whole abdomen in the case of minute flies) the operation

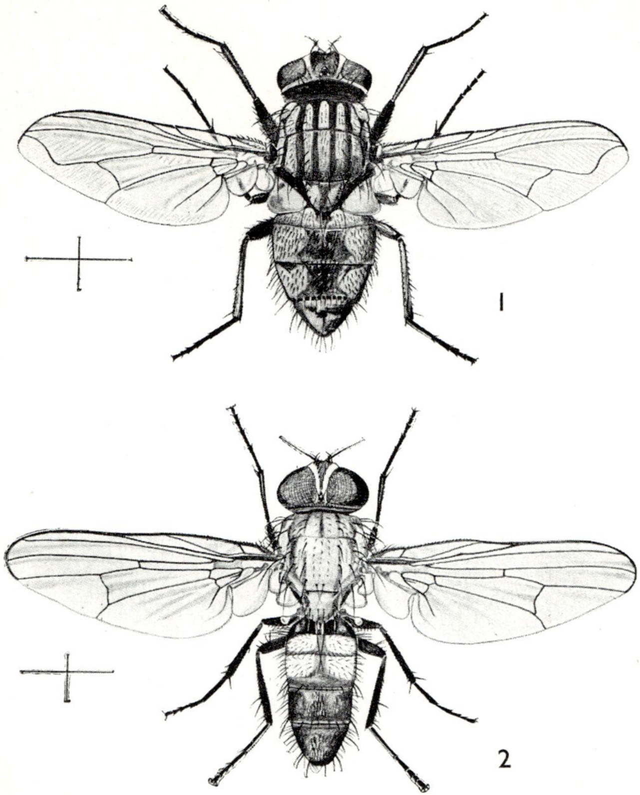

PLATE 49

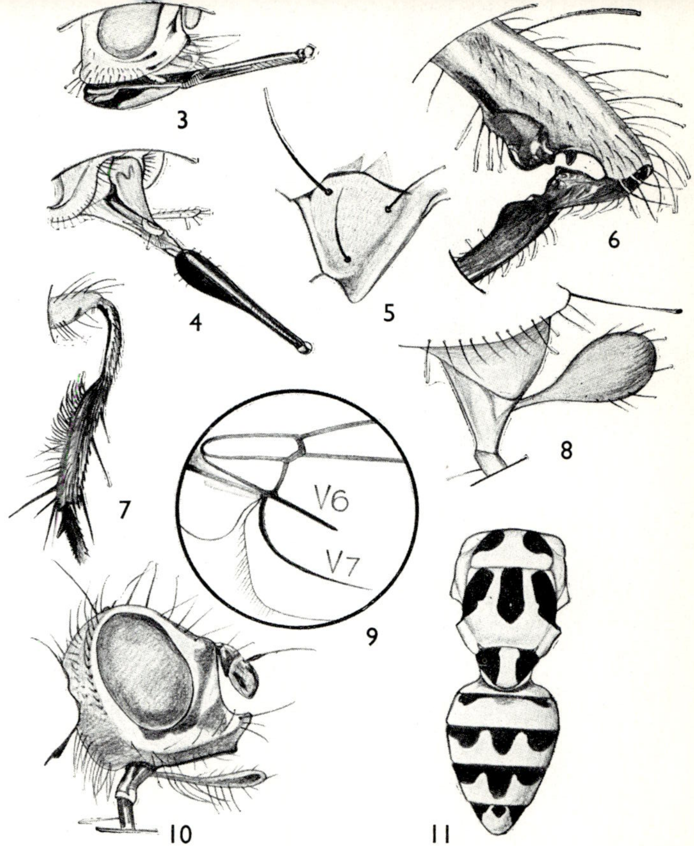

MUSCIDAE-V

1. *Musca domestica* L., ♀, p. 296.
2. *Fannia canicularis* L., ♂, p. 297.

3. *Stomoxys calcitrans* L., proboscis in normal position, p. 301. 4. *Stomoxys calcitrans* L., proboscis extended, p. 301. 5. Coenosiinae, arrangement of sternopleurals, p. 309. 6. *Hydrotaea irritans* Fall., ♂ fore-femur and tibia, p. 303. 7. *Fannia armata* Mg., ♂ mid-tibia, p. 304. 8. *Lispe*, palp, p. 306. 9. *Azelia*, base of wing, p. 307. 10. *Egle muscaria* F., head, p. 308. 11. *Anthomyia pluvialis* L., body markings, p. 308.

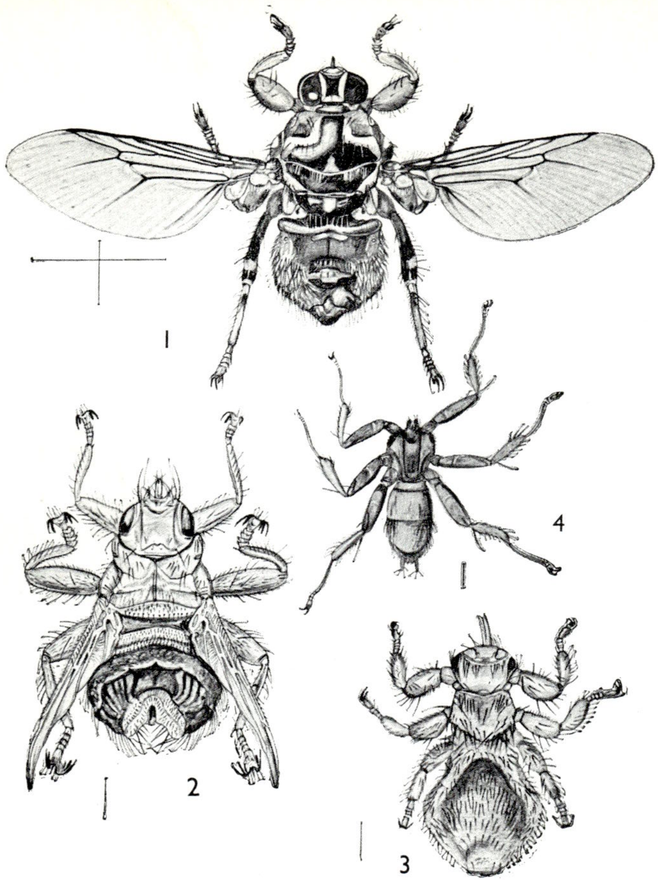

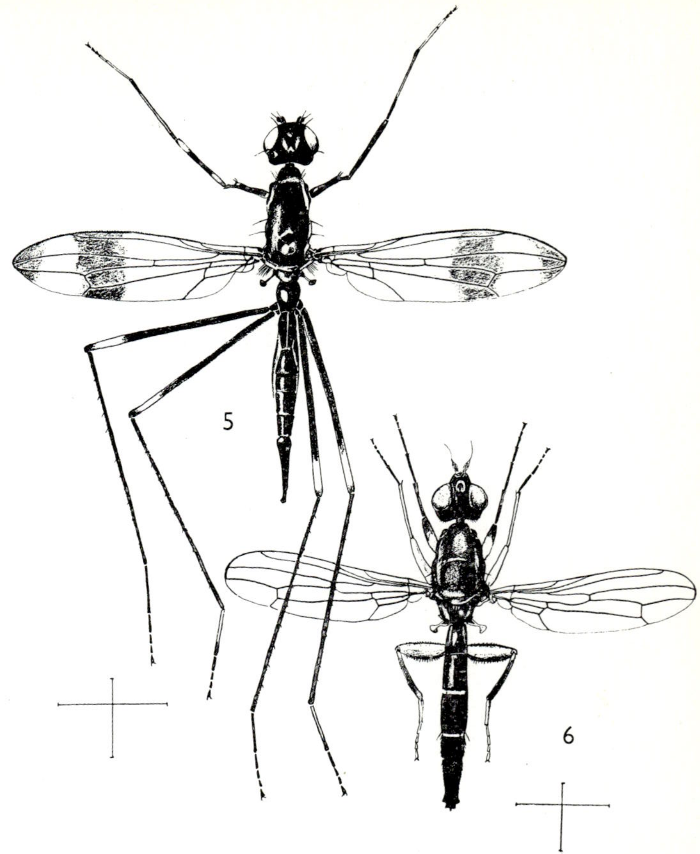

PLATE 50

HIPPOBOSCIDAE-II
1. *Hippobosca equina* L., p. 314. 2. *Stenepteryx hirundinis* L., p. 315.
3. *Melophagus ovinus* L., p. 316.

NYCTERIBIIDAE
4. *Stylidia biarticulata* Hermann, p. 316.

MICROPEZIDAE (=TYLIDAE)-II
5. *Rainieria calceata* Fall., ♀, p. 200.

MEGAMERINIDAE
6. *Megamerina loxocerina* Fall., ♀, p. 199.

is best performed over a large sheet of white paper, so that if the tiny fragment is dropped, there is a reasonable chance of locating it again. The process can be performed also with flies that have been dried and mounted for any length of time, but in such cases, it is a good plan to wet the fine scissors or fine forceps with which the part is removed; in nearly every case it will adhere and thus can be easily transferred to the potash solution. Without this precaution, the tiny fragment is likely to suddenly spring away and be lost. The initial macerating process in potash removes the softer parts and leaves the chitinous structures unchanged; a 10 per cent. solution of Caustic Potash (Potassium Hydroxide) or Caustic Soda (Sodium Hydroxide) may be used, i.e. about 10 gm. in about 100 c.c. of water. If it is desired to speed up this part of the process, and the genitalia are not too delicate (as for instance in very small Nematocera), boiling in a test-tube may be undertaken; the test-tube may be held by means of a twice- or thrice-folded strap of paper round the neck of the tube and the contents gently heated by passing the part of the tube near the *surface* of the liquid back and forth through the flame of a spirit-lamp or Bunsen burner, care being taken to ensure that the liquid shall not effervesce up the tube and leave the specimen stranded on the side. The bottom of the tube should *never* be heated as bubbles of steam will form and probably eject the whole contents of the tube violently—hot potash or soda liquor on exposed skin or the face, or in the eyes would be extremely painful and dangerous. For this reason the mouth of the tube should always be directed away from the person and towards a place where harm will not be occasioned if such an eruption should accidentally happen, e.g. towards a wash-basin or sink. After a minute or two, the specimen may be poured out into a small dish or watch-glass, preferably standing on white paper, so that it can be seen; or one of the little white, glazed earthenware pans used by artists may be employed. While the liquid is still simmering the specimen will probably be near the surface; at a moment when it is at the side which will be lowermost as the tube is tilted, it may be poured out easily with a minimum of liquid accompanying it. If it is allowed to sink to the bottom of the fluid before pouring, it will probably remain there after all the fluid has been drained. Removal of the specimen from one liquid to another may be effected with the aid of a finely-shaved matchstick or a tiny spill of rolled blotting paper or filter-paper. The clearing stage, in clove oil (or oil of wintergreen), may be dispensed with, in the case of temporary or semi-permanent mounts, by mounting direct in lactophenol or one of the Chloral-Gum mountants (*vide* Smart [2]).

LABELLING AND RECORDING.

The extent to which records are maintained by the collector will, of course, be largely governed by the particular object or objects in collecting, but it cannot be too strongly emphasised that the more notes made and preserved with the collection, the greater its scientific value; provided always that the notes are capable of being related to their corresponding specimens and vice versa.

Whilst it is of considerable value to the collector, personally, to allot distinctive numbers to specimens and keep against such specimens, in a register, the fullest possible details, *at least* essential data, such as date and place of capture, and the name of the captor should invariably be with the specimen, since, in time, collections and registers are liable to become separated, especially after the decease of the collector. Moreover, it is imperative that information relating to a particular specimen shall not be capable of being confused with that relating to any other, either now or in the future, so that it is desirable for labels or tickets to be impaled upon the pin bearing the specimen, or its stage, immediately after mounting. Specimens taken in cop., or predators with their prey, should be suitably labelled to denote the fact. If the collection is not to look unsightly, very small labels and neat writing are required ; the top of the label should bear the date of capture, the middle the place of capture, and the bottom the name of the captor (and his serial number if desired), so that these essential particulars can be seen, as far as possible, without removing the specimen from the collection ; other information can be recorded, if desired, on the reverse of the label ; or, if necessary, more than one label can be impaled, in order to accommodate additional data. Place-names should be identifiable on maps in standard use, if necessary, by reference to some feature such as a hill, river or wood, or failing this, by means of an Ordnance Survey Map reference. The county should be stated, in order to avoid confusion with identical place-names in other counties. If in hilly country, it is often of value for the approximate altitude to be recorded : other useful and interesting data include nature of terrain, associated flora and fauna, details of prey taken with predators, or, in the case of parasites, details of the animal or vegetable hosts. It is possible to obtain labels already printed to order, with, say, the captor's name and that of localities frequently visited, so that only the date requires insertion. This saves a lot of time and ensures neatness and legibility.

When specimens are identified, a label bearing the name (and that of the person who identified it, thus : det. A. Smith) may be placed upon the pin and/or the name entered in the register ; simultaneously they may be classified in various ways, if desired, on a card-index or in a loose-leaf book. For example, they may be indexed by species, so that all the data for a particular species may later be referred to ; or by localities, so that all the species taken in a particular locality may be quickly ascertained, and so on. If it is desired to save time at this stage, only the serial number need be entered in the loose-leaf book or card-index ; the necessary details can then be quickly entered, when required, by reference to the register.

PRESERVATION OF THE COLLECTION.

Some thought should be given, at the beginning, by the intending collector, to the manner in which the permanent collection is to be preserved. The elaborate cabinets, with glass-topped drawers and glass-fronted doors, much favoured by lepidopterists, whilst constituting excellent pieces of furniture, are hardly likely to be satisfactory for the dipterist, who will frequently require to remove specimens for examination under lens or microscope, with a minimum of trouble. Moreover, if it is desired to maintain the collection in order of

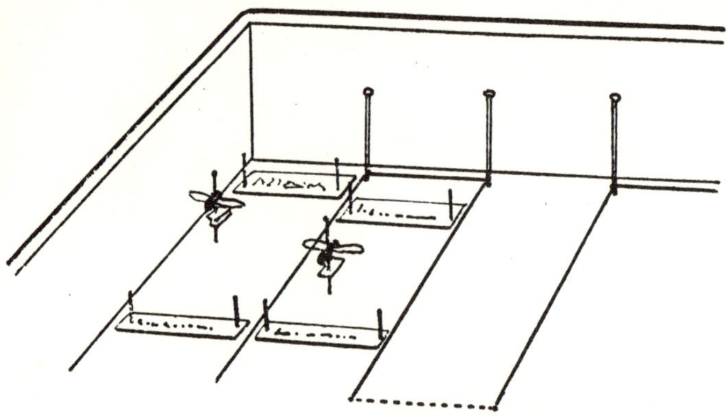

FIG. 16.—A METHOD OF ARRANGEMENT IN THE STOREBOX.

classification, and this is really essential for the serious student, it is difficult, with cabinets, in which drawers are seldom interchangeable, to allow for expansion of the collection once the lay-out has been fixed, without leaving a great deal of space unused, perhaps for a long time. It is generally found more convenient to keep the collection in storeboxes, arranged book-fashion on shelves, so that when one box is full and further accommodation is required for that particular family or section, a fresh box may be inserted without disturbing or rearranging any other part of the collection. Either wooden storeboxes, which are more expensive but more satisfactory, or cardboard cartons may be used—a matter for the collector's pocket. The boxes should be lined with some material to receive the pins, preferably sheet-cork, which should be covered with white paper, and should be deep enough to permit the use of long Continental pins. Neat columns or sections for classifying may be satisfactorily made with the aid of pins and cotton (Fig. 16); this arrangement has the

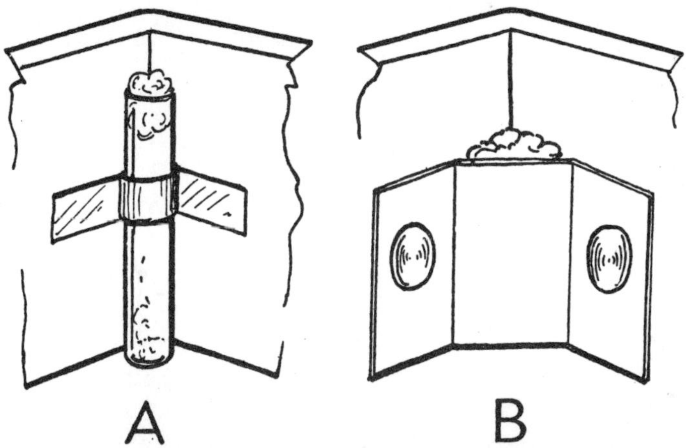

FIG. 17.—IMPROVISED CELLS FOR DISINFECTANTS IN STOREBOXES.
A. Small collecting-tube retained in position with transparent adhesive tape.
B. Card "strap" retained with drawing-pins.

advantage of being easily removable if it is later desired to rearrange the contents of the box in another way. The lids or covers of the boxes or cartons should be tightly fitting to preclude, as far as possible, the entry of pests, which include mites and both adult and larval stages of certain beetles, and to retain the fumes of deterrent substances employed; moreover to ensure that the contents are kept in the dark, thus minimising the fading of colours. There are various ways of deterring pests and preventing the growth of moulds on the specimens. The box may already be fitted with a cell by the maker ; or this may be constructed in the corner with a piece of thin card and drawing-pins (Fig. 17B), paradichlorbenzene or naphthalene being inserted and retained by a small wad of cotton-wool. Paradichlorbenzene, a white crystalline substance, disappears rather rapidly and requires renewal at fairly frequent intervals, but, in our experience, seems to be very effective in keeping away pests. Alternatively, a small collecting-tube, minus its cork, may be secured in the corner of the box with a card strap or, better still, transparent sticky-tape (Fig. 17A) and a wad of cotton-wool inserted therein, which may be impregnated with medicinal creosote and/or glacial carbolic acid. The former should not be used where celluloid mounts are employed since it will turn these brown after a time ; the latter, sold as a white crystalline substance, melts on exposure to air and a little carefully inserted into the bottom of the tube will be soaked up by the cotton-wool and slowly dispersed into the air of the box.

Printed label-lists, such as are available for some other orders, are not obtainable ; to make the collection look neat, labels for families, genera, and species should be carefully written or " printed " by hand or, better still, typewritten. They may be fixed in position with cabinet " points ", or if these are unobtainable, by short points cut from very fine pins, with coarse scissors or a nail-cutter. Family and generic names should be clearly differentiated by being written or typed in block capitals, or in a different colour, or by some such device.

Preservation of immature stages is best effected in 80 per cent. alcohol (p. 338).

Examination of Specimens—Use of Lens and Microscope.

A fair number of our larger flies are recognisable, without much difficulty, in the field, with the naked eye. Generally, however, it is desirable to have at least one good pocket-lens, since there are many instances of closely-similar species, even among the larger flies, which are only separable upon characters not readily visible without one. With the very small species, it will be necessary, at first, to use the lens, in many cases, even to decide the family to which the specimen belongs, although field experience brings, in time, the ability to associate general form, habits, habitats, peculiarities of flight and gait and other characteristics with particular families, so that often a shrewd idea is formed before the lens is actually used.

We have found that a lens with a power of × 10 is very useful for general purposes; it is a good practice to carry it on a lanyard or cord say, around the neck, where it can be dropped into a pocket when not in use, or allowed to hang when both hands are occupied; there is then no fear of it being lost in the excitement of a sudden chase after examining a specimen. It should be realised that it is a good investment to purchase a good-quality lens with a wide, flat field; the satisfaction of using a good lens amply justifies its initial cost. For the examination of very small Diptera in the field, a power of × 20 will prove an asset.

To facilitate the examination of a live specimen, it is advisable to remove it from the net with a collecting-tube which is just large enough, in diameter, to accommodate it without damage; its movements will thus be restricted. Movement along the tube may also be restricted by carefully inserting a small piece of cotton-wool, a small piece of leaf, or by some similar artifice. In the case of very small or minute flies, movement will still be possible even in the smallest collecting-tubes obtainable, but again it may be confined to a very small area, in which the specimens will remain within the focus of the high-powered lens. Such restricting operations should be conducted within the net to lessen the risk of the specimens escaping; likewise removal of the restricting medium or transfer of the specimen or specimens if it is found that they are worthy of retention. The lens should be held close to the eye and the object brought into focus. Wing-venation can best be appreciated, as a rule, if the tube be held up to the light, unless the tube be backed by a piece of white label-paper, as suggested earlier (p. 328). Examination is best conducted out of the glare of the sun.

For critical examination at home, a compound microscope will be necessary. A normal monocular instrument, with the stage removed, can be used fairly satisfactorily, with suitable top-lighting and low-powered objectives, for general examination; and, of course, with the stage and higher-powered objectives, for transparent or semi-transparent features such as the wings or very thin genitalia mounts, capable of being viewed by transmitted light. There is, however, the serious disadvantage of the inverted image, which is, unless the operator be well practised, a particular obstacle for the dipterist who often requires to manipulate his specimens with unsupported fingers, so as to be able to view various aspects in quick succession. Moreover, the image is flat and the inexperienced operator may well miss certain features, assuming them absent, whereas they are merely out of focus.

The most suitable instrument is a binocular of the Greenough pattern (Pl. **52**); here a stereoscopic image is obtained, i.e. the object appears in relief rather than in one plane and, furthermore, is erect instead of inverted, this being accomplished by means of prisms within the instrument and enabling the operator to handle his specimen normally. Special wide-field eyepieces are obtainable which, though expensive, are a great comfort if long spells at the microscope are undertaken. To facilitate manipulation of specimens with the fingers, the pin should be inserted into an ordinary cork of comfortable size. The horseshoe type of stand illustrated allows considerable working-space for the hands. If the upright position of the microscope with this kind of stand is found to cause fatigue, it is not a difficult matter to devise an inclined surface securely fixed to the table or bench, to which

Plate 51

Small-aperture net with long bag, p. 323.

The Sweep-net, p. 326.

PLATE 52

The Suction-bottle or "Pooter", p. 327.

Binocular Microscope (Greenough pattern), p. 348.

the microscope may in turn be secured. In the normal type of inclinable stand, the stage, unless removable, will usually impede manipulation and present difficulties in regard to providing effective top-lighting of the object. Models with inclined eyepieces and horizontal stage are convenient for dissection and wet preparations, as well as for normal observation; an objective-changing turret is a great boon.

Adequate lighting may be provided for low-powered objectives with an ordinary electric bulb in a suitable frame capable of being adjusted for height or direction; the microscope manufacturers' catalogues list a number of inexpensive types. For higher powers, some form of high-intensity lamp capable of being spot-focused, is desirable, although with a certain amount of extra trouble, good results may be obtained with an ordinary lamp and a large bull's-eye condenser on an adjustable stand. It should be emphasised that the lowest powers which will reveal the desired detail should always be used; resolution is more important than mere magnification, and often a slight adjustment to lighting will throw up what is required quite easily. For instance, details of wing-venation viewed against a white-paper background with strong light will be obscured by glare; a reduction of the light, say, by slightly off-directing the beam will often produce the required result; or if the hairs on a compound eye are being examined, they will be better seen against a temporary background of matt black paper, with the light carefully directed on to the object and avoiding the background paper. It should also be remembered that the higher the power used, the smaller the area which can be viewed at one time and the more intense the light-source required; to err in this direction is to incur unnecessary fatigue and to increase the likelihood of inaccurate observation. If the general field is properly comprehended with the lower powers, a special feature of interest may always be investigated on its own, later, with the higher powers.

Packing of Specimens for Transit through the Post, etc.

If exchanging specimens with a fellow-collector, or submitting to a specialist for naming or confirmation, it is generally unsatisfactory to send them unmounted, although very small miscellaneous flies, all of which may not ultimately be preserved, may be sent dry, in numbers, securely packed between thin layers of cellulose wool (not cotton-wool) or fine tissue paper, so as to preclude movement, within a box or container sufficiently stout to withstand the hazards of transit. Such specimens, although dry, may be subsequently mounted by the celluloid cement technique, if this be carefully done. Perhaps a correspondent, not a dipterist, may be sending larger specimens to the collector, and may not have the necessary facilities or faculty for pinning and mounting; in this case, if the transit period be not more than two or three days, the difficulty can be got over by sending the specimens securely positioned in a tube with cellulose wool, lightly damped with ethyl acetate, where they should remain sufficiently relaxed to permit mounting, with care, on receipt. Obviously tubes should be well packed in protective material, such as cotton-wool, sawdust, wood shavings, etc., within a stout wooden or metal box, which in turn should be enclosed within protective material, such as thickly-folded newspapers or corrugated paper, within an outer postal cover.

Pinned or mounted specimens are best sent in one of the cork-lined postal boxes sold by the entomological dealers, substitutes for which can easily be devised by the handyman. It is necessary to ensure that specimens are very securely pinned into the cork, since even one coming adrift will be thrown about and probably ruin all the others. As additional safeguards, heavy or staged specimens may be secured (Fig. 14, B–C), i.e. held in position by long pins, thus minimising the risk of the specimen moving on its pin, or the stage bearing the specimen swinging or working loose, through vibration or jarring; and a piece of cotton-wool may be fastened along the inside edges of the box, so that any parts of flies which may break off in transit may become entangled therein and thus prevented from causing further damage as they are flung about in the box.

Single pinned or mounted specimens may be pinned to the inside of the cork of a collecting-tube, which is then replaced and the tube securely packed in protective material in the usual way.

LITERATURE CITED

1. Edwards, F. E. British Non-Biting Midges. *Trans. ent. Soc. Lond.* **77**: 282. 1929.
2. Smart, J. A Handbook for the Identification of Insects of Medical Importance. 83–4, 100–3, 259–66. British Museum, London. 1943.

Other useful reading:

Eltringham, H. Histological and Illustrative Methods for Entomologists. Clarendon Press, Oxford. 1930.

Audcent, H. Hints on the Mounting of Diptera. *Ent. Rec.* **54**: 69–72. 1942.

Smart, J. Instructions for Collectors, No. 4A. Insects. British Museum, London. 1940.

Day, C. D. British Tachinid Flies. *North Western Naturalist.* **21** (Supp.): 23–5. 1946.

Oldroyd, H. Collecting, preserving and studying insects. 327 pp. Hutchinson, London and New York. 1958.

Buxton, P. A. British Diptera associated with fungi. *Ent. mon. Mag.*, **96**: 61–94. 1961.

Hackman, W. Studies on the Dipterous fauns in burrows of voles (*Microtus, Clethrionomys*) in Finland. *Acta Zool. Fenn.*, **102**: 1–64. 1963

Laurence, B. R. The larval inhabitants of cow pats. *J. Animal Ecol.*, **23**: 234–60. 1954.

Smith, K. G. V. On the Diptera associated with Stinkhorn (*Phallus impudicus* Pers.). *Proc. R. ent. Soc. Lond.* (A) **31**: 49–55. 1956.

Woodroffe, G. E. An ecological study of the insects and mites in the nests of certain birds in Britain. *Bull. Ent. Res.*, **44**: 739–72. 1953.

From p. 60

10. Brindle, A. Notes on the identification of *Limnophila* larvae (Dipt., Tipulidae). *Trans. Soc. Brit. Ent.* **13**: 57–68. 1958.
    Brindle, A. and Bryce, D. The larvae of the British Hexatomini (Dipt., Tipulidae). *Ent. Gaz.* **11**: 207–24. 1961.
    Chiswell, J. R. A taxonomic account of the last instar larvae of some British Tipulinae. *Trans. R. ent. Soc. Lond.* **108**: 409–84. 1954.
11. Laurence, B. R. The British species of *Trichocera* (Dipt., Trichoceridae). *Proc. R. ent. Soc. Lond.* (A). **32**: 132. 1957.
12. Keilin, D. and Tate, P. The early stages of the families Trichoceridae and Anisopodidae (Dipt.). *Trans. R. ent. Soc. Lond.* **90**: 39–62. 1940.
13. Brindle, A. Taxonomic notes on the larvae of British Diptera. 9. The family Ptychopteridae, *Entomologist*, **95**: 212–6. 1962.
14. Dahl, C. Notes on the taxonomy and distribution of Swedish Trichoceridae (Dipt., Nemat.). *Opuscula Entomologica*, **31**: 93–118. 1966.

From p. 100

12. Brindle, A. Taxonomic notes on the larvae of British Diptera. 2. *Xylophagus* Meigen (Rhagionidae). *Entomologist*, **94**: 144–8. 1961. 4. *Atherix* Meigen (Rhagionidae). *Entomologist*, **94**: 218–20. 1961. 12. *Rhagio* F. (Rhagionidae). *Entomologist*, **94**: 311–5. 1961.
13. Hobby, B. M. and Smith, K. G. V. The immature stages of *Chrysopilus cristatus* F. (Rhagionidae). *Ent. mon. Mag.*, **97**: 190–2. 1962.
14. Oldroyd, H. Swarming of Stratiomyidae. *Entomologist*, **70**: 175. 1937.
15. Parmenter, L. The swarming of *Beris morrisii* Dale (Dipt., Stratiomyidae). *Entomologist*, **70**: 213. 1937.

From p. 116

12. Schlinger, E. I. A revision of the genus *Oncodes* Latreille with particular reference to species of the Western Hemisphere. *Proc. U.S. Nat. Mus.* **111**: 227–336. 1960.

From p. 148

Colyer, C. N. Notes on *Megaselia* (*Aphiochaeta*) *tergata* (Lundbeck) with a description of the male. *Broteria Cienc. Nat.* **28**: 19–23. 1959.

Colyer, C. N. Critical notes on the identity of *Megaselia* (*Aphiochaeta*) *spinata* (Wood) and its distinction from *Megaselia* (*A.*) *gemina* Schmitz (Diptera, Phoridae). *Broteria Cienc. Nat.* **29**: 3–6. 1960.

From p. 226

14. Collin, J. E. Some new British Borboridae (Diptera). *J. Soc. Brit. Ent.*, **5**: 172–8. 1956.
15. Laurence, B. R. Two coprophilous Diptera (Borboridae and Sepsidae) new to Britain. *Ent. mon. Mag.*, **88**: 81. 1952.
16. Hackman, W. On the genus *Copromyza* Fall. (Dipt., Sphaeroceridae) with special reference to the Finnish species. *Notul. ent.*, **45**: 33–46. 1965.

From p. 243

14. Collin, J. E. A revision of the British species of *Cetema* Hendel (Dipt., Chloropidae) with two species new to science. *Entomologist*, **99**: 116–20. 1966.
15. Smith, K. G. V. The immature stages of *Gaurax* (= *Botanobia*) *dubius Mcqt.* (Dipt., Chloropidae), with notes on the specific status of *G. fascipes* Becker. *Ent. mon. Mag.*, **100**: 237—9. 1965.

GLOSSARY

of

DIPTEROLOGICAL TERMS IN COMMON USE, ABBREVIATIONS OF AUTHORS' NAMES, ETC.

Abdomen—The hindmost of the three principal divisions of the body (Figs. 1–3).

Acalypterates—A section of Cyclorrhapha Schizophora (Figs. 4, 5 and 12).

Acrostichals—Bristles or hairs on the dorsum, lying between the dorsocentral bristles (Fig. 6B).

Adventitious veins—Veins occurring in abnormal positions.

Aeneous—Brassy, golden-green, bronze-black.

Aestivation—Summer torpor or inactivity, as opposed to hibernation.

Alar squama—A flap-like appendage at the base of the wing, posteriorly (Fig. 7F).

Alula—A lobe of the wing-membrane, near the base posteriorly, lying beyond the alar squama (Fig. 7C, F).

Anal cell—A cell bounded by Veins 5 and 6 (Fig. 7).

Anal gills—Caudal processes near the hind spiracles in certain aquatic larvae, e.g. Tipulidae, p. 48 (Pl. **3**, 1).

Anal lobe—*See* Axillary lobe.

Anal organ (Anal tube)—A caudal process in male Phoridae (Pl. **22**, 3).

Anal tube—*See* Anal organ.

Anal veins—Veins 6 and 7. Vein 6 is often referred to as the first anal and Vein 7, when present, as the second anal, or axillary vein (Fig. 7).

Anchor process (Breastbone, Sternal spatula)—A chitinised process lying within the thoracic region in Cecidomyiid larvae, p. 82.

Annulate(d)—Ringed, but not fully segmented (e.g. certain Brachycerous antennae, Pl. **11**, 2, 4), or banded with colour (e.g. tibiae in Periscelidae, p. 227).

Annulation—A ring-like constriction, or band of colour.

Antennae—The " feelers ", inserted in the front of the head, between the frons and the face (Figs. 1–4).

Antennal foveae—*See* Antennal grooves.

Antennal grooves (Antennal foveae)—Depressions in the face, accommodating the antennae.

Anterior—In front, foremost ; the surface of a structure facing toward the head-end of the body (Fig. 8D).

Anterior cross-vein (Middle cross-vein, Small cross-vein)—A short cross-vein between Veins 3 and 4, usually at the middle of the wing or in the basal half (Fig. 7).

Antero-dorsal—Lying between the anterior and dorsal surfaces (Fig. 8D).

Antero-ventral—Lying between the anterior and ventral surfaces (Fig. 8D).

Aphidagous (Aphidivorous)—Feeding or subsisting on aphids, e.g. the larvae of some species of Syrphidae, p. 156.

Aphidivorous—*See* Aphidagous.

Apical—Lying at the apex ; that part of an appendage lying furthest from the point of attachment to the body (Fig. 8A) ; in the case of the abdomen, the hindmost extremity.

Apical scutellars—The apical marginal bristles of the scutellum (Fig. 6B).

Apical spurs (of the tibiae)—Short, stout bristles at the apex (Fig. 8A).

Appendix (Vein)—A short vein, or extension of a vein, beyond a bend e.g. Vein 4 in some species of Tachinidae (Pl. **36**, 4).

Apterous—Without wings.

Arista—A bristle-like extension of the third antennal segment in Brachycera and Cyclorrhapha—may be visibly segmented, with or without pubescence or cilia (Fig. 4).

Aschiza—A sub-division of the sub-order Cyclorrhapha (Figs. 4A, 12).

Atrophy—A state of becoming or being non-functional ; diminution of a structure through becoming non-functional.

Auxiliary vein—*See* Sub-costa.

Axillary lobe (Anal lobe)—The area behind Vein 6, bearing Vein 7 (axillary vein) when present (Fig. 7).

Axillary vein—Vein 7, the second anal vein ; often vestigial or absent (Fig. 7).

Balancer—*See* Haltere.

Basal—At the base ; that part of a structure nearest the point of attachment to the body (Fig. 8A) ; in the case of the abdomen, the part nearest the thorax.

Basal cells—Cells near the base of the wing, bounded by Veins 3–4 (First basal) and Veins 4–5 (Second basal), lying in front of the anal cell (Fig. 7).

Basicosta—The second scale-like process on the costal margin of the wing, near the base (Fig. 7F).

Basitarsus—*See* Metatarsus.

Bergr.—Bergroth.

Brachycera—One of the three sub-orders of Diptera, linking the primitive Nematocera to the highly-developed Cyclorrhapha (Figs. 2, 12).

Brachycerous—Pertaining to Brachycera.
Branches (of vein)—Two or more arms of a vein, beyond a fork (Fig. 7A, B).
Breastbone—*See* Anchor process.

Callus—A prominence or callosity (cf. Humeral callus, Posterior or Post-alar callus).
Calypter—*See* Squama.
Calypterates—A section of Cyclorrhapha Schizophora (Figs. 4, 5 and 12).
Carina—*See* Keel, Facial ridge.
Carnivorous—Feeding or subsisting on the bodies or body-substances of other animals; in Diptera, usually on other insects or larvae; predatory.
Caudal—Pertaining to the apex of the abdomen; at the "tail".
Cell—An area of the wing bounded by veins (Fig. 7).
Cerci—Paired caudal appendages, often prominent in extruded genitalia of female flies (Fig. 9E).
Chaetotaxy—The arrangement of the bristles; bristle-pattern.
Cheeks (Genae)—A general term applied to the sides of the face (Fig. 4).
Chitinised (Sclerotised)—Hardened or horny, not flexible or membranous.
Cilia—Small hairs, as for instance on costal margin of wing (Pl. **31**, 3), or on the arista (Pl. **38**, 9).
Claspers—Paired lateral appendages of the male genitalia, used to secure the female during mating.
Claws—Curved processes on the fifth, or apical, segment of the tarsus (Fig. 8A, B, C).
Closed cell—A cell entirely bounded by veins (cf. Anal cell, Fig. 7B).
Clubbed—Dilated at the apex; club-like.
Clypeus—A sclerite, or plate, at the base of the proboscis, sometimes visible outside the mouth-opening (cf. Dryomyzidae, p. 205).
Coll.—Collin.
Compound eyes—The large, multi-faceted eyes (Figs. 1–4: Pl. **7**, 3: **7**, 8).
Compressed—Laterally flattened, i.e. narrowed from the dorsal aspect.
Comstock-Needham system—A system of reference to the wing-venation, p. 26.
Conjunctiva—Membrane between sclerites, or plates.
Constricted—Much narrowed or attenuated.
Convergent—Approaching; drawing together.
Coprophagous—Feeding or subsisting on dung.
Corrugated—Wrinkled; with more or less parallel folds on the surface.
Costa (Costal vein)—The stout nervure forming the anterior edge

of the wing ; sometimes denoted by the abbreviation *c* (Fig. 7).
Costal " breaks "—Interruptions in the continuity of the costa, often used as characters for differentiating the Acalypterate families (Fig. 7D).
Costal spine—A spine on the costal margin near the apex of *sc* (Pl. **42**, 6).
Costal vein—See Costa.
Coxa—The basal segment of the leg, attached to the thorax (Fig. 8A).
Cross-veins—Transverse nervures between the longitudinal veins (Fig. 7).
Curt.—Curtis.
Cyclorrhapha—One of the three sub-orders of Diptera, which includes the most highly-developed forms (Figs. 3, 12) ; on emergence of the adult, a " cap " or more or less circular piece of the puparium is thrown off (cf. Orthorrhapha, p. 35).
Cyclorrhaphous—Pertaining to Cyclorrhapha.

Decumbent—Hanging down ; lying down ; pendulous ; depressed.
Decussate—Crossing, e.g. decussate bristles.
Deg.—Degeer.
Depressed—Flattened dorsally and ventrally, i.e. narrowed in profile ; in connection with pubescence or hairs, lying down, not erect.
Dichoptic—A condition in which the compound eyes are widely, or at least clearly, separated by the frons (Pl. **7**, 9: **48**, 2).
Dilated—Widened and flattened ; broadened.
Dimorphism—A state of having two different forms ; *see* Sexual Dimorphism.
Disc—The central area of a structure, e.g. the middle of the squamae, scutellum, or abdominal tergites.
Discal bristles—Bristles occurring on the disc, e.g. abdominal discals on the abdominal tergites.
Discal cell—A cell bounded by Veins 4 and 5 (Fig. 7A, B) ; sometimes confluent with second basal cell (Fig. 7D).
Distal—Furthest from the middle line of the body ; apical.
Divaricate—Widely divergent ; forked.
Divergent—Drawing apart ; separating or separated.
Dorsal—Lying on the upper surface (Fig. 8D).
Dorsocentral bristles—Bristles occurring on the dorsum, lying outside the acrostichals (Fig. 6B).
Dorsum (Notum)—The dorsal surface ; particularly used in connection with the thorax (Figs. 1–3, 6B).
Dsv.—Desvoidy ; or R–D., Robineau-Desvoidy.

Egg.—Egger.
Empodium—A pad, bristle or hair between the pulvilli and claws on the fifth, or apical, segment of the tarsus (Fig. 8B, C).

Entomophagous—Feeding or subsisting on insects.
Excavation—A hollow in one structure which accommodates another, or part of another, e.g. abdominal excavation to accommodate the scutellum (Fig. 3).
Excision—A piece cut out; an indentation.

F.—Fabricius.
Face—Broadly, the area of the head-surface bounded by the antennae, the compound eyes and the anterior edge of the mouth-opening (Fig. 4).
Facets—The divisions, or lenses, of the surface of the compound eyes (Pl. **7**, 3, 8).
Facial bristles—Bristles occurring on the lateral borders of the face, above the vibrissae (Fig. 4B).
Facial ridge—*See* Keel, Carina.
Fall.—Fallén.
Family—A grouping of genera or sub-families to form a sub-division of a sub-order (Fig. 12).
Female—denoted by the sign ♀.
Femur (pl. Femora)—A principal leg-segment; the "thigh", between the trochanter and tibia (Fig. 8A).
First anal vein—Vein 6 (Fig. 7).
Forked cell—A cell bounded by two branches of a forked vein (Fig. 7B).
Frons (Front)—The area of the head-surface bounded by the vertex, the compound eyes and the antennae (Fig. 4).
Front—*See* Frons.
Frontal bristles—Bristles occurring along the inner edge of the orbits, close to the frontal stripe (Fig. 4B).
Frontal stripe (Frontal vitta)—The area of the frons between the orbits in Cyclorrhapha (Fig. 4B).
Frontal vitta—*See* Frontal stripe.
Fronto-orbital bristles—The orbital bristles (Fig. 4B).

Galls—Swellings or deformities in plants caused by flies and other insects, in which the larvae develop.
Genae—*See* Cheeks.
Generic—Pertaining to a genus.
Geniculate—Kneed; bent; jointed.
Genitalia—The sexual or reproductory organs.
Genus (pl. Genera)—A grouping of species to form a sub-division of a family or sub-family (Fig. 12).
Grimshaw's system—A system of reference to the leg-bristles, pp. 30–1 (Fig. 8D).

Habitat—The locality or situation in which an insect normally lives; the medium in which development takes place.

Hal.—Haliday.

Haltere (Balancer)—An appendage, usually knobbed, arising from the hind part of the pleura, present in nearly all Diptera (Figs. 1, 2, 5, 6A).

Head (Head-capsule)—The anterior of the three principal divisions of the body (Figs. 1–3).

Hibernation—Winter torpor or inactivity, as opposed to aestivation.

Hirsute—Hairy.

Holoptic—A condition in which the compound eyes are touching above the antennae, or are not widely or clearly separated by the frons (Pl. **7**, 8: **48,** 1).

Homonyms—Identical names applied to different species, etc.; the later of two homonyms is suppressed in favour of a new name or its earliest synonym.

Humeral bristle(s)—A bristle or bristles occurring on the humerus (Fig. 6B).

Humeral callus—A differentiated area at one of the anterior corners of the dorsum (Fig. 6B).

Humeral cross-vein—A cross-vein at the base of the wing, anteriorly, between the costa and sub-costa and extended to base of Vein 1 (Fig. 7).

Humerus (pl. Humeri)—One of the anterior corners of the dorsum; a " shoulder ".

Hyaline—Clear; transparent; translucent; not pigmented.

Hyperparasite—A parasite living at the expense of another parasite; a parasite within a parasite.

Hypopleural bristles—Bristles occurring on the hypopleura, e.g. as in Calliphoridae and Tachinidae (Fig. 6A) (Pl. **46**, 6).

Hypopleuron (pl. Hypopleura)—A pleurite, or pleural plate, above the hind coxa and posterior to the sternopleuron (Fig. 6A).

Hypopygium—The male genitalia and associated parts of the apical abdominal segments.

Imago—The adult stage, which emerges from the pupa.

In coitu—*See* In copula.

In copula—Usually abbreviated to " In cop."—In the act of mating or pairing.

Insertion (of Antennae)—The point of attachment to the head.

Integument—The outer envelope of the body, or " skin "; the exoskeleton.

Interfrontal bristles—Bristles, often crossed, on the frontal stripe, i.e. lying inside the frontals or fronto-orbitals.

Intra-alar bristles—Bristles on the dorsum posterior to the transverse suture and lying between the supra-alars and dorsocentrals (Fig. 6B).

Iridescent—Displaying brilliant rainbow-like colours ; with brightly-coloured reflections.

Jowl—The general area between the lower margin of the compound eye and the mouth-opening (or the lower margin of the head, viewed in profile) (Fig. 4).

Keel (Carina, Facial ridge)—A ridge or prominence separating the antennal grooves.

L.—Linnaeus, Linné.
Labellum (pl. Labella)—A lobe-like structure at the apex of the proboscis (Fig. 4).
Lamella (pl. Lamellae)—A leaf-like plate ; applied to certain paired appendages of the male genitalia, e.g. in Dolichopodidae, p. 134.
Large cross-vein—See Posterior cross-vein.
Larva—The grub or maggot-stage (Fig. 10), which precedes the pupa.
Larviparous—Bearing larvae.
Lateral—On the sides ; at the side.
Lateral bristles—Bristles occurring on the sides of a structure, e.g. abdominal laterals on the abdominal tergites.
Latr.—Latreille.
Linear—Long and narrow, with parallel sides ; strap-like ; line-like.
Longitudinal veins (Long veins)—The principal nervures, radiating from the base of the wing to the margin (Fig. 7).
Lower Calypter—*See* Thoracic Squama.
Lunule—A crescent-shaped area above the base of the antennae in Cyclorrhapha (Fig. 4) ; a crescent-shaped marking in any place, e.g. on the abdomen in some Syrphidae (Pl. **16,** 2).
Lw.—Loew.

Macq.—Macquart.
Macrochaetae—The large or principal bristles, p. 24.
Macrotrichia—The larger of the microscopic bristles or hairs (if differentiated) on the wing-membrane, p. 26.
Maggot—Usually applied to the typical form of Cyclorrhaphous larva (Fig. 10F).
Male—Denoted by the sign ♂.
Mandibles—Paired appendages of the mouthparts ; the " jaws " of primitive biting-mouthparts in some Diptera.
Marginal—At, or near, the edge or border ; on the fore or hind-margins, as opposed to lateral (on the sides), e.g. marginal bristles on an abdominal tergite.
Marginal scutellar bristles—Bristles occurring on the sides or hind-margin of the scutellum (Fig. 6B).

Median—Along the middle; in the middle line.
Membrane (of wing)—The area of the wing between the nervures.
Mesonotum—The dorsum, comprising the prescutum, scutum, scutellum and postscutellum (if present) (Fig. 6B).
Mesopleural bristles—Bristles occurring on the mesopleura (Fig. 6A).
Mesopleuron (pl. Mesopleura)—A pleurite, or pleural plate, above the sternopleuron and anterior to the pteropleuron (Fig. 6A).
Metapleural bristles—Bristles occurring on the metapleura (Fig. 6A).
Metapleuron (pl. Metapleura)—A pleurite, or pleural plate, posterior to the pteropleuron and above the hypopleuron (Fig. 6A).
Metatarsus—The first or basal segment of the tarsus (more correctly referred to as the basitarsus) (Fig. 8A).
Mg.—Meigen.
Microchaetae—The less prominent bristles; bristles other than the macrochaetae.
Microtrichia—The less prominent of the microscopic bristles or hairs on the wing-membrane, if differentiated, p. 26.
Middle cross-vein—*See* Anterior cross-vein, Small cross-vein.
Mouth-beard—An assembly of prominent hairs on the anterior mouth-edge, or round the oral opening, e.g. Asilidae (Pl. **15,** 4).
Mouth-hooks—Reduced mouthparts in the less primitive forms of Dipterous larvae, in the shape of protrusible hooks (Fig. 10F).
Mouth-opening—*See* Oral opening.

Neck—The narrow structure joining the head to the thorax.
Nematocera—One of the three sub-orders of Diptera, containing the most primitive forms (Figs. 1, 12).
Nematocerous—Pertaining to Nematocera.
Notopleural bristles—Bristles occurring on the notopleura (Fig. 6A).
Notopleuron (pl. Notopleura)—A pleurite, or pleural plate, situated at the extremity of the transverse suture, posterior to the humerus (Fig. 6A).
Notum—*See* Dorsum.

Occipital bristles—Bristles occurring on the occiput.
Occiput—The back of the head, bounded by the vertex, the compound eyes and the neck.
Ocellar bristles—Bristles occurring on the ocellar triangle (Fig. 4B).
Ocellar triangle—An area of the frons, more or less differentiated from the remainder, on which are located the ocelli, or simple eyes (Fig. 4).
Ocellus (pl. Ocelli)—A simple eye (Figs. 3, 4).
Oral opening—The mouth-opening; the aperture from which the proboscis is protruded (Pl. **35,** 12).
Orbit—The eye-margin; often a differentiated strip (Fig. 4B); cf. Vertical plates.

Orbital bristles—Bristles occurring on the orbits; *see also* Fronto-orbital Bristles (Fig. 4B).
Orthorrhapha—A division of the order Diptera, in which emergence of the adult from the puparium takes place through a longitudinal rupture (cf. Cyclorrhapha, p. 35); includes the sub-orders Nematocera and Brachycera (Fig. 12).
Orthorrhaphous—Pertaining to Orthorrhapha.
Ovate—Egg-shaped; ovoid.
Oviparous—Producing eggs.
Oviposition—The act of laying eggs.
Ovipositor—The egg-laying apparatus, or genital tube of female flies, sometimes horny and non-retractile, adapted for piercing (Fig. 1; Pl. **28**, 1, 2).
Ovoid—Egg-like; egg-shaped; ovate.

Palp (pl. Palpi)—Appendages of the proboscis (Fig. 4).
Pand.—Pandellé.
Papillate—Covered with papillae or pimple-like excrescences.
Parasite—An organism living at the expense of another.
Pectinate—Comb-like; combed; feathered.
Pendulous—Hanging downwards; hanging from one end; drooping.
Pennate—Plumed; feathered.
Peristoma—A general term for the region adjacent to the mouth-opening; sometimes used in connection with the anterior oral margin, or the jowls.
Pharynx—The anterior part of the alimentary canal; the part immediately after the opening.
Phytophagous—Feeding or subsisting on plants.
Pictured wings—Wings with prominent markings in various patterns, as in Trypetidae (Pl. **28**).
Pile—Soft down; short, fine, dense hairs.
Pilose—Covered with pile.
Pleurite—A pleural sclerite; a pleuron.
Pleuron (pl. Pleura)—A lateral sclerite of a segment; usually applied to the thoracic segments; in the plural, applied to the pleural plates collectively, or the sides of the thorax.
Plumose—Plumed; feathered; a plumose arista may have cilia on one or both sides.
Pollinose—Covered with "dust", or "bloom".
Porrect—Extending horizontally; not drooping.
Post-alar bristles—Bristles occurring on the post-alar or posterior callus, or posterior corner of the dorsum (Fig. 6B).
Post-alar callus—*See* Posterior callus.
Posterior—At the rear; behind; the surface of a structure facing away from the head-end of the body (Fig. 8D).

Posterior callus (Post-alar callus)—A prominence or callosity at one of the posterior corners of the dorsum ; in Calypterates (Fig. 5), differentiated from remainder of dorsum and usually very pronounced.

Posterior cross-vein (Large cross-vein)—The cross-vein lying between Veins 4 and 5, apically ; the cross-vein closing the discal cell apically (Fig. 7).

Posterior ocellars—Bristles standing on the posterior part of the ocellar triangle (Fig. 4B).

Postero-dorsal—Lying between the posterior and dorsal surfaces (Fig. 8D).

Postero-ventral—Lying between the posterior and ventral surfaces (Fig. 8D).

Posthumeral bristle(s)—A bristle or bristles situated behind the humerus (Fig. 6B).

Postocellar bristles—*See* Postvertical bristles.

Postocular bristles—*See* Postorbital bristles.

Postorbital bristles (Postocular bristles)—Bristles situated on the posterior orbits, i.e. on the eye-margin at the back of the head (Fig. 4).

Postscutellum—A projection under the scutellum.

Postsutural bristles—Bristles occurring posterior to the transverse suture, e.g. postsutural dorsocentrals (Fig. 6B).

Postvertical bristles (Postocellar bristles)—Bristles arising on the occiput, just below the vertex and behind the ocellar triangle (Fig. 4).

Praescutum—*See* Prescutum.

Pre-alar bristle—The anterior supra-alar bristle ; often reduced or absent.*

Preapical bristle (of tibia)—A short dorsal bristle just before the apex ; the presence of a preapical bristle on all or any of the tibiae is a character frequently used in differentiating the Acalypterate families.

Predatory—Preying on other organisms.

Pre-pupa—The first pupal stage, where two stages occur.

Prescutellar bristles—Bristles lying immediately anterior to scutellum, e.g. prescutellar acrostichals (Fig. 6B).

Prescutum (Praescutum)—The part of the dorsum anterior to the transverse suture (Fig. 6B).

Prestomal teeth—Hardened serrate processes surrounding the opening of the pharynx, p. 254 (Pl. **42,** 5).

Presutural bristle—A bristle lying just in front of the point where the transverse suture joins the notopleuron (Fig. 6B).

Proboscis—The mouthparts, excluding the palpi (Fig. 4).

* In fact, not shown in Fig. 6B—if present, would be immediately posterior to the transverse suture.

Profile—The side view.
Pronotum—The dorsal part of the prothorax.
Propleural bristles (Prothoracic bristles)—Bristles occurring on the propleura (Fig. 6A).
Propleuron (pl. Propleura)—The area of the prothorax below the humerus and anterior to the mesopleuron (Fig. 6A).
Prosternum—The underside of the prothorax between the two anterior coxae.
Prothoracic bristles—*See* Propleural bristles.
Prothorax—The first segment of the thorax, from which the anterior legs arise.
Pruinose—" Frosted "; covered with fine " dust "; hoary.
Pseudopod—A false foot; a ventral protuberance or tubercle.
Pseudotracheae—The " collecting channels " of the labella, p. 254.
Pteropleural bristles—Bristles occurring on the pteropleura (Fig. 6A).
Pteropleuron (pl. Pteropleura)—A pleural plate, or pleurite, just below the base of the wing (Fig. 6A).
Ptilinal suture—The groove, in fully-developed Cyclorrhapha Schizophora, indicating where the face has fallen forward when the ptilinum was everted (Fig. 4B; Pl. **46,** 7).
Ptilinum—An inflatable sac above the base of the antennae in Cyclorrhapha Schizophora; used by the escaping fly to rupture the puparium and to penetrate the pupation-medium, p. 277 (Pl. **46,** 7).
Pubescence—Short, fine hair.
Pubescent—Bearing, or covered with pubescence.
Pulvilli—Pads on the fifth, or apical segment of the tarsus (Fig. 8B, C).
Punctate—Pitted; covered with small holes, dots or pits.
Punctures—Small round surface depressions; small pits.
Pupa—The stage, often quiescent, which precedes the imago or adult stage (Fig. 11).
Puparium—The pupal integument or " shell ".
Pupipara—A sub-division of the sub-order Cyclorrhapha (Fig. 12), in which the females deposit larvae which immediately pupate.
Pustulate—Covered with blister-like prominences or swellings.
Pz.—Panzer.

R-D.—Robineau-Desvoidy.
Receding (face)—In profile, sloping backward from the base of the antennae toward the mouth-edge (Pl. **31,** 9).
Recurrent—Running backward; returning in direction of origin.
Respiratory appendages—External processes connected with breathing in immature stages, especially aquatic larvae and pupae, e.g. Culicidae, Chironomidae (Figs. 10, 11).

Respiratory horns (Respiratory trumpets)—Elongated processes, connected with the anterior spiracles, on the thorax of certain pupae, especially Nematocera, e.g. Tipulidae, p. 47.
Respiratory siphon—A caudal breathing process in larvae of Culicidae (Fig. 10D).
Respiratory trumpets—*See* Respiratory horns.
Retractile—Capable of being withdrawn, or drawn inwards.
Rond.—Rondani.

Saphrophagous—Feeding or subsisting on dead and decaying organic matter.
Sc—Sub-costa ; auxiliary vein (Fig. 7).
Schizophora — A sub-division of the sub-order Cyclorrhapha, comprising the Calypterates and the Acalypterates (Figs. 4, 5, 12).
Sclerite—A chitinised or sclerotised plate.
Sclerotised—*See* Chitinised.
Scop.—Scopoli.
Scutellar marginals—*See* Marginal scutellar bristles.
Scutellum—The shield-shaped part of the dorsum lying posterior to the scutum (Figs. 1–3, 6B).
Scutum—The part of the dorsum posterior to the transverse suture and anterior to the scutellum (Fig. 6B).
Second anal vein—Vein 7, or axillary vein ; *see* Anal veins (Fig. 7).
Sensu lato—In the wide sense ; comprehensively ; for example, the sub-genera making up the genus *Psila* may be collectively referred to as *Psila* (sensu lato) or *Psila* sens. lat.
Sensu stricto—In the narrow or strict sense ; for example, the sub-genus *Psila* of the genus *Psila* may be referred to thus : *Psila* (sensu stricto) or *Psila* s.s.
Sexual Dimorphism—Differences of form as between the sexes of a species (Pl. **28**, 5, 6: **45**).
Simple—Without modification or adornment ; not complex.
Simple eyes—*See* Ocelli.
Sinuate—Waved ; wavy ; doubly bent.
Sinuous—Sinuate ; bending in and out.
Small cross-vein—*See* Anterior cross-vein, Middle cross-vein.
Spatulate—Broadened or dilated at the apex ; spoon-shaped ; shaped like a spatula.
Species—Individuals with common characteristics and called by a common name (Fig. 12).
Specific—Pertaining to a species.
Spinose—*See* Spinous.
Spinous (Spinose)—Armed with spines.
Spiracle—An external opening of the tracheae or breathing-tubes (Fig. 6A, 10A, E, F).

Spiracular disc—The anal area bearing the hind spiracles in Dipterous larvae (Fig. 10A: Pl. **2**, 1, 2).

Squamae (Calypters)—Flap-like appendages at the base of the wing, or attached to the thorax close to the base of the wing (Figs. 2, 3, 7F).

Stalked cell—A closed cell, bounded by two long veins which join or fuse before the margin (Fig. 7C: Pl. **46**, 1, 2).

Sternal spatula—*See* Anchor process.

Sternite—The ventral plate, or sclerite, of a segment.

Sternopleural bristles—Bristles occurring on the sternopleura (Fig. 6A).

Sternopleuron (pl. Sternopleura)—A pleural plate, or pleurite, above the middle coxa (Fig. 6A).

Stigma—A thickened or darkened area of the wing, lying behind the costa in the region of the apex of Veins 1 and 2 (Pl. **7**, 10); the term is sometimes applied to a spiracle.

Striate—Bearing impressed lines or scratches; lined; bearing parallel lines.

Style—A thick terminal portion of the third antennal segment in some flies, usually tapered, p. 20 (Pl. **19**, 5).

Sub-apical—Just before the apex; almost apical.

Sub-apical cell—The cell at or near the apex of the wing, bounded by Veins 3 and 4.

Sub-apterous—With reduced or vestigial wings; not fully winged.

Sub-basal—Near the base; almost basal.

Sub-costa (Auxiliary vein)—The vein between the costa and Vein 1, sometimes vestigial or absent (Fig. 7); often abbreviated to *sc*.

Sub-family—A grouping of genera to form a sub-division of a family (Fig. 12).

Sub-genus—A grouping of species to form a sub-division of a genus (Fig. 12); expressed thus: *Psila* (*Pelethophila*) *fimetaria* L., i.e. the species *fimetaria* of the sub-genus *Pelethophila* of the genus *Psila*.

Supplementary veinlets—Small veinlets springing from the long veins in abnormal positions; adventitious veins.

Supra-alar bristles—Bristles near the sides of the dorsum, posterior to the transverse suture and above the wing-bases (Fig. 6B).

Synonyms—Different names for the same species, genus, etc.

Tarsus (pl. Tarsi)—The foot, or apical part of the leg, usually composed of five segments (Fig. 8A).

Tegula—The first scale-like process on the costal margin of the wing, near the base (Fig. 7F).

Teratology—Abnormality of structure, e.g. venation-teratology, abnormality of vein-arrangement (Pl. **9**, 1).

Tergite—The dorsal plate, or sclerite, of a segment.

Terminal—At the apex; at the end.
Tessellate—Chequered.
Thoracic squama (Lower calypter)—A flap-like appendage of the thorax, near the base of the wing (Fig. 7F).
Thorax—The second, or middle of the three principal divisions of the body, between the head and the abdomen (Figs. 1–3).
Thorn—A stout thorn-like spur or bristle.
Tibia—The fourth segment of the leg, or " shin ", between the femur and tarsus (Fig. 8A).
Tomentum—Short, soft pubescence.
Tracheae—The breathing-tubes, opening at the spiracles, conveying oxygen to the blood in the body-cavity.
Transverse—Across the longitudinal axis.
Transverse suture (of dorsum)—A suture or groove dividing the prescutum from the scutum; sometimes incomplete (Fig. 6B).
Trapezoid—With four unequal sides, none of them parallel.
Trivial name—The specific name, or name of a species.
Trochanter—The small second segment of the leg, between the coxa and the femur (Fig. 8A).
Truncate—Abruptly cut off, transversely; with the apex cut off in a more or less straight line.
Tubercle—A small prominence or elevation, more or less rounded.

Unicolorous—Of one colour.
Upper Calypter—*See* Alar squama.
Uterus—The vagina; the womb; usually applied to larviparous flies, e.g. *Sarcophaga*, p. 276.

V. or var.—*See* Variety.
Variety—A form of a species, other than the typical form, usually indicated thus: *Scopeuma stercorarium* L. v. (or var.) *merdarium* F.
v.d. Wulp—van der Wulp.
Veinlet—A small vein; a small vein not ending in another vein or the wing-margin, i.e. ending free.
Veins—The nervures of the wings (Fig. 7).
Vena spuria—A false vein, as in Syrphidae (Fig. 7C).
Venation—The arrangement of the wing-veins.
Ventral—Lying on the lower surface; the underside (Fig. 8D).
Vertex—The top of the head, bounded by the frons, the compound eyes and the occiput (Fig. 4).
Vertical bristles—Bristles occurring on the vertex (Fig. 4).
Vertical plates—Downward extensions of the vertex, differentiated from the frons; cf. Orbits.
Vibrissae—The large bristles arising from the angles at the sides of the mouth in many Cyclorrhaphous flies (Fig. 4B).

Vill.—Villeneuve.
Vitta—A stripe; usually applied to the part of the frons between the orbits, viz. frontal vitta.
Viviparous—Producing young in an active state, as opposed to oviparous.

Whlbg—Wahlberg.
Wied.—Wiedemann.

Zett.—Zetterstedt.

♂ = Male.
♀ = Female.

USEFUL REFERENCES

Henderson, I. F. A Dictionary of Scientific Terms. Oliver & Boyd, London. 1920.
Jardine, F. E. S. The Dictionary of Entomology. R. B. Janson & Sons, London.
Grimshaw, P. H. Introduction to the Study of Diptera, with a key for the Identification of Families. *Proc. roy. Phys. Soc. Edin.* **22**: 187–215. 1934.
Curran, C. H. The Families and Genera of North American Diptera. 480–93. New York. 1934.
Torre-Bueno, J. R. de la A. A Glossary of Entomology. Brooklyn, N.Y. 1950.

INDEX

The heavy figures denote principal page references.